# IUTAM SYMPOSIUM ON VARIATIONS OF DOMAINS AND FREE-BOUNDARY PROBLEMS IN SOLID MECHANICS

SOLID MECHANICS AND ITS APPLICATIONS
Volume 66



*Aims and Scope of the Series*

The fundamental questions arising in mechanics are: *Why?*, *How?*, and *How much?* The aim of this series is to provide lucid accounts written by authoritative researchers giving vision and insight in answering these questions on the subject of mechanics as it relates to solids.

The scope of the series covers the entire spectrum of solid mechanics. Thus it includes the foundation of mechanics; variational formulations; computational mechanics; statics, kinematics and dynamics of rigid and elastic bodies; vibrations of solids and structures; dynamical systems and chaos; the theories of elasticity, plasticity and viscoelasticity; composite materials; rods, beams, shells and membranes; structural control and stability; soils, rocks and geomechanics; fracture; tribology; experimental mechanics; biomechanics and machine design.

The median level of presentation is the first year graduate student. Some texts are monographs defining the current state of the field; others are accessible to final year undergraduates; but essentially the emphasis is on readability and clarity.

*For a list of related mechanics titles, see final pages.*

IUTAM Symposium on

# Variations of Domain and Free-Boundary Problems in Solid Mechanics

Proceedings of the IUTAM Symposium
held in Paris, France,
22–25 April 1997

Edited by

P. ARGOUL

*Laboratory of Materials and Structures of Civil Engineering,*
*Public Works Research Laboratory and CNRS,*
*Champs-sur-Marne, France*

M. FRÉMOND

*Laboratory of Materials and Structures of Civil Engineering,*
*Public Works Research Laboratory and CNRS,*
*Champs-sur-Marne, France*

and

Q. S. NGUYEN

*Laboratory of Solid Mechanics,*
*École Polytechnique,*
*Palaiseau, France*

SPRINGER SCIENCE+BUSINESS MEDIA, B.V.

A C.I.P. Catalogue record for this book is available from the Library of Congress.

ISBN 978-94-010-5992-3 ISBN 978-94-011-4738-5 (eBook)
DOI 10.1007/978-94-011-4738-5

---

*Printed on acid-free paper*

# CONTENTS

# PREFACE

This proceedings volume contains 39 papers presented at the IUTAM Symposium on Variations of Domains and Free Boundary Problems in Solid Mechanics, held in Paris from April 22nd to 25th 1997, at the Ecole des Mines and the Ecole Polytechnique.

This symposium offered an opportunity for researchers from all engineering disciplines and applied mathematics to review the state of the art and to identify new trends and new features in the field. Mechanical modelling, mathematical discussion and numerical resolution have been the primary goals of the meeting.

Principal subjects of discussion concerned ground freezing, shape memory alloys, crystal growth, phase change in solids, piezo-electricity, wavelets, delamination, damage, fracture mechanics, polymerization, adhesion, friction, porous media, nucleation, plasticity, inverse problems, and topological optimization.

More than 80 scientists of different nationalities participated in this symposium. Efforts of many people made this symposium possible. We would like to thank all the authors and participants for their contributions and the members of the Scientific Committee for their patronage and assistance in selecting papers. The effectiveness of the Organizing Committee is acknowledged. We are pleased to thank all the involved members of the two Laboratories : Laboratoire de Mécanique des Solides and Laboratoire des Matériaux et des Structures du Génie Civil, specially Valérie François and Michelle Valenti for their strong contributions in the preparation and organization.

We are furthermore indebted to the following sponsors by their financial and material supports :

– Centre National de la Recherche Scientifique
– Commissariat à l'Energie Atomique
– Ecole des Mines de Paris
– Ecole Polytechnique
– Electricité de France
– European Science Foundation (Free Boundary Problem Program)
– International Union of Theoretical and Applied Mechanics (IUTAM)
– Kluwer Academic Publishers BV
– Laboratoire Central des Ponts et Chaussées

– The European Union (Human Capital and Mobility Program "Phase Transition and Surface Tension")

Pierre Argoul
Michel Frémond
Quoc Son Nguyen

Paris, France
July 1998

## International Scientific Committee

- M. FREMOND (France)
- P. GERMAIN (France)
- M.E. GURTIN (USA)
- K.C. HWANG (China)
- H. KITAGAWA (Japan)
- Z. MROZ (Poland)
- Q.S. NGUYEN (France)
- N.V. NIKITIN (Russia)
- O. RICHMOND (USA)
- B. STORAKERS (Sweden)

## Local Organizing Committee

- P. ARGOUL (Secretary)
- M. FREMOND (Co-Chairman)
- Q.S. NGUYEN (Co-Chairman)

# STEADY STATE SOLUTIONS TO A MULTI-DIMENSIONAL PHASE CHANGE PROBLEM IN GROUND FREEZING

JOSÉ FRANCISCO RODRIGUES
*CMAF/Univ. Lisboa, Av. Prof. Gama Pinto, 2*
*1699 Lisboa Codex, Portugal*

## 1. Introduction

In spite of the importance to applications, the mathematical analysis of the conduction-convection free boundary that arises in the freezing or thawing of saturated porous media is not yet completely treated. A coherent model must take into account both thermal and hydraulic aspects that are strongly coupled according to basic continuum mechanics principles (see [1]).

It is well-known (see, for instance [7] and its bibliography) that the effect of seepage flow on the heat transfer with change of phase in a porous medium cannot be neglected in many engineering problems, such as ground stabilization in maintenance and construction sites, freezing of food stuff, or geological applications, such as geothermal reservoirs or migration of magma.

From the mathematical point of view, one of the simplest models of this very complex phenomenon may be treated as a Stefan problem with convection only in the frozen region. This has been done in two-dimensional cases by Elliott, DiBenedetto and Friedman, in [3] and [4]. The main difficulty in their approach lies in the existence of a separating line that is *a priori* unknown and must be found as a level set of the temperature field, which, in general, is not a smooth function and may degenerate into a mushy region. This case was considered in [5], by Frémond, Ghidouche and Point, where a time dependent model of the cryogenic suction on the frost line was studied as a Darcy–Stefan type problem, in which the mechanical effects, stresses in the solid region and convection in the liquid one, were neglected. On the other hand, it is also known that physical instabilities can arise in this solidification process due to the buoyance-driven convection in the melted region, as a linear instability analysis work by McKay [8] has confirmed.



*P. Argoul et al. (eds.),*
*IUTAM Symposium on Variations of Domains and Free-Boundary Problems in Solid Mechanics,* 1–7.

In this work we consider a steady state problem, for which we propose a new weak formulation for a Stefan-type problem with convection governed by Darcy's law with the Boussinesq approximation for the buoyancy force. For this formulation, which has the advantage of not requiring the continuity of the temperature that restricted the preceding works [3] and [4] to two-dimensional particular cases, we shall give an existence result with general assumptions and, for small Péclet numbers, also an uniqueness theorem.

Our approach can be extended to the time dependent problem and to more general situations (see [10]).

## 2. Mathematical Formulations and Results

We assume the porous medium occupies a bounded domain $\Omega \subset \mathbb{R}^n$ ($n = 2, 3$), with a Lipschitzian boundary $\partial\Omega$, which is divided by an unknown free boundary $\Gamma$, where the dimensionless temperature $\vartheta = \vartheta(x)$ is zero, into two sets $\Omega_+ = \{x \in \Omega\colon \vartheta(x) > 0\}$ and $\Omega_- = \{x \in \Omega\colon \vartheta(x) < 0\}$, representing, respectively, the melted and the frozen regions.

Denoting by $\mathbf{v} = \mathbf{v}(x)$ and $\pi = \pi(x)$, respectively, the dimensionless velocity and hydraulic potential, the mathematical problem arising from the steady state convection–conduction and continuity equations and Darcy's law, consists in finding $\{\vartheta, \mathbf{v}, \pi, \Gamma\}$, such that ($\nabla\cdot = \mathrm{div}$):

$$-\Delta\vartheta + \lambda\mathbf{v}\cdot\nabla\vartheta = 0 \quad \text{in } \Omega\backslash\Gamma = \Omega_+ \cup \Omega_- , \tag{1}$$

$$\nabla\cdot\mathbf{v} = 0 \;\text{ and }\; \mathbf{v} = \nabla\pi + \mathbf{b}(\vartheta) \quad \text{in } \Omega_+ , \tag{2}$$

$$\mathbf{v} = 0 \quad \text{in } \Omega_- , \tag{3}$$

with the following boundary conditions

$$\vartheta = \vartheta_{\mathrm{D}}(x) \quad \text{on } \partial\Omega , \tag{4}$$

$$\mathbf{v}\cdot\mathbf{n} = 0 \quad \text{on } \partial\Omega_+ . \tag{5}$$

Here $\lambda > 0$ is a prescribed physical constant, essentially the Péclet number (see [7]) or the Grashoff number in the dimensionless model of [4], $\mathbf{b} = \mathbf{b}(\vartheta)$ is a buoyance force which, according to the Boussinesq approximation, can be a linear or nonlinear function of the temperature (see [8] and [4] and their references) and $\vartheta_{\mathrm{D}}$ is a prescribed temperature on $\partial\Omega$ that may take positive and negative values.

Since the densities of the frozen and unfrozen fluids are supposed to be the same, we consider there is no transfer of mass across $\Gamma$ and by (5) the free boundary stationary conditions for the temperature and for the jump of its normal derivative

$$\vartheta = [\nabla\vartheta\cdot\mathbf{n}]_-^+ = 0 \quad \text{on } \Gamma \tag{6}$$

are implicitly assumed (here $\mathbf{n}$ denotes the unit normal to $\Gamma$). That also implies, together with (2) and (3) that the velocity field $\mathbf{v}$ is divergent free in the sense of distributions $\mathcal{D}'(\Omega)$ in all $\Omega$. Then it is easily seen that (1), with (6)and (3) can be rewritten globally in the form

$$-\Delta\vartheta + \lambda\nabla\cdot(\vartheta\,\mathbf{v}) = 0 \quad \text{in } \mathcal{D}'(\Omega)\ . \tag{7}$$

On the other hand, the hydraulic potential from (2) and (5) should solve the Neumann problem

$$-\Delta\pi = \nabla\cdot\mathbf{b}(\vartheta) \ \text{ in } \Omega_+, \qquad -\partial\pi/\partial n = \mathbf{b}(\vartheta)\cdot\mathbf{n} \ \text{ on } \partial\Omega_+ \tag{8}$$

in the a priori unknown domain $\Omega_+ = \{\vartheta > 0\}$, which is the main problem and requires regularity properties on the temperature. Imposing the continuity of $\vartheta$ and using a natural approach of [2], where the velocity field is governed by Navier–Stokes equations (see also [9], where that was extended to a more general framework), a weak formulation was introduced in [3] (see also [4]). This did not take into account the boundary condition (8) on $\partial\Omega_+ \cap \Gamma$, and only yields mathematical results for two-dimensional problems.

The variational formulation we propose here, keeps the definition of the velocity field $\mathbf{v}$ verifying (2) and (3) and is based on solving the Neumann problem (8) in the whole domain $\Omega$, instead of only $\Omega_+$, and on the simple remark

$$\vartheta\,\mathbf{v} = \mathbf{v}\,\vartheta^+ = [\nabla\pi + \mathbf{b}(\vartheta)]\,\vartheta\chi_{\{\vartheta>0\}} \qquad \text{a.e. in } \Omega\ , \tag{9}$$

where $\vartheta^+ = \max(\vartheta, 0)$ is the positive part of the temperature, i.e. $\vartheta^+ = \vartheta\,\chi_{\{\vartheta>0\}}$, where $\chi_{\{\vartheta>0\}} = 1$ if $\vartheta(x) > 0$ and $\chi_{\{\vartheta>0\}} = 0$ if $\vartheta(x) \le 0$.

**Definition.** We say that the triple $\{\vartheta, \pi, \mathbf{v}\}$ is a weak solution of (1)–(5), if

$$\vartheta \in H^1(\Omega)\cap L^\infty(\Omega), \quad \vartheta = \vartheta_D \qquad \text{on } \partial\Omega\ , \tag{10}$$

$$\pi \in H^1(\Omega), \qquad \mathbf{v} = [\nabla\pi + \mathbf{b}(\vartheta)]\,\chi_{\{\vartheta>0\}} \ \in\ \mathbf{L}^2(\Omega)\ , \tag{11}$$

$$\int_\Omega [\nabla\vartheta - \lambda\,\vartheta\,\mathbf{v}]\cdot\nabla\varphi\,dx = 0, \qquad \forall\varphi \in H_0^1(\Omega)\ , \tag{12}$$

$$\int_\Omega [\nabla\pi + \mathbf{b}(\vartheta)]\cdot\nabla\psi\,dx = 0, \qquad \forall\psi \in H^1(\Omega)\ . \tag{13}$$

**Remark 1.** We do not require *a priori* the continuity of the temperature, as in [3] and [4], at the cost of obtaining in (13) an artificial hydraulic potential in the whole domain $\Omega$, although with the definition (11) only its values in the measurable subset $\{\vartheta > 0\}$ are relevant for the convection–conduction problem (12).

**Remark 2.** In the weak formulation of [3] the condition $\nabla \cdot \mathbf{v} = 0$ globally in $\Omega$ was demanded, $\pi$ was required to belong only to $H^1(\{\vartheta > 0\})$, since by the continuity of $\vartheta$ the subset $\{\vartheta > 0\}$ was open, and therefore the corresponding variational equation (13) was restricted only to the $\psi \in H^1(\Omega)$ such that $\operatorname{supp}\psi \subset \{\vartheta > 0\}$.

**Remark 3.** On the contrary, (13) with the definition (11) contains in a generalized sense the boundary condition (5), i.e., $\pi$ solves the Neumann problem (8) in a weak sense. Indeed, if the free boundary $\Gamma \equiv \{\vartheta = 0\} \cap \Omega = \partial\{\vartheta > 0\} \cap \Omega$ is regular (for instance, of class $C^1$) and $\nabla\vartheta$ and $\nabla\pi + \mathbf{b}(\vartheta)$ are continuous in a neighbourhood of $\Gamma$, its normal vector $\mathbf{n}_\Gamma$ is parallel to $\nabla\vartheta|_\Gamma$. Hence, for $\delta > 0$ sufficiently small, by defining

$$H_\delta(t) = 1 \quad \text{if } t \geq \delta, \quad H_\delta(t) = t/\delta \quad \text{if } 0 \leq t \leq \delta \quad \text{and} \quad H_\delta(t) = 0 \quad \text{if } t \leq 0\,,$$

and choosing $\psi = H_\delta(\varphi\vartheta)$ in (13), where $\varphi \in C^1(\overline{\Omega})$, $\varphi > 0$ in $\overline{\Omega}$, is arbitrary, we obtain

$$\int_{\{0<\varphi\vartheta<\delta\}} \left[\nabla\pi + \mathbf{b}(\vartheta)\right] \cdot \nabla(\varphi\vartheta)\, dx = 0\,.$$

Dividing by $\operatorname{meas}\{0 < \varphi\vartheta < \delta\}$ and letting $\delta \to 0$, we find

$$\int_{\{\vartheta=0\}\cap\Omega} \left[\nabla\pi + \mathbf{b}(\vartheta)\right] \cdot \nabla(\varphi\vartheta)\, d\Gamma = \int_\Gamma \varphi\left[\nabla\pi + \mathbf{b}(\vartheta)\right] \cdot \nabla\vartheta\, d\Gamma = 0 \qquad (14)$$

first for positive $\varphi$. Hence (14) holds also for any $\varphi \in C(\Gamma)$, and we may conclude that $[\nabla\pi + \mathbf{b}(\vartheta)] \cdot \mathbf{n}_\Gamma = 0$ everywhere in $\Gamma$.

**Remark 4.** Under the regularity conditions of the preceding remark, if we assume also $\mathbf{b}(\vartheta) = \mathbf{0}$ for all $\vartheta \leq 0$, it is clear that from (13), $\pi$ also solves the (homogeneous) Neumann problem in $\Omega_- = \{\vartheta < 0\}$ and consequently also $\nabla\pi = 0$ in $\Omega_-$, and our weak solution is not only a classical one, i.e., satisfies all the conditions (1)–(5), but is also a weak solution in the sense of [3] or [4].

**Theorem 1.** *Assume $\mathbf{b} \in C^0(\mathbb{R}; \mathbb{R}^n)$, i.e. $\mathbf{b}$ is continuous, and $\vartheta_D \in H^1(\Omega) \cap L^\infty(\Omega)$ are given functions. Then there exists at least a solution $\{\vartheta, \pi, \mathbf{v}\}$ of (10)–(13).*

**Theorem 2.** *Assume $\mathbf{b} \in C^{0,1}(\mathbb{R}; \mathbb{R}^n)$, $\vartheta_D \in W^{1,q}(\Omega)$ for some $q > n$ and $\partial\Omega$ is of class $C^{1,1}$. Then any solution of (10)–(13) has the regularity*

$$\{\vartheta, \pi, \mathbf{v}\} \in \left[W^{1,q}(\Omega) \cap C^{0,\alpha}(\overline{\Omega})\right] \times \left[W^{2,q}(\Omega) \cap C^{1,\alpha}(\overline{\Omega})\right] \times \mathbf{L}^\infty(\Omega)$$

*with* $\alpha = 1 - n/q > 0$. *Moreover, there exist a* $\lambda_* > 0$ *(*$\lambda_*$ *depends only on* $\Omega$, $\|\vartheta_D\|_{W^{1,q}(\Omega)}$ *and* $\|\mathbf{b}\|_{C^{0,1}}$*) such that the solution of (10)–(13) is unique provided* $|\lambda| < \lambda_*$.

**Notations.** We denote by $W^{m,p}(\Omega)$ and $C^{m-1,\alpha}(\overline{\Omega})$, $m = 1, 2$, $1 \leq p \leq \infty$, $0 < \alpha \leq 1$, respectively, the usual Sobolev and Hölder spaces ($\alpha = 1$ corresponding to the Lipschitz condition). Similarly we set $H^m(\Omega) = W^{m,2}(\Omega)$ and $W_0^{1,p}(\Omega)$ denotes the closure of the $C^\infty$-functions with compact support in $\Omega$ in $W^{1,p}(\Omega)$ ($W_0^{1,2}(\Omega) = H_0^1(\Omega)$) and $\mathbf{L}^q(\Omega) = [L^q(\Omega)]^n$, $1 \leq q \leq \infty$.

## 3. Proofs and Further Remarks

### 3.1. PROOF OF THEOREM 1

The proof of Theorem 1 can be done by elementary methods of elliptic equations (see [6]) and reduced to an application of the Schauder fixed point theorem in $L^2(\Omega)$. Set

$$\mathbf{K}_M = \left\{ \varphi \in L^2(\Omega) \colon \|\varphi\|_{L^\infty(\Omega)} \leq M \right\} ,$$

where we fix $M = \|\vartheta_D\|_{L^\infty(\Omega)}$. For each $\tau \in \mathbf{K}_M$, we can solve, up to constants, the linear Neumann problem

$$p_\tau \in H^1(\Omega) \colon \int_\Omega \left[ \nabla p_\tau + \mathbf{b}(\tau) \right] \cdot \nabla \psi \, dx = 0, \qquad \forall \psi \in H^1(\Omega) \,, \tag{15}$$

which satisfies the following estimate independently of $\tau \in \mathbf{K}_M$:

$$\|\nabla p_\tau\|^2_{\mathbf{L}^2(\Omega)} \leq \|\mathbf{b}(\tau)\|^2_{\mathbf{L}^2(\Omega)} \leq B_M = \max_{|t| \leq M} |\mathbf{b}(t)|^2 \int_\Omega 1 \,. \tag{16}$$

Defining $\mathbf{w}_\tau = \nabla p_\tau + \mathbf{b}(\tau) \in \mathbf{L}^2(\Omega)$ and letting $T_M(\omega) = \min(M, \max(\omega, 0))$, $\omega \in \mathbb{R}$, we solve now the quasi-linear elliptic problem

$$u - \vartheta_D \in H_0^1(\Omega) \colon \int_\Omega \left[ \nabla u - \lambda T_M(u)\, \mathbf{w}_\tau \right] \cdot \nabla \varphi \, dx = 0, \quad \forall \varphi \in H_0^1(\Omega) \,. \tag{17}$$

Since $0 \leq T_M(\omega) \leq M$ and $T_M$ is Lipschitz continuous there exists at most one solution to (17) (see [6]) and, by letting $\varphi = u - \vartheta_D$ in (17), the a priori estimate

$$\int_\Omega |\nabla u|^2 \leq 4 \int_\Omega |\nabla \vartheta_D|^2 + 4\lambda^2 M^2 \int_\Omega |\mathbf{w}_\tau|^2 \equiv C_M, \qquad \forall \tau \in \mathbf{K}_M \,, \tag{18}$$

assures easily the existence of the solution $u$. By letting $\varphi = (u - M)^+ \in H_0^1(\Omega)$ and $\varphi = (M - u)^+ \in H_0^1(\Omega)$ in (17) and remarking that from (15) we have

$$\int_\Omega T_M(u)\,\mathbf{w}_\tau\cdot\nabla(u-M)^+ = \int_{\{u>M\}} M\,\mathbf{w}_\tau\cdot\nabla u$$
$$= M\int_\Omega \mathbf{w}_\tau\cdot\nabla(u-M)^+ = 0$$

and since $T_M(u)=0$ if $u\le 0$,

$$\int_\Omega T_M(u)\,\mathbf{w}_\tau\cdot\nabla(M-u)^+ = -\int_{\{u<-M\}} T_M(u)\,\mathbf{w}_\tau\cdot\nabla u = 0\ ,$$

we conclude, respectively, that $u\le M$ and $u\ge -M$ a.e. in $\Omega$.

Consequently, the unique solution $u$ to (17) is in $\mathbf{K}_M$ and therefore $T_M(u)=u^+$. By the continuous dependence result of the elliptic problems and the compactness of the imbedding $H^1(\Omega)\subset L^2(\Omega)$, we have defined a completely continuous mapping $\mathcal{S}:\ \tau\mapsto\mathbf{w}_\tau\mapsto u=\mathcal{S}(\tau)$, satisfying $\mathcal{S}(\mathbf{K}_M)\subset\mathbf{K}_M$. The Schauder fixed point $\vartheta=\mathcal{S}(\vartheta)$, with $\pi=p_\vartheta$ and $\mathbf{v}=\chi_{\{\vartheta>0\}}\,\mathbf{w}_\vartheta$ clearly satisfy (10)–(13) and yields the conclusion of Theorem 1.

**Remark 5.** In fact, Theorem 1 still holds without any regularity assumption on $\partial\Omega$, i.e., for any open bounded subset $\Omega\subset\mathbb{R}^n$. It is sufficient to rewrite the Dirichlet boundary condition (10) in the form $\vartheta-\vartheta_D\in H^1_0(\Omega)$, recalling that the compactness of $H^1_0(\Omega)\subset L^2(\Omega)$ still holds and repeating the proof for $\widetilde{\vartheta}=\vartheta-\vartheta_D$ with the convex subset $\widetilde{\mathbf{K}}_M=\vartheta_D+\mathbf{K}_M\subset L^2(\Omega)$.

**Remark 6.** The preceding existence result can also be extended to more general cases like in [3], with a non-homogeneous right-hand side $f\in W^{-1,q}(\Omega)$, $q>n$, in (12) and $\mathbf{b}=\mathbf{b}(x,\vartheta)$, such that $\sup_{|\tau|\le M}|\mathbf{b}(\cdot,\tau)|\in L^2(\Omega)$. We can also consider a mixed boundary condition for the temperature $\vartheta$, as in [9], provided it preserves the weak maximum principle and the $L^\infty$-estimate.

### 3.2. PROOF OF THEOREM 2

The proof of Theorem 2 is a consequence of the $L^q$-estimates for elliptic boundary value problems. Indeed, from the regularity of the Neumann problem when $\partial\Omega$ is at least of class $C^1$ (see [12]), since $\mathbf{b}(\vartheta)\in\mathbf{L}^\infty(\Omega)$, from (13) we have $\nabla\pi\in\bigcap_{1<p<\infty}\mathbf{L}^p(\Omega)$ and, by (11), also $\mathbf{v}\in\bigcap_{1<p<\infty}\mathbf{L}^p(\Omega)$. Then, by the $L^q$-regularity for the Dirichlet problem (see [11]) from (10) and (12) it follows $\vartheta\in W^{1,q}(\Omega)\cap C^{0,\alpha}(\overline{\Omega})$ with $\alpha=1-n/q>0$. Since $\mathbf{b}$ is Lipschitz continuous, from (13) we have that $-\Delta\pi=\mathbf{b}'(\vartheta)\cdot\nabla\vartheta\in\mathbf{L}^q(\Omega)$ and by the regularity of the Neumann problem it follows $\pi\in W^{2,q}(\Omega)\cap C^{1,\alpha}(\overline{\Omega})$, and also $\mathbf{v}\in\mathbf{L}^\infty(\Omega)$.

In order to show the uniqueness consider two solutions $\{\vartheta_i, \pi_i, \mathbf{v}_i\}$, $i = 1, 2$, and denote by $\overline{\vartheta} = \vartheta_1 - \vartheta_2$ and $\overline{\pi} = \pi_1 - \pi_2$. Let $B' = \sup_{|\tau|\leq M} |\mathbf{b}'(\tau)|$ and taking $\psi = \overline{\pi}$ in (13) we obtain first

$$\|\nabla\overline{\pi}\|_{\mathbf{L}^2(\Omega)} \leq B' \, \|\overline{\vartheta}\|_{L^2(\Omega)} \ . \tag{19}$$

Setting $\mathbf{f}_i = \vartheta_i \, \mathbf{v}_i = \vartheta_i^+[\nabla\pi_i + \mathbf{b}(\vartheta_i)]$ and taking $\varphi = \overline{\vartheta}$ in (12) we obtain

$$\|\nabla\overline{\vartheta}\|_{\mathbf{L}^2(\Omega)} \leq |\lambda| \, \|\mathbf{f}_1 - \mathbf{f}_2\|_{\mathbf{L}^2(\Omega)} \ . \tag{20}$$

Denoting $M_1 = \|\nabla\pi_1 + \mathbf{b}(\vartheta_1)\|_{\mathbf{L}^\infty(\Omega)}$ and by $C_0$ the Poincaré constant (i.e., $\|\varphi\|_{L^2(\Omega)} \leq C_0\|\nabla\varphi\|_{\mathbf{L}^2(\Omega)}$, $\forall\, \varphi \in H_0^1(\Omega)$), the right-hand side of (20) can be estimated using (19), with the help of $\pm\vartheta_2^+[\nabla\pi_1 + \mathbf{b}(\vartheta_1)]$, by

$$\begin{aligned} \|\mathbf{f}_1 - \mathbf{f}_2\|_{\mathbf{L}^2(\Omega)} &\leq M_1\|\vartheta_1^+ - \vartheta_2^+\|_{L^2(\Omega)} + 2MB'\|\overline{\vartheta}\|_{L^2(\Omega)} \\ &\leq C_0(M_1 + 2MB') \, \|\nabla\overline{\vartheta}\|_{\mathbf{L}^2(\Omega)} \ . \end{aligned}$$

Hence, from (20), the uniqueness follows for $|\lambda| < [C_0(M_1 + 2MB')]^{-1}$.

**Acknowledgement.** Partially supported by FCT (JNICT), Praxis XXI and Praxis /2/2.1/MAT/125/94 projects.

## References

1. Bear, J. (1972) *Dynamics of Fluid in Porous Media.* Elsevier, New York.
2. Cannon, J.R., DiBenedetto, E. and Knightly, G.H. (1980) The steady state Stefan problem with convection, *Arch. Rational Mech. Anal.*, **73**, pp. 79–97.
3. DiBenedetto, E. and Elliott, C.M. (1985) Existence for a problem in ground freezing, *Nonlinear Analysis, Th. Meth. Appl.*, **9**, pp. 953–967.
4. DiBenedetto, E. and Friedman, A. (1986) Conduction–convection problems with change of phase, *J. Diff. Eqs.*, **62**, pp. 129–185.
5. Frémond, M., Ghidouche, H. and Point, N. (1985) Freezing a porous medium with water supply coupled Stefan problem, *J. Math. Anal. Appl.*, **108**, pp. 371–402.
6. Gilbarg, D. and Trudinger, N.S. (1983) *Elliptic Partial Differential Equations of Second Order.* Springer-Verlag, 2nd ed..
7. Goldstein, M.E. and Reid, R.L. (1978) Effect of fluid flow on freezing and thawing of saturated porous media, *Proc. R. Soc. London*, **A364**, pp. 45–73.
8. McKay, G. (1996) Patterned ground formation and convection in porous media with phase change, *Continuum Mech. Thermodyn.*, **8**, pp. 189–199.
9. Rodrigues, J.F. and Urbano, J.M. (1997) On the stationary Boussinesq–Stefan problem with constitutive power-laws, *Int. J. Non-Linear Mechanics*, **29**.
10. Rodrigues, J.F. and Urbano, J.M. (to appear) *On the Darcy–Stefan problem.*
11. Simader, C.G. (1972) On Dirichlet's boundary value problem, *Lect. Notes in Math.*, **268**, Springer-Verlag, Heidelberg.
12. Simader, C.G. and Sohr, H. (1992) *A new approach to the Helmholtz decomposition and the Neumann problem in $L^q$-spaces for bounded and exterior domains*, in pp. 1–35 of "Mathematical Problems Relating to the Navier–Stokes Equations", G.P. Galdi (Editor). World Scientific, Singapure.

# EVOLUTION OF PHASES IN A SHAPE MEMORY ALLOY BY LOCAL AND GLOBAL MINIMISATIONS OF A NON CONVEX POTENTIALS

O. MAISONNEUVE, S. PAGANO, P. ALART AND F. LEBON
*LMGC Université de Montpellier II - CNRS*
*UMR 5508 Place Eugène Bataillon*
*34095 Montpellier Cedex 05*

## 1. Introduction

A Shape Memory Alloy (SMA) is a mixture of martensite and austenite or only martensite or austenite. Austenite-martensite phase transformation leads to variations of composition. This transformation can appear by mechanical or thermal action. The characteristic pseudo-elastic behaviour appears from this modification in the material over certains conditions for which observe an hysteresis loop on the stress-strain relationship. This hysteresis can be influenced by several factors : thermal effect, dissipation, structural effect, existence of geometrical defect, etc..

Here, we use a model taking the temperature, the deformation and the volumic proportions of austenite and martensite as state variables. We adopt, here, Fremond's modelling in the non dissipative case which gives a non monotonic stress-strain relationship. Some experiments (Chrysochoos *et al.*, 1995) show that during room temperature test, dissipative effects are strongly dominated by effects resulting from temperature variations of material associated to the latent heat. In our case, after elimination of the volumic proportions of martensite, the bulk energy density is a non convex function.

For a given problem, it is classical to look for the mechanical equilibrium by a global minimization of the "total" energy (potential energy). Moreover, the energy associated at each problem is not quasiconvex. Existence of absolute minima is not garanteed in all the case. It is possible to pose displacement boundary-value problems for which the total energy has no minimizer and then no absolute minimum ; the minimizing sequences are produced by displacements with fine-scale irregularities. Consequently, we



*P. Argoul et al. (eds.),*
*IUTAM Symposium on Variations of Domains and Free-Boundary Problems in Solid Mechanics*, 9–16.

introduce two relaxation procedures (convexification and quasiconvexification) for which existence of minimizer is garanteed in order to compare in the three cases, displacement solutions and evolutions of phases. The algorithm to find global minimum is based on an augmented Lagrangian method (decomposition-coordination method).

From a mechanical point of view, there is no reason to exclude local minima as possible equilibrium states. So, the evolution of phases may be different from the loading process to the unloading one. In order to fine local minima, we write the non convex potential as a difference of two convex functions (D.C.). Then we can formulate a problem of two variational inequalities in term of displacements and stresses. In the finite dimensional case, this problem is equivalent to search solutions which locally minimize the "total" energy. By applying partial convex duality, an inf-inf problem can be formulated with respect to an appropriately defined Lagrangian function.

We present a theoretical and numerical comparison between these two points of view in focusing out the evolution of phases.

## 2. Constitutive law

Here, we adopt the Fremond modelling (Fremond, 1987) with only two variants of martensite, and we denote by $\beta_1$ and $\beta_2$ their volumic proportions. We assume that the volumic free energies of the phases (respectively the austenitic one and the two martensitic ones) have the following form involving the small strain tensor $\mathbf{e}$,

$$W_0(\mathbf{e},T) = \tfrac{1}{2}\mathbf{eKe} - c(T) - C_0 T log T$$
$$W_-(\mathbf{e},T) = W_0(\mathbf{e},T) + \alpha(T)\tau : \mathbf{e} + c(T)$$
$$W_+(\mathbf{e},T) = W_0(\mathbf{e},T) - \alpha(T)\tau : \mathbf{e} + c(T)$$

with $\alpha(T) = \begin{cases} -a(T-T_c) & if T \leq T_c \\ 0 & otherwise \end{cases}$ and $c(T) = \frac{L}{T_0}(T - T_0)$

The phases are assumed to have the same elastic tensor $\mathbf{K}$, and the same heat capacity C. The austenite-martensite phase transformation is associated with a residual stress tensor $\alpha\tau$ which is be chosen identically for the two variants. L is the latent heat of phase transformation. Fremond writes the total volumic free energy $\Psi$ as a convex combination of the three energies: the volumic proportions are constrained to stay in the simplex C via the addition of the indicator function of this simplex :

$$\begin{aligned} \Psi(\mathbf{e},T,\beta_1,\beta_2) &= \beta_1 W_-(\mathbf{e},T) + \beta_2 W_+(\mathbf{e},T) \\ &+ (1-\beta_1-\beta_2)W_0(\mathbf{e},T) + T I_C(\beta_1,\beta_2) \end{aligned}$$

where $I_C(\beta_1,\beta_2)$ is the indicator function of the convex set :

$$\mathcal{C} = \{(\gamma_1, \gamma_2) \quad | \quad 0 \leq \gamma_i \leq 1 \quad ; \quad \gamma_1 + \gamma_2 \leq 1\}.$$

We introduce the formalism of generalized standard materials and we neglect dissipation (Chrysochoos *et al.*, 1995): we can eliminate the internal variables $(\beta_1, \beta_2)$. Then, a new potential can be defined as the minimum of three convex functions. Note that this potential is not convex:

$$\begin{aligned} W(\mathbf{e}, T) &:= \min\{\Psi(\mathbf{e}, T, \beta_1, \beta_2), (\beta_1, \beta_2) \in C\} \\ &= \min\{W_0(\mathbf{e}, T), W_-(\mathbf{e}, T), W_+(\mathbf{e}, T)\} \end{aligned}$$

With the three-well energy W, which does not depend on $\beta_1, \beta_2$, the material constitutive law is now written :

$$\sigma \in \bar{\partial}_{\mathbf{e}} W(\mathbf{e}(\mathbf{u}), T) \tag{1}$$

where $\bar{\partial}$ denotes the subdifferential operator of Clarke.

We consider an extension test and we represent the energy W at different temperatures in Figure 2 : low temperature $T \leq T_0$, intermediate temperature $T_0 \leq T \leq T_c$ and high temperature $T \geq T_c$.

*Figure 1.* Energy W for $T \leq T_0$, $T_0 \leq T \leq T_c$ and $T \geq T_c$

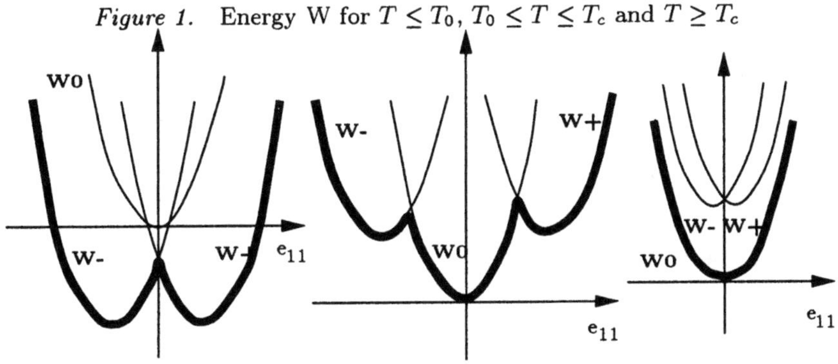


With the constitutive law (1), we etablish a classical formulation of the mechanical problem.

## 3. Mechanical problem at fixed temperature

Consider a boundary problem on a domain $\Omega$ of boundary $\Gamma$. Let $\mathbf{f}$ denote the density of body force applied on $\Omega$, $\mathbf{F}$ the density of surfacic force applied on the part $\Gamma_0$ of $\Gamma$, and $\mathbf{U}$ a displacement given on the complementary part $\Gamma_1$.

The mechanical problem at fixed temperature consists in finding the displacement field $\mathbf{u}$ and the stress field $\sigma$ verifying the following equations :

$$\begin{aligned} \mathbf{div}\sigma + \mathbf{f} = 0 &\quad \text{in } \Omega \\ \mathbf{e}(\mathbf{u}) = \nabla_{\mathbf{s}}\mathbf{u} &\quad \text{in } \Omega \\ \textit{Constitutive law}: \sigma \in \bar{\partial}_{\mathbf{e}} W(\mathbf{e}(\mathbf{u})) &\quad \text{in } \Omega \\ \mathbf{u} = \mathbf{U} &\quad \text{on } \Gamma_0 \\ \sigma.\mathbf{n} = \mathbf{F} &\quad \text{on } \Gamma_1 \end{aligned}$$

In the following, the set of kinematically admissible displacement fields with finite energy of the structure will be defined by :
$\mathbf{U}_{ad} = \left\{\mathbf{v} \in \left[\mathrm{H}^1(\Omega)\right]^3, \mathbf{v} = \mathbf{U} \text{ on } \Gamma_0\right\}$ and the stress spaces $\mathbf{H} = \left[\mathrm{L}^2(\Omega)\right]^6$.

## 4. Global minimum

### 4.1. THE THREE-WELL ENERGY

At first, we search for a global minimum of the three-well potential energy. To solve this problem, we consider an augmented Lagrangian. It is a function of a multiplier $\mu$ and of two independent variables $\mathbf{v}$,$\mathbf{q}$ corresponding to the displacement and strain fields.

$$\mathrm{L_r}(\mathbf{v}, \mathbf{q}, \mu) = \int_\Omega W(\mathbf{q})\, dx - l(\mathbf{v}) + \int_\Omega \mu : (\mathbf{e}(\mathbf{v}) - \mathbf{q})\, dx + \frac{r}{2}\int_\Omega ((\mathbf{e}(\mathbf{v}) - \mathbf{q}))^2\, dx$$

with $l(\mathbf{v}) = \int_\Omega \mathbf{f}.\mathbf{v}\, d\Omega + \int_{\Gamma_1} \mathbf{F}.\mathbf{v}\, d\Gamma$ and $r \geq 0$.

We extend this non-convex density energy, the classical strategy of searching for the saddle points of the augmented Lagrangian i.e. the $(\mathbf{u}, \mathbf{p}, \lambda)$ solutions of the problem :

$$\mathrm{L_r}(\mathbf{u}, \mathbf{p}, \mu) \leq \mathrm{L_r}(\mathbf{u}, \mathbf{p}, \lambda) \leq \mathrm{L_r}(\mathbf{v}, \mathbf{q}, \lambda), \forall v \in \mathbf{U}_{ad}, \forall \mathbf{q} \in \mathbf{H}, \forall \mu \in \mathbf{H}$$

To solve the initial problem, we use an algorithm inspired of by Uzawa method (LeTallec *et al.*, 1989) which involves a two-step iteration : a linear global problem and a non linear local problem.

### 4.2. THE "RELAXED" ENERGY

To profit by an existence theorem, we calculate the quasiconvexification QW of the previous potential by using the approach of Pipkin (Pipkin, 1990), and the convexification CW of W. To determine the quasiconvexification, we take a unit cube B, we use the definition of QW (Pipkin, 1990)

$$QW(\mathbf{e}) := \liminf_{\mathbf{u}_N \to 0 \text{ uniformly}} \int_B W(\mathbf{e} + \nabla_s \mathbf{u}_N)\, dx, \tag{2}$$

and consider a sequence of displacement fields $\mathbf{u}_N$ for which the strain varies rapidly and periodically between two alternate constant values $e_1$ and $e_2$. The details of the calculus are given in the paper (Pagano *et al.*, to appear). $\mathbf{u}_N$ is a periodic function of the variable $\mathbf{n.x}$, of period $\frac{1}{N}$.

Pipkin defines this sequence on one period $[0, \frac{1}{N}]$ by

$$\mathbf{u}_N := \begin{cases} 2\theta(\mathbf{n.x})\mathbf{u} & \text{on}[0, \frac{1-\theta}{N}] \\ -2(1-\theta)(\mathbf{n.x} - \frac{1}{N})\mathbf{u} & \text{on}[\frac{1-\theta}{N}, \frac{1}{N}] \end{cases}$$

and $\mathbf{e} + \nabla_s \mathbf{u}_N = \begin{cases} e_1 = e + \theta(\mathbf{u}\otimes\mathbf{n} + \mathbf{n}\otimes\mathbf{u}) \\ e_2 = e + (1-\theta)(\mathbf{u}\otimes\mathbf{n} + \mathbf{n}\otimes\mathbf{u}) \end{cases}$

where $\mathbf{u}$ and $\mathbf{n}$ are two constant vectors and $0 < \theta < 1$. We can prove that the density energy QW is :

$$QW(\mathbf{e}) = \min_{\mathbf{u},\mathbf{n},0\leq\theta\leq 1} [(1-\theta)W(\mathbf{e}_1) + \theta W(\mathbf{e}_2)]. \tag{3}$$

This minimization can be performed by discrimination of all the possible cases for $W(\mathbf{e}_1)$ and $W(\mathbf{e}_2)$. We obtain a similar form, noted RW, for the convexification and the quasiconvexification of the density energy W. for $c + \frac{\mathcal{G}}{2} \geq 0$ :

$$RW(\mathbf{e}) = \begin{cases} W_-(\mathbf{e}) & \text{if } \alpha\tau:\mathbf{e} \leq -c + \frac{\mathcal{G}}{2} \\ W_0(\mathbf{e}) + \frac{1}{2\mathcal{G}}(\alpha\tau:\mathbf{e} + c + \frac{\mathcal{G}}{2})^2 & \text{if } -c + \frac{\mathcal{G}}{2} \leq \alpha\tau:\mathbf{e} \leq -c - \frac{\mathcal{G}}{2} \\ W_0(\mathbf{e}) & \text{if } \alpha|\tau:\mathbf{e}| \leq c + \frac{\mathcal{G}}{2} \\ W_0(\mathbf{e}) + \frac{1}{2\mathcal{G}}(\alpha\tau:\mathbf{e} - c - \frac{\mathcal{G}}{2})^2 & \text{if } c + \frac{\mathcal{G}}{2} \leq \alpha\tau:\mathbf{e} \leq c - \frac{\mathcal{G}}{2} \\ W_+(\mathbf{e}) & \text{if } \alpha\tau:\mathbf{e} \geq c - \frac{\mathcal{G}}{2} \end{cases} \tag{4}$$

The functional RW is equal to the convexification of W if $\mathcal{G} = \mathcal{G}_c$ and to the quasiconvexification for $\mathcal{G} = \mathcal{G}_q$ with :

$$\mathcal{G}_c = 2\left[\min_{\mathbf{E}\in M_s^{3x3}} W_-(\mathbf{E}) - c\right] = -\alpha^2\tau:\mathbf{K}^{-1}\tau$$

$$\text{and} \quad \mathcal{G}_q = 2\left[\min_{\mathbf{u},\mathbf{n}} W_-(\mathbf{u}\otimes\mathbf{n} + \mathbf{n}\otimes\mathbf{u}) - c\right]$$

In particular, $\mathcal{G}_q = \mathcal{G}_c$ if $\tau$ has the form $\mathbf{a}\otimes\mathbf{b} + \mathbf{b}\otimes\mathbf{a}$ where $\mathbf{a}$ and $\mathbf{b}$ are two vectors. If we restrict our attention to the case of isotropic Hooke's law with Lamé coefficients $\lambda$ and $\mu$, and if we suppose that $\tau = \mathbf{I}$, we can calculate the coefficients $\mathcal{G}_q$ and $\mathcal{G}_c$ to obtain the following results :

$$\mathcal{G}_c = \frac{-\alpha^2}{2\lambda + \mu} \quad \text{and} \quad \mathcal{G}_q = \frac{-\alpha^2}{\lambda + 2\mu}$$

The optimal values of $\theta$ which minimize (3), noted $\theta_-$ and $\theta_+$ may be interpreted as two internal variables which represent the "proportions of martensite" of the mixture. These values are the following :

$$(\theta_-,\theta_+) = \begin{cases} (1,0) & \text{if } \alpha\tau : \mathbf{e} \le -c + \frac{\mathcal{G}}{2} \\ (\frac{\alpha\tau : \mathbf{e} + c + \frac{\mathcal{G}}{2}}{\mathcal{G}}, 0) & \text{if } -c + \frac{\mathcal{G}}{2} \le \alpha\tau : \mathbf{e} \le -c - \frac{\mathcal{G}}{2} \\ (0,0) & \text{if } \alpha|\tau : \mathbf{e}| \le c + \frac{\mathcal{G}}{2} \\ (0, \frac{-\alpha\tau : \mathbf{e} + c + \frac{\mathcal{G}}{2}}{\mathcal{G}}) & \text{if } c + \frac{\mathcal{G}}{2} \le \alpha\tau : \mathbf{e} \le c - \frac{\mathcal{G}}{2} \\ (0,1) & \text{if } \alpha\tau : \mathbf{e} \ge c - \frac{\mathcal{G}}{2} \end{cases} \tag{5}$$

For the case $c + \frac{\mathcal{G}}{2} \le 0$ we obtain, also, similar expressions for QW and CW (analogous to the Pipkin's result for a two-well energy). We see that QW and CW are piecewise quadratic and $C^1$. From a well known result of Dacarogna, any solution of this relaxed problem is a limit of a minimizing sequence of the potential associated to the initial problem (Dacorogna, 1985). To solve the relaxed problem, we use the same method as for the initial problem.

## 5. Local minimum

The potential W, at fixed temperature, can appear as the difference of two convex functions (d.c.) (Pagano *et al.*, to appear) $\varphi_1$ and $\varphi_2$ :

$$\varphi_1(\mathbf{e}) = W_0(\mathbf{e}) \quad \text{and} \quad \varphi_2(\mathbf{e}) = \max(\alpha|tr\mathbf{e}| - c)$$

Based on this decomposition, we formulate the mechanical problem as two variational inequalities. In the finite dimensional case, this problem is equivalent to finding local minima of the potential energy. The decomposition leads to the introduction of a Lagrangian $L_{II}$, in the terminology of Auchmuty (1989). It depends on the displacement field corresponding to a stress field, and it is function of the energy $\Phi_1$ and the Fenchel convex conjugate function $\Phi_2^*$. This Lagrangian is separately convex in each of the dual variables, but the convexity of $L_{II}$ is not guaranteed in general. An algorithm is derived to reach *∂-critical points* of the Lagrangians $L_{II}$. These points are defined as follows :

$(\mathbf{u}, \sigma_2) \in \mathbf{U}_{ad}\mathrm{x}\mathbf{H}$ is a *∂-critical point* of $L_{II}$ if :

$$\begin{aligned} L_{II}(\mathbf{u}, \sigma_2) &\le L_{II}(\mathbf{v}, \sigma_2), \quad \forall \mathbf{v} \in \mathbf{U}_{ad} \\ L_{II}(\mathbf{u}, \sigma_2) &\le L_{II}(\mathbf{u}, \tau), \quad \forall \tau \in \mathbf{H} \end{aligned}$$

The relationship between critical points of the potential energy and those of $L_{II}$ is somewhat more delicate. If $(\mathbf{u}, \sigma_2) \in \mathbf{U}_{ad}\mathrm{x}\mathbf{H}$ is a ∂-critical point of the Lagrangian then $\mathbf{u}$ is a critical point of the potential energy.

This convex difference algorithm constructs minimizing sequence of $L_{II}$. It permits us to characterize a $\partial$-critical point of $L_{II}$. In practice, it determines local minima of the Lagrangian. It is also split into two steps, (global and local steps).

We remark that the "relaxed" energy can, also, be split into the difference of two convex functions $\varphi_1$ and $R\varphi_2$. The same numerical method can be applied to the resolution of the relaxed problem.

$$(R\varphi_2) = \begin{cases} -\alpha\tau : \mathbf{e} - c & \text{if } \alpha\tau : \mathbf{e} \leq -c + \frac{\mathcal{G}}{2} \\ \frac{-1}{2\mathcal{G}}(\alpha\tau : \mathbf{e} + c + \frac{\mathcal{G}}{2})^2 & \text{if } -c + \frac{\mathcal{G}}{2} \leq \alpha\tau : \mathbf{e} \leq -c - \frac{\mathcal{G}}{2} \\ 0 & \text{if } \alpha|\tau : \mathbf{e}| \leq c + \frac{\mathcal{G}}{2} \\ \frac{-1}{2\mathcal{G}}(\alpha\tau : \mathbf{e} - c - \frac{\mathcal{G}}{2})^2 & \text{if } c + \frac{\mathcal{G}}{2} \leq \alpha\tau : \mathbf{e} \leq c - \frac{\mathcal{G}}{2} \\ \alpha\tau : \mathbf{e} - c & \text{if } \alpha\tau : \mathbf{e} \geq c - \frac{\mathcal{G}}{2} \end{cases} \tag{6}$$

## 6. Numerical simulations at fixed temperature

In this part, we compare the three points of view. We assume that $\tau = \mathbf{I}$ and we consider a flat sample with a geometrical defect submitted to a strain controlled loading-unloading process. For symmetry reasons, only a quarter of the sample is represented [Figure 3].

*Figure 2.* Flat sample

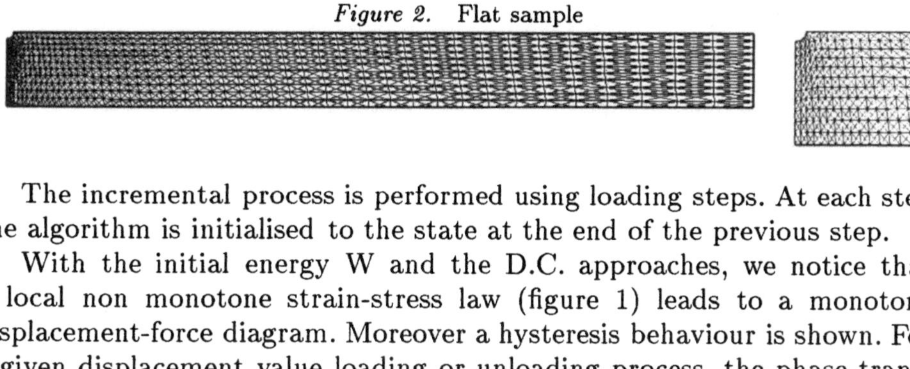

The incremental process is performed using loading steps. At each step the algorithm is initialised to the state at the end of the previous step.

With the initial energy W and the D.C. approaches, we notice that a local non monotone strain-stress law (figure 1) leads to a monotone displacement-force diagram. Moreover a hysteresis behaviour is shown. For a given displacement value loading or unloading process, the phase transformation in the flat is not the same. The first picture in figure 4 represents the binary phases distribution in the flat. According to the chosen modelling, the proportions are either zero or one almost everywhere. Roughly speaking the phase transition is locally instantaneous.

With the global minimization, (quasiconvexified or not), the hysteresis vanishes. The distribution of phases, issuing from the minimisation of the quasiconvexification, appears as the "homogenization" of the phase distribution obtained for the non convex potential W. Some differences are obtained about the spatial repartition of phases for the quasiconvexification (figure 4-2) and the convexification (figure 4-3). With CW, the zone where the volumic proportion of martensite lies between zero and one is more important than for relaxed energy QW.

**Aknowledgement :** We thank Doctor Licht for his very interesting and helpful comments.

## References


Pipkin, A.C. (1990) Elastic material with two preferred states, *Q. J. Mech. appl. Math.*Vol. no **3**,pp. 1–15

Pagano, S., Alart, P., Lebon, F. and Maisonneuve, O. (to appear) Evolution of phases in a shape memory alloys by local and global minimisations of a non convex potential

Fremond, M. (1987) Matériaux à mémoire de forme, *R. Acad. Sci. Paris*Vol. no **304**,pp. 239–244

LeTallec, P. and Vidrascu, M. (1989) Une méthode numérique pour les problèmes d'équilibre de corps hyperélastiques incompressibles en grandes déformations, *Numer. Math.*Vol. no **43**,pp. 199–224

Dacorogna, F. (1989) Direct methods in the Calculus of Variations, *Berlin : Springer-verlag*

Auchmuty, G. (1989) Duality algorithms for nonconvex variational principles, *Num. Func. An. & Opt.*Vol. no **10**,pp. 211–264

Chrysochoos, A., Pham, H., and Maisonneuve, O. (1995) Couplages thermomécanique du comportement d'alliages Cu-Zn-Al et Ni-Ti, *C. R. Acad. Sci. Paris*Vol. no **320**,pp. 217–223


*Figure 3.* Results with energies W(4-1), CW(4-2) and QW(4-3)

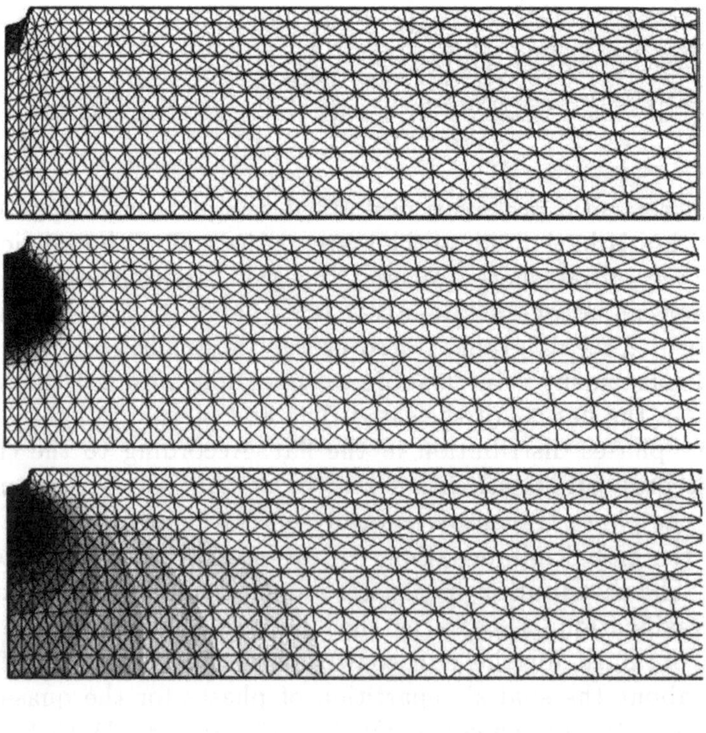

# MICROMECHANICAL MODELLING OF THE THERMOMECHANICAL BEHAVIOR OF SHAPE MEMORY ALLOYS

E. PATOOR, M. BERVEILLER
*Laboratoire de Physique et Mécanique des Matériaux (UMR CNRS 7554)*
*Institut Supérieur de Génie Mécanique et Productique*
*ENSAM, 4 rue Augustin Fresnel, 57070 METZ, France.*

**Abstract.**

This paper uses modern scale transition methods to establish the thermomechanical behavior of shape memory alloys. In section 2, kinematics, kinetics and thermodynamical aspects related to the martensitic transformation are described from a continuum point of view using moving boundary concepts. Relations so obtained are applied in section 3 using a discrete internal variables description. Polycrystalline behavior so compute shows excellent agreement with experimental measurements.

## 1. Introduction

Phase transition problems deal with complex microstructure at different length scales. During the martensitic transformation several interfaces are produced inside the material (between the parent and product phases and between the different variants of martensite). In this work, the transformation strain related to the phase change is considered as an eigenstrain and jumps of this transformation strain field across these interfaces are accounted. Moving boundaries concepts allow one to characterize the two inelastic strain mechanisms observed in shape memory alloys (SMA) and to define the driving forces associated with interface motions. Relations so obtained are applied by using a discrete internal variables description. Two characteristic length scales are considered, the granular structure responsible for the intergranular internal stress field and the martensite multivariant pattern inside the grain. Two models are then developed. The first one derives the constitutive equations governing the behavior of a grain inside a polycrystalline material. Each variant of martensite is represented by its volume fraction and many internal variables are introduced. The second model deals with the intergranular interaction. A scale transition from the grain level to the polycrystalline material is performed using a self consistent approximation. This approach is used to compute the overall behavior of a polycrystalline copper-based SMA.

## 2. Continuous Description

Progress of the transformation occurs by nucleation of new domains, growing of existing martensitic domains by interface motion or (and) exchange between variants. Considering a austenite-martensite Representative Volume Element (RVE) moving



*P. Argoul et al. (eds.),*
*IUTAM Symposium on Variations of Domains and Free-Boundary Problems in Solid Mechanics*, 17–24.

boundaries framework seems well adapted to describe such mechanisms. Definition of the driving forces involved requires the calculation of the free energy and the dissipation associated to the transformation.

## 2.1. KINEMATICS AND KINETICS OF MARTENSITIC PHASE TRANSITION

Martensitic phase transformation is a diffusionless first order phase transition [1]. Chemical composition remains constant but atoms are organized in a new crystallographical lattice. An inelastic lattice strain called Bain strain describes this geometrical transformation. Since this strain field is usually strongly incompatible, a lattice invariant strain occurs to relax (at least partially) high internal stresses produced by these incompatibilities. The transformation strain $\varepsilon^T$ related to a volume element is composed of these two strains. This transformation strain is assumed to be a constant well defined quantity and considered as a stress free or inelastic strain in the sense of Eshelby [2] or Kröner [3].

During the transformation of some part of a RVE of volume V, the total deformation field $\varepsilon(r)$ observed inside the material results from elastic, thermal and transformation contributions. Considering adequate boundaries conditions the overall deformation E is equal to the mean value over volume V of this total deformation field. In the small strain framework with homogeneous elastic compliances and thermal expansion coefficients, the global transformation strain $E^T$ over the RVE is simply defined as:

$$E_{ij}^T = \frac{1}{V}\int_V \varepsilon_{ij}^T(r)\,dV \tag{1}$$

In addition, transformation field $\varepsilon^T(r)$ is assumed to be piecewise uniform having a well-defined value, denoted by $\varepsilon^{TI}$ inside each variant of martensite of volume $V_I$.

$$\varepsilon_{ij}^T(r) = \sum_I \varepsilon_{ij}^{TI}\theta^I(r) \quad \text{with} \quad \theta^I(r) = \begin{cases} 1 & \text{if } r \in V_I \\ 0 & \text{if } r \notin V_I \end{cases} \tag{2}$$

From Eqns. (1) and (2), one gets:

$$E_{ij}^T = \sum_I f^I \varepsilon_{ij}^{TI} \tag{3}$$

where parameter $f^I$ (= $V_I/V$) represents the volume fraction of a variant I. Progress of the transformation involves motion of interfaces. In the small strain approximation, the time derivative of Eqn. (1) is given by:

$$\frac{dE_{ij}^T}{dt} = \frac{1}{V}\int_V \frac{\partial\varepsilon_{ij}^T}{\partial t}(r,t)\,dV - \frac{1}{V}\int_\Sigma \left[\varepsilon_{ij}^T\right] w_\alpha\, n_\alpha\, d\Sigma \tag{4}$$

Notation [x] means the jump of a quantity x across an interface $\Sigma$ having a normal velocity $w_\alpha\, n_\alpha$. The internal interface $\Sigma$ is composed with austenite-martensite interfaces $\Sigma_{AM}$ and martensite-martensite interfaces $\Sigma_{MM}$, so that Eqn. (4) becomes:

$$\frac{dE^T_{ij}}{dt} = \frac{1}{V}\int_{\Sigma_{AM}} \varepsilon^T_{ij} w_\alpha n_\alpha d\Sigma - \frac{1}{V}\int_{\Sigma_{MM}} \left[\varepsilon^T_{ij}\right] w_\alpha n_\alpha d\Sigma \tag{5}$$

Two strain mechanisms involved in the behavior of SMA appear in Eqn. (5). The first integral is related to the transformation itself. The second integral which corresponds to deformation without phase transformation, describes the mechanism of reorientation of the martensite variants yet formed.

## 2.2. FREE ENERGY AND DISSIPATION

### 2. 2. 1. *Free Energy Related to the Martensitic Transformation of the RVE*

Total potential energy of the RVE in a two phase state is composed of an elastic strain energy with density w(r), a crystallographic (usually called chemical) free energy with density φ(r) and an interfacial energy between all the constituents. This last contribution can be neglected for shape memory alloys. Other sources of energy (dislocations, point defects) are also neglected in the present approach. The stress-free austenitic state is taken as reference configuration. It is assumed that the temperature remains uniform over the RVE. For a unit volume V, the free energy Φ is given by:

$$V\Phi = \int_V \varphi(r)dV + \int_V w(r)dV \tag{6}$$

Around the thermodynamical equilibrium temperature $T_0$, chemical free energy may be linearized by:

$$\frac{1}{V}\int_V \varphi(r)dV = B\left(T - T_0\right)f \tag{7}$$

where f denotes the global amount of martensite. Determination of the elastic strain energy requires the knowledge of the local stress state σ. These stresses are related to the overall stresses Σ and also to the internal ones τ. Since transformation strain field $\varepsilon^T(r)$ presents some incompatibilities, elastic energy W may be expressed as :

$$W = \int_V w(r)dV = \frac{1}{V}\int_V \sigma_{ij}(r)\varepsilon^e_{ij}(r)dV = \frac{1}{2}E_{ij}\Sigma_{ij} - \frac{1}{2V}\int_V \sigma_{ij}(r)\varepsilon^T_{ij}(r)dV \tag{8}$$

Considering the internal stress field must be in equilibrium and having its mean value equal to zero, we can rewritte Eqn. (8) as:

$$W = \frac{1}{2}\left(E_{ij} - E^T_{ij}\right)C_{ijkl}\left(E_{kl} - E^T_{kl}\right) - \frac{1}{2V}\int_V \tau_{ij}(r)\varepsilon^T_{ij}(r)dV \tag{9}$$

where E denotes the total strain of the RVE, and C is the fourth order elastic tensor. In this approach, internal stresses depend only on the incompatibilities of the transformation strain field. Calculation of this internal stress field may be performed using Green function techniques [4, 5]. To account with the volume $V_I$ and the morphology of the different transformed regions, an exact calculation requires the spatial description of the transformation field this greatly complicates the problem.

2.2.2. *Dissipation and Diving Forces for Martensitic Transformation and Reorientation*

For quasi-static evolutions and isothermal approximation, intrinsic dissipation D of a system of volume V is deduced from the comparison between the external power (denoted by $P_{ext}$) and the change in free energy.

$$D = P_{ext} - \frac{d\phi}{dt} \tag{10}$$

Due to discontinuity across the moving internal boundaries $\Sigma$, one gets:

$$\frac{d\phi}{dt} = \frac{1}{V}\int_V \frac{\partial\big(w(r)+\varphi(r)\big)}{\partial t} dV - \frac{1}{V}\int_\Sigma [w+\varphi] w_\alpha n_\alpha d\Sigma \tag{11}$$

The chemical energy density and the transformation strain $\varepsilon^{TI}$ are assumed to be time independent quantities, the first integral in Eqn. (11) reduces to:

$$\frac{1}{V}\int_V \frac{\partial\big(w(r)+\varphi(r)\big)}{\partial t} dV = \frac{1}{V}\int_V \sigma_{ij}(r)\,\dot{\varepsilon}_{ij}(r)dV \tag{12}$$

Jump of the chemical energy is equal to zero across martensite-martensite interfaces and is uniform and equal to $(\varphi_A - \varphi_M)$ across martensite-austenite interfaces. This gives:

$$\frac{1}{V}\int_\Sigma [\varphi]\, w_\alpha\, n_\alpha\, d\Sigma = (\varphi_A - \varphi_M)\frac{1}{V}\int_\Sigma w_\alpha\, n_\alpha\, d\Sigma \tag{13}$$

Assuming uniform elasticity and using the symmetry properties of the elastic complience tensor, one gets:

$$[w] = \frac{1}{2}\left(\sigma_{ij}^{+} + \sigma_{ij}^{-}\right)\left[\varepsilon_{ij} - \varepsilon_{ij}^{T}\right] \tag{14}$$

where $\sigma^+$ and $\sigma^-$ denote the stress state on each side of the interface considered. Setting Eqns. (12) to (14) into (11) defines the change of free energy as:

$$\frac{d\phi}{dt} = \frac{1}{V}\int_V \sigma_{ij}\,\dot{\varepsilon}_{ij}\, dV - \frac{1}{2V}\int_\Sigma \left(\sigma_{ij}^{+} + \sigma_{ij}^{-}\right)\left[\varepsilon_{ij} - \varepsilon_{ij}^{T}\right] w_\alpha\, n_\alpha\, d\Sigma$$
$$- (\varphi_A - \varphi_M)\frac{1}{V}\int_\Sigma w_\alpha\, n_\alpha\, d\Sigma \tag{15}$$

The external power $P_{ext}$ results from the boundary conditions applied on external boundary S of normal n. If there are no volume forces applied in the VER:

$$VP_{ext} = \int_S \sigma_{ij}\, n_j\, v_i\, dS \tag{16}$$

where $v_i$ denotes the velocity of particles at external boundary S. Using the divergence theorem applied on a volume V containing some moving discontinity surfaces $\Sigma$, we get:

$$V\,P_{ext} = \int_V \left(\sigma_{ij}\, v_i\right)_{,j} dV + \int_\Sigma \left[\sigma_{ij}\, n_j\, v_i\right] d\Sigma \tag{17}$$

Hadamard [6] conditions on [v] give:

$$\left[v_i\right] = -\left[u_{i,j}\right] n_j w_\alpha n_\alpha \tag{18}$$

From Eqn. (18) and the continuity conditions on $\sigma$ and the displacement u at the moving surfaces $\Sigma$, Eqn. (17) becomes:

$$V\,P_{ext} = \int_V \sigma_{ij}\,\dot{\varepsilon}_{ij}\, dV - \frac{1}{2}\int_\Sigma \left(\sigma_{ij}^{+} + \sigma_{ij}^{-}\right)\left[\varepsilon_{ij}\right] w_\alpha\, n_\alpha\, d\Sigma \tag{19}$$

Finally, the dissipation follows from Eqns. (15) and (19):

$$D = \frac{1}{V}\int_\Sigma \left\{\frac{1}{2}\left(\sigma_{ij}^{+} + \sigma_{ij}^{-}\right)\left[\varepsilon_{ij}^{T}\right] + [\varphi]\right\} w_\alpha n_\alpha d\Sigma \tag{20}$$

In Eqn. (20) the term $F = \frac{1}{2}\left(\sigma_{ij}^{+} + \sigma_{ij}^{-}\right)\left[\varepsilon_{ij}^{T}\right] + [\varphi]$ defines the driving force for interface motion [2]. Relations determined in this section theorytically allow us to describe the thermomechanical behavior of a material undergoing a thermoelastic martensitic transformation. In these relations, the internal stress field produced by incompatibilities of the transformation field plays a very important role (see Eqn. (9) or (11)). Complete determination of these stresses implies a very accurate description of the microstructure and leads to an inextricable problem, approximations must be made.

A numerical resolution using finite element method is one of these solutions, but due to the complexity of the microstructure, a real 3-D problem will be non realistically time consuming. Other solution is to neglect the internal stress field, but this extreme solution leads only to some basic result, such as the Patel and Cohen criterion [7]. A more realistic way is to use a discrete internal variables framework.

## 3. Discrete Internal Variables Description

To deal with the different length scales appearing in polycrystalline SMA, two models are developed. The first one concerns the behavior of the grains of a polycrystalline material, and the second uses a self consistent approximation to compute the intergranular interaction. This crystallographic approach introduces many internal variables having a strong physical meanning. The only internal variables required are the volume fraction of the different variants of martensite for each grain.

### 3.1 CONSTITUTIVE EQUATIONS FOR A GRAIN INSIDE A POLYCRYSTAL

Grain is considered as single crystal and taken as VER for this modeling. In this approach, all the martensitic plates having the same crystallographic orientation are considered to belong to the same variant. Application of Eqn. (3) gives the transformation strain $E^T$ for a grain.

$$E_{ij}^{T} = \sum_{n} \varepsilon_{ij}^{n} \; f^{n} \tag{21}$$

In this section, the constant transformation strain occuring inside a variant of volume fraction $f^n$ is denoted by $\varepsilon^n$. Eqns. (6) and (7) show that chemical and elastic energies are functions of control parameters (applied stress $\Sigma$ and temperature T) and a set of internal variables related to the transformation strain field itself. Micromechanical analysis [7] had shown that the contribution related to the internal stress field in Eqn. (9) can be represented using an interaction matrix $H^{nm}$. This allows us to consider the $f^n$ as the only internal variables.

$$W_{int} = -\frac{1}{2V} \int_{V} \tau_{ij}(r) \, \varepsilon_{ij}^{T}(r) \, dV = \frac{1}{2} \sum_{n,m} H^{nm} \, f^{n} \, f^{m} \tag{22}$$

Substituting Eqn. (22) into Eqn. (9) one express the complementary free energy defined from Eqn. (6) by (where $M = C^{-1}$ denotes the elastic stiffness tensor):

$$\Psi(\Sigma_{ij}, T, f^{n}) = \frac{1}{2} \Sigma_{ij} M_{ijkl} \Sigma_{kl} + \Sigma_{ij} \sum_{n} \varepsilon_{ij}^{n} f^{n} - B(T-T_0) \sum_{n} f^{n} - \frac{1}{2} \sum_{n,m} H^{nm} f^{n} f^{m} \tag{23}$$

Due to the physical meanning of the internal variables used, the following kinematical constraints must be considered:

$$f^{n} \geq 0 \qquad \text{and} \qquad \sum_{n} f^{n} \leq 1 \tag{24}$$

To account for these kinematical constraints, Lagrange multipliers $\lambda_0$ and $\lambda_n$ are introduced [8]. Thermodynamical forces acting on each variant are obtained in the form:

$$F^{n} = \frac{\partial \Psi}{\partial f^{n}} - \lambda_0 + \lambda_n = \Sigma_{ij} \varepsilon_{ij}^{n} - B(T-T_0) - \sum_{m} H^{nm} f^{m} - \lambda_0 + \lambda_n \tag{25}$$

The transformation proceeds until the thermodynamical forces reach critical values. This allows us to define a local transformation criterion on each variant. When both this criterion and constraints (24) are satisfied, the constitutive equation at the grain level are obtained using the consistency condition:

$$\dot{E}_{ij}^{T} = \sum_{n} \varepsilon_{ij}^{n} \sum_{m} (H^{nm})^{-1} \left[ \varepsilon_{kl}^{m} \dot{\Sigma}_{kl} - B \dot{T} \right] \tag{26}$$

These relations are used as local relationships in the polycrystalline model.

## 3.2. SELF-CONSISTENT APPROXIMATION

In the polycrystalline case intergranular stresses must be evaluated. This is done by assuming that the global behavior is expressed using effective macroscopic thermomechanical tangent moduli ($L^{eff}$ and $M^{eff}$). Local and overall stress and strain rate are linked together by concentration tensors A(r) and a(r):

$$\dot{\varepsilon}_{ij}(r) = A_{ijkl}(r) \dot{E}_{kl} + a_{ij}(r) \dot{T} \tag{27}$$

Let us consider a homogeneous reference medium characterized by a set of uniform tangent moduli ($L^o$ and $M^o$) and subject to a uniform strain $E^o$. Using the Green tensor method we define a localization expression with an integral form :

$$\dot{\varepsilon}_{mn}(r) = \dot{E}^o_{mn} + \int_V \Gamma^o_{mnij}(r-r')\,[\delta l^o_{ijkl}(r')\,\dot{\varepsilon}_{kl}(r') - \delta m^o_{ij}(r')\,\dot{T}]\,dV' \qquad (28)$$

where $\Gamma^o$ denotes the modified Green tensor. $\delta l^o(r)$ and $\delta m^o(r)$ are the deviations of the local tangent moduli $l(r)$ and $m(r)$ from $L^o$ and $M^o$. In this work, a self-consistent approximation is chosen to determine piecewise uniform concentration tensors having constant values in each grain of volume $V^N$. Knowledge of these quantities allows us to determine the effective behavior of the polycrystal [7].

$$L^{eff}_{ijkl} = \sum_N l^N_{ijkl} A^N_{klmn} \frac{V^N}{V} \quad \text{and} \quad M^{eff}_{ij} = -\sum_N \{l^N_{ijkl} a^N_{kl} - m^N_{ij}\} \frac{V^N}{V} \qquad (29)$$

where $l^N$ and $m^N$ are the local piecewise uniform tangent moduli derived from Eqn. (26). This framework is applied to compute the superthermoelastic behavior of a Cu-Al-Be SMA [9]. Polycrystalline aspects are described by taking into account grain shape and crystallographical orientation. Transformation strain $\varepsilon^n$ is defined from measurement of the lattice parameters in the two phases. Parameter B in Eqn. (26) is experimentally obtained from tensile tests performed on a single crystal having the same chemical composition. Interaction matrix $H^{mn}$ is deduced from micromechanical considerations [7]. This description is successfully applied to simulate both the superelastic behavior (Figure 1) and the anisothermal creep on the same alloy (Figure 2).

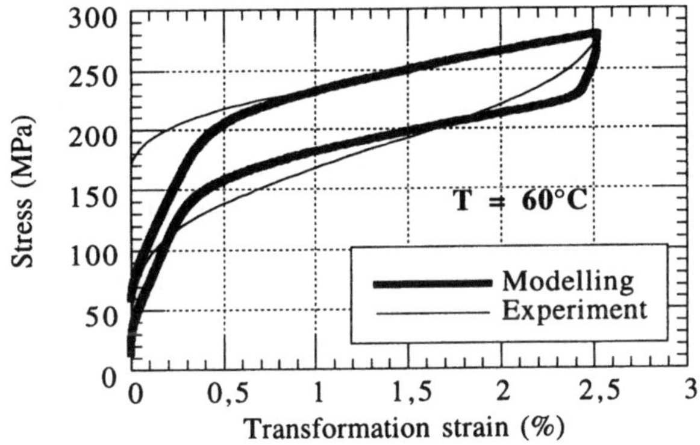


*Figure 1.* Simulation of the superelastic behavior of a polycrystalline Cu-Al-Be SMA. Comparison with experimental data ($M_s$ = - 40°C)

## 4. Conclusion

This approach leads to excellent agreement with experimental measurements without any fitting parameter. Variables involved in this scheme can be experimentally determined from experiments performed on a single crystal having the same chemical composition. Due to the three dimensional description of the transformation used in this framework, the asymmetry experimentally observed during a tensile-compression test, as well as the behavior obtained during multiaxial loading, are described well [7].

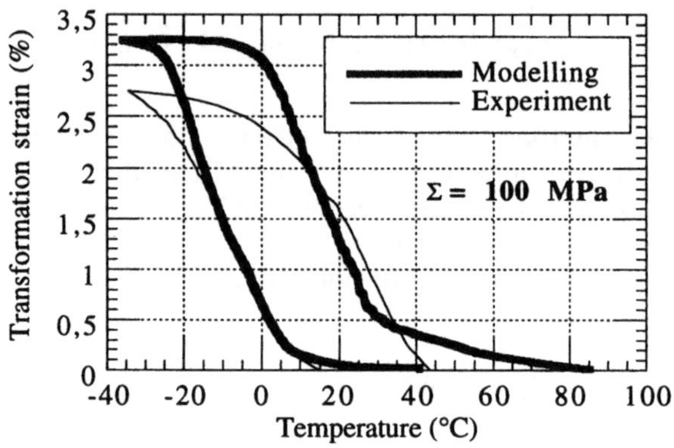


*Figure 2.* Simulation of the anisothermal creep of the same polycrystalline Cu-Al-Be SMA. Comparison with experimental data ($M_S$ = - 40°C)

## 5. References

1. Delaey L. (1991) Diffusionless transformations, in R.W. Cahn, P. Haasen and E.J. Kramen (eds), *Materials Science and Technologies, Vol. 5 : Phase Transformations in Materials*, VCH Publishers, pp. 339-404.
2. Eshelby J.D. (1961) Elastic inclusions and inhomogeneities, in I.N. Sneddon and R. Hill (eds), *Progress in Solids Mechanics,Vol. 2*, North-Holland, Amsterdam, pp. 87-140.
3. Kröner E. (1961) Zur plastischen verformung des vielkristalls, *Acta Metall.*, **9**, 155-161.
4. Kröner E. (1989) Modified Green functions in the theory of heterogeneous and/or anisotropic elastic media, in G. J. Weng, M. Taya and M. Abe (eds), *Micromechanics and Inhomogeneity, Vol. 2*, Springer Verlag.
5. Berveiller M. and Zaoui A. (1984) Modelling of the plastic behavior of inhomogeneous media, *J. Engng. Mat. and Technology*, **106**, pp. 295-299.
6. Hadamard J. (1903) *Leçons sur la propagation des ondes et les équations de l'hydrodynamique*, Cours du Collège de France, Librairie Scientifique A. Hermann, Paris.
7. Patoor E. and Berveiller M. (1997) Micromechanical modelling of the thermomechanical behavior of shape memory alloys, in M. Berveiller and F. D. Fischer (eds) *Mechanics of Solids with Phase Changes*, CISM Courses and lectures n° 368, Springer Wien New York, pp. 121-188.
8. Moumni Z. and Nguyen Q. S. (1996) A model of materials with phase change and applications, Coll. C1, J. de Physique IV, C1-335-345.
9. Entemeyer, D. (1996) *Etude micromécanique du comportement thermomécanique des alliages à mémoire de forme*, Thèse de Doctorat, Université de Metz, France.

# MICROMECHANICS CONSTITUTIVE MODELING FOR POLYCRYSTALLINE SMA

G-Q SONG
*Institute of Engineering Mechanics,*
*Nanchang University, Nanchang, P.R.CHINA*

AND

Q-P SUN & K-C HWANG
*Department of Engineering Mechanics,*
*Tsinghua University, Beijing, P.R.CHINA*

**Abstract.** A micromechanics constitutive model for martensite transformation of polycrystalline shape memory alloys is proposed. The model is established on the following bases: (1) the transformation theory of unconstrained single crystal shape memory alloys, (2) the proper simulation for the evolution of microstructure, i.e.,nucleation and growth of martensite variants, (3) the effect of orientation distribution on transformation process. Theoretical predictions of this model are in agreement with the available experimental observations.

## 1. Introduction

The constitutive response of shape memory alloys (abbreviated to SMA) is characterized by the process of thermoelastic martensite transformation. The study on the constitutive models of the SMA materials attracts interest of a lot of researchers. Some models that can simulate the main constitutive behavior have been presented, but there is still a great need for describing well the martensite transformation process, especially the evolution of microstructure in SMA materials.

For polycrystalline SMA materials, micromechanics is an effective method by which the macrobehaviour can be derived from internal microstructure evolution. Patoor *et al.* [1] obtained the constitutive relations of polycrystalline SMA from those of monocrystal by applying refined self-consistent



*P. Argoul et al. (eds.),*
*IUTAM Symposium on Variations of Domains and Free-Boundary Problems in Solid Mechanics*, 25–33.

method, but Sun and Hwang [2, 3] derived the relations directly by using micromechanics under the assumption that microscopic martensite transformation in a grain is completed instantaneously. In Sun *et al.*'s model, a phenomenological linear hardening factor was introduced in the evolution equation of internal variables in order to simulate the hardening behavior in macroscopic stress-strain relation.

The present paper is primarily concerned with micromechanics investigation of transformation induced plasticity of polycrystalline SMA materials, incorporating microstructures, crystallography and micromechanics into the continuum formulation of macroscopic constitutive behavior. In section 2, the martensite forward and reverse transformation conditions of unconstrained single crystal SMA are given first. Based on these results, those conditions of the bulk polycrystalline materials are obtained by means of Mori-Tanaka's mean-field theory which is used to calculate internal stresses and interaction of microstructures in section 3. Two internal variables of different levels are introduced to describe transformation process: the microstrain $g$, as an order parameter, shows the microscopic phase structure of material, and the martensite volume fraction $f$ indicates martensite transformation macro-process. In section 4, the constitutive equations for polycrystalline SMA are derived. Two kinds of hardening mechanisms are taken into account: the change of martensite variant orientations from the most favorable to the most unfavorable and the morphology of martensite variant turning from thin to thick oblate spheroid. They can simulate well the microstructural evolution during transformation. Results of numerical computation and comparison with experiments will be given in a separate paper.

For simplicity we will neglect the transformation lattice volume change, which is appropriate for most SMA materials.

## 2. The martensite transformation theory of unconstrained single crystal SMA

According to martensite transformation crystallographic theory, transformation strain is an invariant plane strain, and the deformation gradient tensor of martensite transformation $\mathbf{F}^p$ can be written as

$$\mathbf{F}^p = \mathbf{I} + \Delta g \mathbf{d} \otimes \mathbf{n} \tag{1}$$

where $\mathbf{d}$, $\mathbf{n}$ are unit vectors specifying the direction of displacement, the normal to the invariant plane, $\Delta g$ gives the magnitude of transformation strain, and $\mathbf{I}$ is a unit tensor. Following Landau transformation theory, the whole of strain $g$ in the direction of $\mathbf{d}$ can be known as an order parameter which indicates the state of phase. If $g$ jumps from $g_p$ to $g_m$ during the forward transformation and from $g_{mr}$ to $g_{pr}$ during reverse transformation, we assume

$$\Delta g = g_m - g_p = g_{mr} - g_{pr} = \text{const} \tag{2}$$

Considered as small deformation, the transformation strain $\varepsilon^p$ can be given as

$$\varepsilon^p = \pm\frac{1}{2}\Delta g(\mathbf{d}\otimes\mathbf{n}+\mathbf{n}\otimes\mathbf{d}) = \pm\Delta g\mathbf{R} \tag{3}$$

for forward(+) and reverse(-) transformation. Here $\mathbf{R}$ is called the orientation of the martensite variant. In the light of the crystallographic theory of martensite transformation, there are $N$ kinds of possible martensite variants with different orientations $\mathbf{R}$ and identical $\Delta g$.

The phase structure of an unconstrained single crystal of SMA material is either parent phase or one of $N$ kinds of martensite variants.A simplified Landau-Devonshire theory in which free energy functions are segmentally parabolic is appropriate to model the transformation behavior of unconstrained single crystal SMA [4, 5].

Assuming for simplicity that the parent phase and all martensite variants have the same elastic property, and neglecting thermal expansion and specific heat, with strain $\varepsilon$ specialized in the direction of orientation $\mathbf{R}(\varepsilon = g\mathbf{R})$, the Helmholtz free-energy $H(g,T)$ has the forms

$$H(g,T) = \tfrac{1}{2}\eta(T)g^2 \qquad \text{for parent phase} \tag{4}$$

$$H(g,T) = \tfrac{1}{2}\eta(T)(g-\Delta g)^2 - \Theta(1-\frac{T}{T_1}) \qquad \text{for a martensite variant} \tag{5}$$

where $\Theta$ represents the latent heat of the parent martensite transition at the equilibrium temperature $T_1$. The stress response function $\sigma(g,T)$ associated with $H(g,T)$ is

$$\sigma(g,T) = \eta(T)g \qquad \text{for parent phase} \tag{6}$$

$$\sigma(g,T) = \eta(T)(g-\Delta g) \qquad \text{for martensite} \tag{7}$$

where $\sigma = \sigma : \mathbf{R}$ is the component of stress $\sigma$ in the direction of the orientation $\mathbf{R}$. By using the same method as in Abeyaratne *et al.*'s paper [5], we can derive the following transformation conditions of forward and reverse phase transformation at temperature $T$

$$\text{forward} \qquad \sigma \geq \frac{\Theta}{T_1\Delta g}(T-M^0) \tag{8}$$

$$\text{reverse} \qquad \sigma \leq \frac{\Theta}{T_1\Delta g}(T-A^0) \tag{9}$$

where $M^0, A^0$ are respectively the transition temperatures of martensite and parent phase at stress-free state. The energy differences between parent phase and martensite in forward and reverse phase transformation are expressed by

$$\Delta H^{p\to m} = H(g_m, T) - H(g_p, T) \tag{10}$$

$$\Delta H^{m\to p} = H(g_{pr}, T) - H(g_{mr}, T) \tag{11}$$

They are caused by structure change during transformation, and known also as the differences of structural chemical free energy.

## 3. The transformation criterion of polycrystalline SMA bulk body

As done in micromechanics, a representative material sample, which is known as the constitutive element, is taken from the polycrystalline SMA bulk body as the subject of study. A temperature $T$ is uniformly distributed in the element and the external macroscopic stress $\Sigma$ or strain $\mathbf{E}$ is applied on the boundary. Let $f$ denote the volumetric fraction of martensite variants in constitutive element. As shown by a lot of micrographs, martensite variants appear in a shape of plate or blade, each variant is considered as an oblate spheroid inclusion surrounded by parent matrix with short axis normal to the invariant plane. For stress induced martensite transformation, microscopic transformation strain is uniform and equals $\epsilon^p$ in a martensite inclusion.

Denoting the volume of the constitutive element, parent matrix and martensite inclusions by $V$, $V_M$, $V_I$, volume average by $< >$, and neglecting the thermal expansion and difference in elastic constants between parent phase and martensite, we can decompose the energy changes during martensite transformation into the following terms with the stress-free parent phase of the constitutive element taken as reference state. The boundary surface energy can be neglected.

(*a*) *Elastic strain energy.* Accompanying martensite transformation at a given temperature, the total elastic energy stored in a unit volume of the constitutive element is the sum of $W_1$ and $W_2$, where $W_1$ is caused directly by the external macroscopic stress

$$W_1 = \tfrac{1}{2}\Sigma : \mathbf{C}^{-1}(T) : \Sigma \tag{12}$$

and $W_2$ is caused by the internal stress of transformation. Taking account of the interaction between the stress field produced by individual variant, and making use of Eshelby's solution [6] and the Mori-Tanaka self-consistent theory [7, 8], we have the self-consistent derivation [9]

$$W_2 = -\tfrac{1}{2}f\langle\varepsilon^p : \mathbf{C}(T) : (\mathbf{S}-\mathbf{I}_4) : \varepsilon^p\rangle_{V_I} + \tfrac{1}{2}f^2\langle\varepsilon^p\rangle_{V_I} : \mathbf{C}(T) : \langle(\mathbf{S}-\mathbf{I}_4) : \varepsilon^p\rangle_{V_I} \tag{13}$$

where $\mathbf{C}(T)$and $\mathbf{C}^{-1}(T)$ are the macroscopic elastic stiffness and compliance tensors of the constitutive element, respectively, $\mathbf{S}$ is Eshelby's tensor, and $\mathbf{I}_4$ is identity tensor of rank 4. The macroscopic strain $\mathbf{E}$ in the constitutive element can be expressed as

$$\mathbf{E} = \mathbf{E}^e + \mathbf{E}^p = \mathbf{C}^{-1}(T) : \Sigma + f\langle\varepsilon^p\rangle_{V_I} \tag{14}$$

(*b*) *Structural chemical free energy.* The variation of structural chemical free energy in a unit volume of the constitutive element has the following expression

$$\Delta H = \Delta H^{p\to m} f \tag{15}$$

To sum up (a)-(b), the Gibbs free energy $G$ of the constitutive element per unit volume can be formulated as

$$G(\Sigma, T, f, \varepsilon^p) = W_1 + W_2 + \Delta H - \Sigma : \mathbf{E} \tag{16}$$

Under the condition of $\Sigma$, $T$, $f$, $\varepsilon^p$, let a martensite inclusion be nucleating and growing or shrinking and disappearing. Assuming that the inclusion transforming has volume $dV$, volumetric fraction $df$ and orientation $\mathbf{R}$, we can give the change rate of Gibbs free energy density of the constitutive element with respect to the growth or shrinkage of the martensite inclusion

$$\begin{aligned} G_f(\Sigma, T, f, \varepsilon^p; \Delta g) &= \frac{d}{df}G(\Sigma, T, f, \varepsilon^p) = -\Delta g\Sigma : \mathbf{R} \\ &\quad -\mu(T)f\Delta g\mathbf{R} : (2\langle\varepsilon^p\rangle_{V_I} - \langle\mathbf{S} : \varepsilon^p\rangle_{V_I} - \langle\varepsilon^p\rangle_{V_I} : \mathbf{S}) \\ &\quad +\tfrac{1}{2}\mu(T)(\Delta g)^2(1 - 2S_{1313}) + \Delta H^{p\to m} \end{aligned} \tag{17}$$

where $S_{1313}$ is the component of $\mathbf{S}$ on the local Cartesian coordinate system in the martensite inclusion with axis 1 and axis 3 parallel to $\mathbf{d}$ and $\mathbf{n}$ of the variant, $\mu(T)$is the shear modulus of the constitutive element. Here we assume that the elasticity is isotropic for simplicity. Substituting $\Delta g = g_m - g_p$ and (10) into (17), we can define a energy function $G_f(\Sigma, T, f, \varepsilon^p; g_m)$ of order parameter $g_m$. The energy function means the variation with $g_m$ of the Gibbs free energy density of the constitutive element. Then the equilibrium of the constitutive element is ensured by the following equation

$$\frac{\partial}{\partial g_m}G_f(\Sigma, T, f, \varepsilon^p; g_m) = 0 \tag{18}$$

By the definition $\sigma = \partial H(g,T)/\partial g$, we can obtain the generalized force $\sigma$ conjugated with variable $g_m$ from (18)

$$\sigma = \Sigma : \mathbf{R} - \mu(T)(\Delta g)(1 - 2S_{1313}) + \mu(T) f \mathbf{R} : (2\langle \varepsilon^p \rangle_{V_I} - \langle \mathbf{S} : \varepsilon^p \rangle_{V_I} - \langle \varepsilon^p \rangle_{V_I} : \mathbf{S}) \tag{19}$$

where $\sigma$ measures the action at the direction of orientation $\mathbf{R}$ on the inclusion that is nucleating, including external stress, interaction of inclusions and constraint of surrounding matrix. If we consider this inclusion as an unconstrained body, $\sigma$ is equivalent to external force acting on the inclusion as stated in section 2. Consequently, the forward transformation condition of bulk body can be inferred from that of unconstrained body

$$\Sigma : \mathbf{R} + \mu(T) f \mathbf{R} : (2\langle \varepsilon^p \rangle_{V_I} - \langle \mathbf{S} : \varepsilon^p \rangle_{V_I} - \langle \varepsilon^p \rangle_{V_I} : \mathbf{S}) \geq k(T - M') \tag{20}$$

$$k = \frac{\Theta}{T_1 \Delta g} \quad , \quad M' = M^0 - \frac{1}{k}\mu(T)(\Delta g)(1 - 2S_{1313}) \tag{21}$$

Similarly, the reverse transformation condition of bulk body can be also obtained

$$\Sigma : \mathbf{R} + \mu(T) f \mathbf{R} : (2\langle \varepsilon^p \rangle_{V_I} - \langle \mathbf{S} : \varepsilon^p \rangle_{V_I} - \langle \varepsilon^p \rangle_{V_I} : \mathbf{S}) \leq k(T - A') \tag{22}$$

$$A' = A^0 - \frac{1}{k}\mu(T)(\Delta g)(1 - 2S_{1313}) \tag{23}$$

The equalities in (20) and (22) express the critical conditions, and they are also known as yield conditions of transformation induced plasticity. For each given $\mathbf{R}$, the yield surface defined by (20) or (22) is a hyperplane. If we assume that the orientations of all possible martensite variants are uniformly distributed for macroscopically isotropic polycrystalline SMA body, the initial yield surface of the polycrystal is a closed smooth envelop of the yield hyperplanes corresponding to all the orientations. From (17) it is clear that a flat variant minimizes the increase of free energy, so we can conclude that the shape of martensite variants trends to be flat spheroid (i.e. the ratio $\rho$ of short axis to long axis equals zero) at nucleation.

## 4. Transformation flow rules of polycrystalline SMA bulk body

When yield conditions (20) and (22) are satisfied, if we change the loading parameters stress $\Sigma$ and temperature $T$, the transformation may advance. The subsequent forward transformation process is composed of two parts: the variants that have formed continue to grow and the transformation extends form the favorable to unfavorable directions. For given orientation $\mathbf{R}$

of martensite variants, the orientations of grains that can crystallographically transform to those variants can be determined. Denoting by $V_{\mathbf{R}}$ the volume distribution of those grains, and by $f_{\mathbf{R}}$ the volume fraction distribution of martensite variants with orientation $\mathbf{R}$ with respect to $V_{\mathbf{R}}$. $V_{\mathbf{R}}$ can be assumed to be a constant for isotropic polycrystal. Then we have the expression of macroscopic plastic strain

$$\mathbf{E}^p = f\langle\varepsilon^p\rangle_{V_I} = \frac{V_{\mathbf{R}}}{V}\int_{\Omega} f_{\mathbf{R}}\varepsilon^p d\omega \tag{24}$$

where $\Omega$ is the region composed of all the possible orientations $\mathbf{R}$ in polycrystal, and $d\omega$ is a differential measure of orientation, $\int_{\Omega} d\omega = 1$. The $f$ is related to $f_{\mathbf{R}}$ by

$$f = \frac{V_{\mathbf{R}}}{V}\int_{\Omega} f_{\mathbf{R}} d\omega \tag{25}$$

For $f_{\mathbf{R}}$ equals one in (25), $f$ must equal $N$, so

$$\frac{V_{\mathbf{R}}}{V} = N \tag{26}$$

The relations (20) and (22) show that under the condition of given $\Sigma, T, f, \mathbf{R}, V_I$ the transformation yield is dependent only on the shape parameter $\rho$ of the variant, *i.e.*, all the inclusions in each orientation have identical shape parameter. According to experiment observations that the inclusions grow independently at the initial stage of transformation and they will merge with each other to form the larger inclusions, we can assume phenomenologically that the shape parameter $\rho_{\mathbf{R}}$ of inclusions with orientation $\mathbf{R}$ is related to $f_{\mathbf{R}}$ by $n$th order power function

$$\rho_{\mathbf{R}} = z(f_{\mathbf{R}})^n \tag{27}$$

where $z$ is a factor depending on the shape of grain. When the grain has the shape of sphere, we can let $z = 1$.

With aid of (24), the incremental stress-strain relation is

$$\dot{\mathbf{E}} = \dot{\mathbf{E}}^e + \dot{\mathbf{E}}^p = \mathbf{C}^{-1}(T) : \dot{\Sigma} + \frac{V_{\mathbf{R}}}{V}\int_{\Omega} \dot{f}_{\mathbf{R}}\varepsilon^p d\omega \tag{28}$$

where $\dot{f}_{\mathbf{R}}$ can be determined by the consistency condition. For forward and reverse transformation yield conditions (20) and (22), the consistency equation has the same form

$$F(f,\varepsilon^p)S'_{1313}\dot{\rho}_{\mathbf{R}} + 2\mu(T)\mathbf{R} : \frac{V_{\mathbf{R}}}{V}[(S_{1313}-1)\int_{\Omega} \dot{f}_{\mathbf{R}}\varepsilon^p d\omega$$
$$+ \int_{\Omega}(S_{1313}\dot{f}_{\mathbf{R}} + S'_{1313}\dot{\rho}_{\mathbf{R}})\varepsilon^p d\omega] = \dot{\Sigma} : \mathbf{R} - \vartheta(T)\dot{T} \tag{29}$$

$$F(f,\varepsilon^p) = -2\mu(T)\Delta g + 2\mu(T)\mathbf{R} : \frac{V_{\mathbf{R}}}{V}\int_{\Omega} f_{\mathbf{R}}\varepsilon^p d\omega \tag{30}$$

$$\vartheta(T) = k - [f\mathbf{R} : (2\langle\varepsilon^p\rangle_{V_I} - \langle\mathbf{S} : \varepsilon^p\rangle_{V_I} - \langle\varepsilon^p\rangle_{V_I} : \mathbf{S}) - \Delta g(1 - 2S_{1313})]\frac{d\mu(T)}{dT} \tag{31}$$

where $S'_{1313} = dS_{1313}(\rho_{\mathbf{R}})/d\rho_{\mathbf{R}}$ and we have used the equality

$$\mathbf{S} : \mathbf{R} = \mathbf{R} : \mathbf{S} = 2S_{1313}\mathbf{R} \tag{32}$$

The consistency condition (29) is an integral equation. Under the constrained conditions:

(1) (*forward transformation*) the present loads $\Sigma, T, f, V_I$, satisfy the forward transformation yield condition (20), and

$$\sum_{\mathbf{R}_1\in\mathbf{R}'} f_{\mathbf{R}_1} < 1, \quad (\mathbf{R}' = \mathbf{Q}\mathbf{R}\mathbf{Q}^T, \mathbf{Q} \in \mathfrak{g}) \tag{33}$$

then $\dot{f}_{\mathbf{R}} \geq 0$;

(2)(*reverse transformation*) the present loads $\Sigma, T, f, V_I$, satisfy the reverse transformation yield condition (22), and $f_{\mathbf{R}} > 0$, then $\dot{f}_{\mathbf{R}} \leq 0$,
the solution $\dot{f}_{\mathbf{R}}$ of (29) gives the evolution rate of martensite volume fraction distribution at loading step $\dot{\Sigma}, \dot{T}$. In (33), a rotation group $\mathfrak{g}$ is introduced to denote the mutual relations of all the possible orientations $\mathbf{R}$ in a single crystal.

## 5. Conclusion

Based on micromechanics, crystallographic theory and nucleation and growth criterion of variant, a constitutive model to describe thermoelastic martensitic transformation for polycrystalline SMA is proposed. Compared with the existing models the microstructural change of the material during transformation is characterized by variables $f$, $f_{\mathbf{R}}$, $\rho_{\mathbf{R}}$, which have clear physical meaning and corresponding microstructural reality.

## Acknowledgments

The authors are sincerely grateful to the National Natural Science Foundation of China, Special Research Facilities for Ph. D. Programs of the State Education Commission of China and Province Natural Science Foundation of Jiangxi for their support.

## References

1. Patoor, E., Eberhardt, A. and Berveiller, M. (1988), *Arch. Mech.*, **40**, 775-794.

2. Sun, Q. P. and Hwang, K. C.(1993), *J. Mech. Phys. Solids,* **41**, 1-33.
3. Sun, Q. P. and Hwang, K. C. (1994), in *Advances in Applied Mechanics*, Vol. **31**, Academic Press, New York 249-298.
4. Muller, I and Xu, H. (1991), *Acta Metall. Mater.* **39**, 263-271.
5. Abeyaratne, R. and Knowles, J.K. (1993), *J. Mech. Phys. Solids*, **41**, 541-574.
6. Eshelby, J.D. (1957), *Proc. R. Soc. London*, Ser. A, **241**, 376-396.
7. Mori, T. and Tanaka, K. (1973), *Acta metall.*, **21**, 571-574.
8. Mura, T. (1987), Micromechanics of Defects in Solids, Martinus Nijhoff, Dordrecht.
9. Song, G.-Q. (1996), Ph.D. Thesis, Tsinghua University, Beijing.

# DAMPING OF A VIBRATING SMA ROD THROUGH PHASE TRANSFORMATION

EDUARD ROMAN OBERAIGNER
*Institute of Mechanics*
*Montanuniversität Leoben, Austria*
KIKUAKI TANAKA
*Department of Aerospace Engineering*
*Tokyo Metropolitan Institute of Technology, Hino/Tokyo, Japan*
AND
FRANZ DIETER FISCHER
*Institute of Mechanics*
*Montanuniversität Leoben, Austria*

## 1. Introduction

Vibration damping is an important task in structural engineering. Several ways exist to accomplish it. One possibility is to 'weaken' the material by introducing a phase transformation through cooling or heating and to force the material to dissipate its mechanical energy during the transformation. The vibration damping due to phase transformation can be formulated mathematically by taking into account a micromechanical model of the authors [1, 2, 3] on the behavior of shape memory alloys (SMA's). This model incorporates a kinetic law to describe the stress-temperature-transformed volume fraction-relation and a constitutive law (stress-strain-temperature-transformed volume fraction-relation). These are nonlinear ordinary differential equations, which are coupled with two partial differential equations, the heat conduction equation and the wave equation.
The four mentioned differential equations can be combined to give an integral-equation for the second derivative of the displacement. This gives in further steps the displacement itself, the strain-, and stress-distributions and the temperature and the transformed volume-fraction as functions of position and time. Side results are the positions of moving boundaries be-



*P. Argoul et al. (eds.),*
*IUTAM Symposium on Variations of Domains and Free-Boundary Problems in Solid Mechanics*, 35–44.

tween the inelastic and the elastic domains in the rod.
In former approaches of the authors, the set of the four relevant differential equations was solved for special cases (adiabatic approximation) [4]. In this approach no analytical approximation is used. Only some numerical approximation is applied.

The main result is the mechanical energy as function of time which shows the intensity of the damping.

## 2. Mathematical model

In this section the mathematical model to describe the vibration damping of a freely vibrating SMA rod through phase transformation will be discussed.

The model has to contain the wave equation, the heat conduction equation (since in the present paper phase transformation is induced through cooling or heating), a kinetic law to describe the stress-temperature-transformed volume fraction-relation, and finally a constitutive law. The later is also a relation between stress, strain, temperature and transformed volume fraction.
In the following the four mentioned equations will be discussed in more detail.
Let $\Sigma$ be the stress within the considered rod, $T$ the temperature, and $\epsilon$ the total strain. $\xi$ is the transformed (martensitic) volume fraction. $u$ denotes the displacement and $\rho$ the density of the material. All the introduced quantities depend on the positon $x$ and the time $t$.

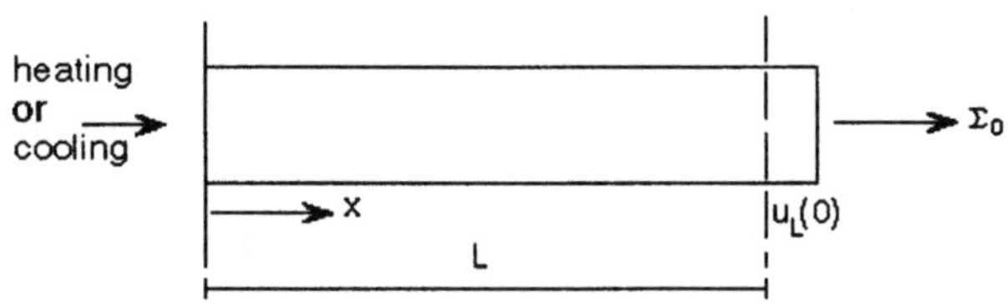


*Figure 1.* Elastic rod under initial stress $\Sigma_0$.

The following situation will be considered (see Fig. 1.):
At the beginning (time $t = 0$) the rod is under the stress $\Sigma_0$. $u_L(0)$ is the corresponding initial elongation at the position $L$. The rod is further homogeneous and consists only of the austenitic phase. $\Sigma_0$ is assumed to be below the yield stress, so that no prior plastification exists, and lies further on the right side of the $M_s$-line in the stress-temperature diagramm (see

Fig. 2).

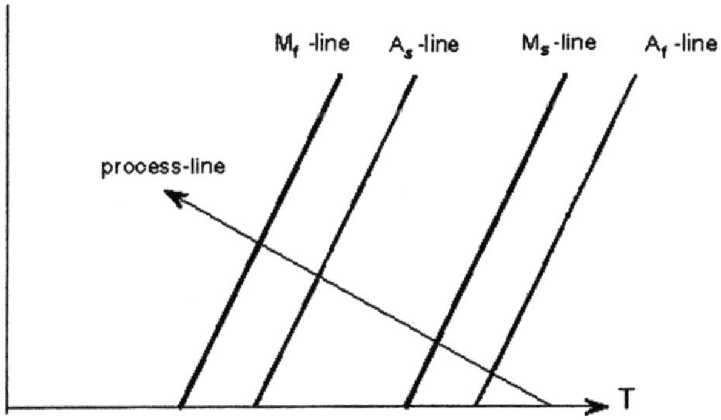


*Figure 2.* Stress-temperature plane.

The $M_s$-line is the martensite-start-line, the $M_f$-line the martensite-finish-line, whereas the $A_s$-line and the $A_f$-line denote the austenite-start and finish-line, resp., for the reverse transformations. Inclination and position of these lines are assumed to be given by experiments.
At the starting time of the vibration (i.e., $t = 0^+$) the rod is released, and at the same moment it is heated or cooled on the fixed end in order to induce a phase transformation. Through the heat conduction and the accompanying phase transformation the vibration will be 'weakened'. It will come to a damping of the rod by dissipation of the mechanical energy. After some time one will have the situation shown in figure 3.

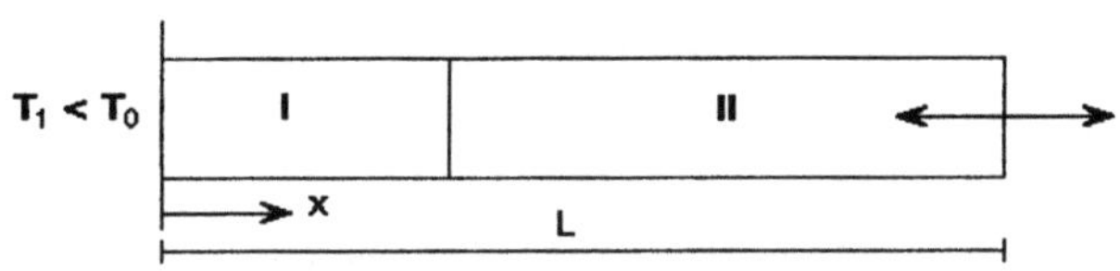


*Figure 3.* Elastic and inelastic domain on the damped rod.

There might be two (or even more) domains. In an early phase of the process the domain $I$ will be the two phase domain with a martensitic volume fraction between 0 and 1, the domain $II$ will still be a one phase (i. e. austenitic) domain. Domain $I$ is inelastic, domain $II$ still elastic.
At a later time, when the heat propagation has continued, domain I may split into the domains $I_a$ and $I_b$, where $I_a$ is already totally transformed, and $I_b$ only partially. Then $I_a$ will be elastic. (See Fig. 4).

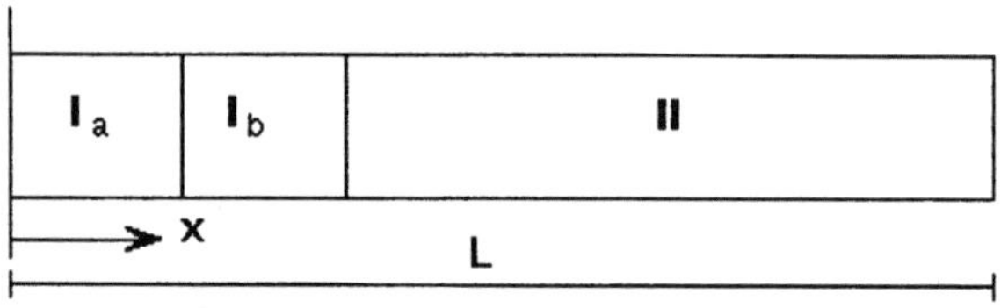


*Figure 4.* Appearence of a new elastic domain.

Figure 5 shows schematically the hysteresis used in the numerical calculations.

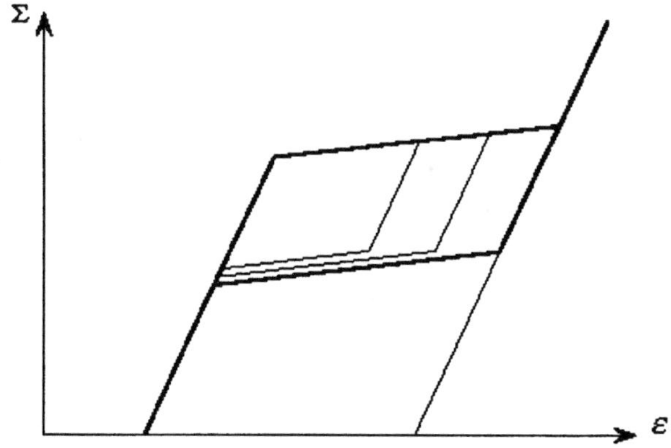


*Figure 5.* Schematic stress-strain curve.

The stress depends via a constitutive law on the temperature, the total strain, and the martensitic volume fraction. In differential form this can be written as

$$d\Sigma = E(d\epsilon - \Omega d\xi - \Theta dT) \tag{1}$$

$\epsilon^{tr} = \Omega d\xi$ is the strain due to phase transformation, $\epsilon^{th} = \Theta dT$ the thermal strain, and, finally, $d\epsilon^{el} = \frac{1}{E}d\Sigma$ the elastic strain. The differential of the total strain is the sum of these

$$d\epsilon = d\epsilon^{el} + d\epsilon^{tr} + d\epsilon^{th} \tag{2}$$

From equation (1) we can see immediately the relations

$$\frac{\partial \Sigma}{\partial \epsilon} = E, \quad \frac{\partial \Sigma}{\partial \xi} = -E\Omega, \quad \frac{\partial \Sigma}{\partial T} = -E\Theta, \tag{3}$$

which will be used in the wave equation later on. $\Omega$ itself is a function of $\xi$ and $T$, or can be written equivalently as a function of stress instead of $\xi$. Even $E$ and $\Theta$ might be considered as functions. The kinetic law to describe the progress of the transformation is of the form

$$d\xi = ad\Sigma + bdT \tag{4}$$

$a$ and $b$ are functions of stress and temperature and are taken together with $\Omega$ from a former publication of the authors [2]. The wave equation for a rod reads

$$\frac{\partial \Sigma}{\partial x} = \rho \frac{\partial^2 u}{\partial t^2}. \tag{5}$$

Due to

$$\Sigma = \Sigma(T, \epsilon, \xi) \tag{6}$$

one has

$$\frac{\partial \Sigma}{\partial x} = \frac{\partial \Sigma}{\partial T}\frac{\partial T}{\partial x} + \frac{\partial \Sigma}{\partial \epsilon}\frac{\partial \epsilon}{\partial x} + \frac{\partial \Sigma}{\partial \xi}\frac{\partial \xi}{\partial x} \tag{7}$$

Insertion of the kinetic law, Eq. (4), into the constitutive law, Eq. (1), yields

$$\begin{aligned} d\epsilon &= \frac{1}{E} d\Sigma + \Omega(ad\Sigma + bdT) + \Theta dT \\ &= (\frac{1}{E} + a\Omega)d\Sigma + (b\Omega + \Theta)dT \end{aligned} \tag{8}$$

or rearranged

$$d\Sigma = \frac{E}{1 + aE\Omega} d\epsilon - \frac{E(b\Omega + \Theta)}{1 + aE\Omega} dT \tag{9}$$

This means, with the help of the kinetic law the dependence of the stress on $\xi$ was eliminated. Of course, $\Sigma = \Sigma(\epsilon, T)$ possesses another dependence on $\epsilon$ and $T$ than $\Sigma = \Sigma(\epsilon, T, \xi)$ from above.
For the further calculations it is better to use the form $\Sigma = \Sigma(\epsilon, T)$ and to avoid, therefore, in Eq. (7) the last term. Insertion into Eq. (5) gives now

$$\frac{\partial \Sigma}{\partial T}\frac{\partial T}{\partial x} + \frac{\partial \Sigma}{\partial \epsilon}\frac{\partial \epsilon}{\partial x} = \rho \frac{\partial^2 u}{\partial t^2} \tag{10}$$

or more specific

$$-\frac{E(b\Omega + \Theta)}{1 + aE\Omega}\frac{\partial T}{\partial x} + \frac{E}{1 + aE\Omega}\frac{\partial^2 u}{\partial x^2} = \rho \frac{\partial^2 u}{\partial t^2} \tag{11}$$

Due to

$$\frac{E}{1 + aE\Omega} = E\left(1 - \frac{aE\Omega}{1 + aE\Omega}\right) \tag{12}$$

Eq.(11) can be rewritten into the form

$$\left(\frac{\rho}{E}\right)\frac{\partial^2 u}{\partial t^2}-\frac{\partial^2 u}{\partial x^2}=-\frac{aE\Omega}{1+aE\Omega}\frac{\partial^2 u}{\partial x^2}-\left(\frac{b\Omega+\Theta}{1+aE\Omega}\right)\frac{\partial T}{\partial x} \tag{13}$$

The left side contains the terms of the elastic wave equation. The first term on the right side can be considered as a perturbation term and the second term on the right side as a source term.
Due to $0 \leq aE\Omega < \infty$ the coefficient of the perturbation term ranks between zero and one. In most cases it is small compared to one. The coefficient of the source term lies between $\Theta$ and 0.
The boundary conditions for Eq. (13) are

$$u(x=0,t)=0 \tag{14}$$

$$\Sigma(x=L,t)=0 \Rightarrow \frac{\partial u}{\partial x}(x=L,t)=0 \tag{15}$$

The initial conditions are

$$u(x,t=0)=\frac{x}{L}u(x=L,t=0)=\frac{x}{L}u_L(0) \tag{16}$$

$$\dot{u}(x,t=0)=0 \tag{17}$$

Equation (13) can be solved with a perturbation method using Green's functions. One has formally

$$u(x,t)=\int_0^t\int_0^L G(x,\bar{x};t,\bar{t})w(\bar{x},\bar{t})d\bar{x}d\bar{t} \tag{18}$$

For fixed boundary conditions as in the present case one has time-transformation-invariance of $G$. According to [5] one gets for the Green's function

$$\begin{aligned} G(x,\bar{x};t,\bar{t}) \;&= \frac{4}{\pi v_s}\sum_{k=0}^{\infty}\frac{1}{2k+1}\quad \sin\left(\frac{(2k+1)\pi x}{2L}\right)\sin\left(\frac{(2k+1)\pi\bar{x}}{2L}\right) \\ &\qquad \sin\left(\frac{v_s(2k+1)\pi(t-\bar{t})}{2L}\right) \end{aligned} \tag{19}$$

and for the standardising function

$$w(\bar{x},\bar{t})=-v_s^2\left(A\frac{\partial^2 u(\bar{x},\bar{t})}{\partial\bar{x}^2}+B\frac{\partial T(\bar{x},\bar{t})}{\partial\bar{x}}\right)+\frac{\bar{x}}{L}U_L(0), \tag{20}$$

with the abbreviations

$$A:=\frac{aE\Omega}{1+aE\Omega},\quad B:=\frac{b\Omega+\Theta}{1+aE\Omega}, \tag{21}$$

$$v_s := \sqrt{\frac{E}{\rho}}. \tag{22}$$

Performing the second derivative of $u(x,t)$ with respect to $x$ in Eq. (18) deliveres an integral equation for $u''(x,t) = \frac{\partial u^2}{\partial x^2}$ :

$$u''(x,t) = -\frac{\pi}{v_s L^2} \sum_{k=0}^{\infty} (2k+1) \int_0^t \int_0^L \sin\left(\frac{(2k+1)\pi x}{2L}\right) \sin\left(\frac{(2k+1)\pi \bar{x}}{2L}\right)$$
$$\sin\left(\frac{v_s(2k+1)\pi(t-\bar{t})}{2L}\right) \cdot \{-v_s^2[Au''(\bar{x},\bar{t}) + BT'(\bar{x},\bar{t})] + \frac{\bar{x}}{L} u_L(0)\} d\bar{x} d\bar{t} \tag{23}$$

This equation has to be solved together with the heat conduction equation

$$\frac{\partial(\rho c_p T)}{\partial t} - \frac{\partial}{\partial x}\left(\kappa \frac{\partial T}{\partial x}\right) = q(x,t) = (k_L + k_D)\dot{\xi} \tag{24}$$

The source term has a contribution from the latent heat, namely $k_L\dot{\xi}$, and a further contribution from heat due to dissipation of mechanical energy, namely $k_D\dot{\xi}$. $k_L$ and $k_D$ are considered here as constants. However, in the process of a phase transformation $c_p$ may not be constant (see e.g. [6]). The boundary conditions are

$$T(x=0,t) = T_1 < T_0 \quad \text{(i.e. cooling)} \tag{25}$$

$$\frac{\partial T}{\partial x}(x=L,t) = 0 \quad \text{(i.e. perfect insulation)} \tag{26}$$

The initial condition is

$$T(x, t=0) = T_0 \tag{27}$$

This means, at $t = 0$ the temperature jumps at the fixed end of the rod from $T = T_0$ to the lower value $T = T_1$. $\dot{\xi}$ cames from Eq. (4).

## 3. Numerical procedure

There is no chance to solve the relevant set of equations (Eqs. (1), (4), (23), (24)) analytically. The problem is attacked numerically in the following way. Starting from the inital situation $t = 0$ the set of equations is solved for each position $x_i = i\Delta x$ for each time step $t_n = n\Delta t$.
If the integral equation (23) is written formally

$$u''(x,t) = f(x,t) + \lambda \int_0^t \int_0^L K(x,t;\bar{x},\bar{t}) u''(\bar{x},\bar{t}) d\bar{x} d\bar{t}, \tag{28}$$

then one has approximatively for the time-step $n$

$$\begin{aligned} u''(x,t_n) &= f(x,t_n) + \lambda \int_0^{t_{n-1}} \int_0^L K(x,t;\bar{x},\bar{t}) u''(\bar{x},\bar{t}) d\bar{x} d\bar{t} \\ &+ \lambda \Delta t \int_0^L K(x,t_n;\bar{x},\bar{t}) u''(\bar{x},t_n) d\bar{x} \end{aligned} \tag{29}$$

Now the following further approximations are made

$$f(x,t_n) \approx f(x,t_{n-1}) \tag{30}$$

$$K(x,t_n;\bar{x},\bar{t}) \approx K(x,t_{n-1};\bar{x},\bar{t}) \tag{31}$$

The kernel $K(x,t;\bar{x},\bar{t})$ can be simply gained by comparison of Eq. (28) with Eq. (23). Both are nonlinear. However, through the above approximations one gets a linear integral equation of Fredholm type in one dimension. $u''(\bar{x},\bar{t})$ has to be known only up to $t_{n-1}$ in the first integral in Eq. (29). Therefore one has to solve now the following reduced integral equation

$$u''(x,t_n) = \tilde{f}(x,t_{n-1}) + \lambda \Delta t \int_0^L K(x,t_{n-1};\bar{x},t_{n-1}) u''(\bar{x},t_n) d\bar{x} \tag{32}$$

$\tilde{f}(x,t_n)$ follows from comparison with formula (29):

$$\begin{aligned} \tilde{f}(x,t_{n-1}) = f(x,t_{n-1}) &+ \lambda \int_0^{t_{n-1}} \int_0^L K(x,t_{n-1};\bar{x},\bar{t}) u''(\bar{x},\bar{t}) d\bar{x} d\bar{t} \\ \cong f(x,t_{n-1}) + \lambda \textstyle\sum_{m=0}^{n-1} \Delta t &\int_0^L K(x,t_{n-1};\bar{x},t_m) u''(\bar{x},t_m) d\bar{x} \end{aligned} \tag{33}$$

Equation (32) can be handled in the usual way described in text books on integral equations (see e.g. [7], [8]). Here a Neumann series algorithm is used. Due to space limitations it is not explained here. Parallel to the integral equation the heat conduction equation has to be solved step by step.

The source term stems always from the former time step. $c_p$ and $\rho$ are considered as constant. It is however, no problem to extend the numerical algorithm to non-constant heat capacities and densities. From $T(x;t_{n-1})$ the derivatives $T'$ and $\dot{T}$ can be gained and inserted into the integral equation and into the discretized kinetic law:

$$\Delta\xi = a\Delta\Sigma + b\Delta T = a\dot{\Sigma}\Delta t + b\dot{T}\Delta t \tag{34}$$

(One has to check, however, if in the current state one has a progress of the transformation or not.)

$\dot{\Sigma}$ can be derived from Eq. (9), where $\epsilon$ follows from $u''$ via integration over $x$.
A characteristic time of the system is the period of one elastic vibration:

$$t_r := \frac{2L}{v_s}. \tag{35}$$

In order to fulfill the Courant-Friedrichs-Lewy-condition [9] it is necessary that

$$\Delta t \ll t_r. \tag{36}$$

The following material parameters are used: $E = 7000MPa$, $\Theta = -0.07K^{-1}$, $\rho = 5g/cm^3$, $c_p = 500J/kg/K$, $\kappa = 10W/m/K$, $M_s = 240K$, $A_s = 236K$, $M_f = 235K$, $A_f = 241K$.
The computational parameters are as follows: $T_0 = 250K$, $\Sigma_0 = 10MPa$, $T_1 = 235K$, $L = 1m$, $\Delta x = 0.01m$, $\Delta t = t_r/500$.

## 4. Discussion of the results

Figure 6 shows the main result, namely the decrease of the normalized mechanical energy (kinetic + strain energy divide by the initial strain energy) in dependence on time.

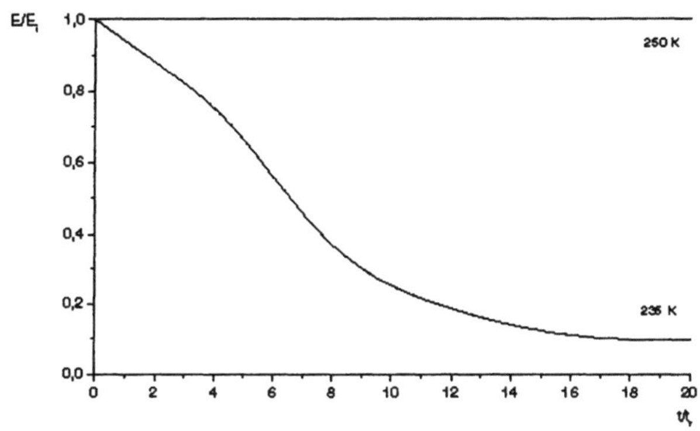


*Figure 6.* Mechanical energy.

The curve shows, that nearly total vibration damping occurs already within a short time. This result is close to that of former calculations, when the adiabatic approximation (i.e., no stress waves) had been used. To get this result, it was necessary to calculate first $u'', \epsilon, \Sigma, \xi, \dot{\Sigma}, \dot{T}$, etc., each at the positions $x_i$ and time steps $t_n$. Out of space limitations these results cannot be presented here. It should only be mentioned, that for these quantities the result differs somehow from the results of the older approximative approach. In a further step parameter studies should be carried out to show firstly the influence of the starting temperature and secondly of the material parameters.

## References

1. Oberaigner, E. R., Fischer, F. D. and Tanaka, K.:
A new micromechanical formulation of Martensite kinetics driven by temperature and/or Stress, *Arch. Appl. Mech.*, 63, 522-533, 1993.
2. Oberaigner, E. R., Fischer, F. D. and Tanaka, K.:
The influence of transformation kinetics on Stress-Strain Relations of Shape Memory Alloys in thermomechanical processes, *J. Intell. Mat. Sys. Struct.*, 5, 474-486, 1994.
3. Tanaka, K., Nishimura, F., Tobushi, H., Oberaigner, E. R. and Fischer, F. D.:
Thermomechanical behavior of an fe-based shape memory Alloy: transformation conditions and hystereses, *Journal de Physique IV*, Colloque C8, 5, C8-463 – C8-468, 1995.
4. Oberaigner, E. R., Fischer, F. D. and Tanaka, K.:
Investigation of the damping behavior of a vibrating shape memory alloy rod using a micromechanical Model, *in Mathematics and Control in Smart Structures* (Varadan, V. V., ed.), SPIE-The International Society for Optical Engineering, Bellingham, 2442, 249-359, 1995 and *Smart Mater. Struct.*, 3, 456-463, 1996.
5. Butkovskiy, A. G.: *Greens Functions and Transfer Functions Handbook*, Ellis Harwood Limited Publishers, Chicester, England, 1982.
6. Bhattacharyya, A., Lagoudas, D. C., Wang, Y., Kinra, V.: On the Role of Thermoelectric Heat Transfer in the Design of SMA Activators: Theoretical modeling and experiment, *Smart Mater. Struct.* 4, 252-263, 1995.
7. Pipkin, A. C.:
*A Course on Integral Equations Texts in Applied Mathematics 9*, Springer Verlag, New York, 1991.
8. Golberg, M. A.:
Editor, *Numerical Solution of Integral Equations, Plenum Press*, New York 1990.
9. William, H. P., Flannery, B. P. (ed.):
*Numerical Recipes*, Cambridge University Press, Cambridge, England, 1986.

# THE EQUILIBRIUM MOTION OF THE MARTENSITIC INTERFACE IN THICK-WALLED INFINITE AUSTENITIC PLATE


B.RANIECKI
*Institute of Fundamental Technological Researches PAS*
*Świętokrzyska 21, 00-049 Warsaw, Poland*

AND

CH. LEXCELLENT
*Laboratoire de Mécanique Appliquée, associée au C.N.R.S.,*
*Faculte des Sciences, 25030 Besançon Cedex, France*


## Notation

$$\boldsymbol{X}\boldsymbol{Y} \rightarrow X_{ij}Y_j \text{ or } X_{ijkl}Y_{kl} \text{ or } H_{ijkl}Y_{klmn} \; ; \; \boldsymbol{X} \cdot \boldsymbol{Y} \rightarrow X_iY_i \text{ or } X_{ij}Y_{ij} \; ;$$

$$X_{ij}^T = X_{ji} \; ; \; \boldsymbol{X} \otimes \boldsymbol{Y} \leftrightarrow X_iY_j \; ; \; \overset{-1}{\boldsymbol{X}} \text{ - inverse of } \boldsymbol{X} \; ; \; \mathbf{1} \text{ - unit tensor}$$

## 1. Preliminaries

**i)** Consider the microregion $R(t)$ subdivided (in actual configuration) by strong discontinuity interface $S(t)$ into parent-phase portion $R_1(t)$ (high-temperature phase-austenite ($\mathcal{A}$)) and product phase portion (low temperature phase-martensite ($\mathcal{M}$)) $R_2$ . $\mathcal{A} \rightarrow \mathcal{M}$ phase transition (p.t.) is regarded as the coherent one, i.e., displacement and temperature $T$ are continuous, whereas deformation gradient $\boldsymbol{F}$, velocity $\boldsymbol{v}$, and Cauchy's stress $\boldsymbol{\sigma}$ experience jump discontinuities at $S(t)$ . The jump in some physical quantity $\psi$ at the interface is denoted by $[\![\psi]\!] = \psi_1 - \psi_2$ , where indexes 1 and 2 indicate a property of parent phase and product phase, respectively. When, at generic instant $t$ , the reference configuration is so chosen that the parent phase particles are identified by their position in the current configuration at that time (region $R_1$) then the geometric and




*P. Argoul et al. (eds.),*
*IUTAM Symposium on Variations of Domains and Free-Boundary Problems in Solid Mechanics*, 45–52.

kinematical compatibilities relations become

$$\boldsymbol{F}_1 = \mathbf{1}\ ; \quad \boldsymbol{F}_2 = \mathbf{1} + \boldsymbol{\lambda} \otimes \boldsymbol{n}\ , \quad [\![\boldsymbol{v}]\!] = c\boldsymbol{\lambda} \tag{1}$$

and, for a quasistatic situation, the equation of balance of internal energy $u$ , mechanical equilibrium equation, and Clausius-Duhem inequality at the interface $S$ can be written in the following form [1]

$$\rho_1 c[\![u]\!] = [\![\boldsymbol{q} \cdot \boldsymbol{n} - \boldsymbol{v} \cdot \boldsymbol{\sigma n}]\!]\ ; \quad [\![\boldsymbol{\sigma}]\!]\boldsymbol{n} = 0 \tag{2}$$

$$T\Delta^s = \rho_1 c\Sigma^f \geq 0\ ; \quad \Sigma^f = [\![\phi]\!] + \boldsymbol{\lambda} \cdot \boldsymbol{\sigma}_1 \boldsymbol{n}/\rho_1 \tag{3}$$

Here $\boldsymbol{\lambda}$ is a vector representing the "amplitude" of actual phase distortion, $\boldsymbol{n}$ is outward unit normal to $S(t)$ pointing into $R_1$, $c$ is the local relative speed of the interface $S(t)$ measured by an observer moving with the parent-phase particle in front of the interface, $\boldsymbol{q}$ is the heat flux, $\rho_1$ denotes the actual mass density of the parent phase, $\phi = u - Ts$ is the specific free energy, $s$ is the specific entropy, and $\Delta^s$ denotes the interfacial entropy production. The driving force $\Sigma^f$ of the interface occurring in $(3)_1$ can be expressed in several different forms (cf. e.g. [1-3]) which will not be discussed here.

**ii)** As it is seen from $(3)_2$ the driving force $\Sigma^f$ depends on the properties of the materials adjacent to the interface as well as on the normal $\boldsymbol{n}$ of the interface. To get the more explicit expression for $\Sigma^f$ suppose that $\boldsymbol{\lambda}$ is infinitesimal and both $\mathcal{A}$ and $\mathcal{M}$ phases are linear elastic. Assume that the properties of the phase mixture are described by the "double well" free energy function

$$\phi(T, \boldsymbol{\varepsilon}) = \min_{\boldsymbol{\varepsilon}}\{\phi_1(T, \boldsymbol{\varepsilon});\ \phi_2(T, \boldsymbol{\varepsilon})\} \tag{4}$$

$$\begin{aligned} \phi_\alpha &= \frac{1}{2\rho_0}(\boldsymbol{\varepsilon} - \boldsymbol{\Gamma}_\alpha) \cdot \boldsymbol{L}(\boldsymbol{\varepsilon} - \boldsymbol{\Gamma}_\alpha) + \phi^0_\alpha(T)\ ; \quad \boldsymbol{\Gamma}_\alpha = \boldsymbol{\Gamma}^0_\alpha + \boldsymbol{\alpha}(T - T_0) \\ \phi^0_\alpha &= C_w[T - T_0 - T ln(T/T_0)] + \overset{*}{u}{}^0_\alpha - T\,\overset{*}{s}{}^0_\alpha \qquad \alpha = 1, 2 \end{aligned} \tag{5}$$

where $C_w, \boldsymbol{L}\ (\boldsymbol{M} = \boldsymbol{L}^{-1})$ and $\boldsymbol{\alpha}$ are specific heat, elastic moduli and thermal expansion coefficients, respectively, $\{T_0, \rho_0,\ \overset{*}{u}{}^0_\alpha,\ \overset{*}{s}{}^0_\alpha\}$ are reference temperature, mass density of $\mathcal{A}$, internal energies and entropies of individual phases in a reference stress-free state, $\boldsymbol{\Gamma}^0_\alpha$ are eigenstrains of individual phases in the reference state, such that $\Delta\boldsymbol{\Gamma} \equiv \boldsymbol{\Gamma}^0_2 - \boldsymbol{\Gamma}^0_1$ is the transformation strain at stress free state. Using the relations of linear elastostatic following from $\boldsymbol{\sigma} = \rho_0 \partial\phi/\partial\boldsymbol{\varepsilon}$ , changing the strain measure from $\boldsymbol{\varepsilon}$ into $\boldsymbol{\varepsilon}' = \boldsymbol{\varepsilon} - \boldsymbol{\Gamma}_1 - \boldsymbol{M}\boldsymbol{\sigma}_1 = [(\boldsymbol{F} + \boldsymbol{F}^T)/2] - \mathbf{1}$ , and applying the method detaily described in [1] one finds eventual form of expressions for $\boldsymbol{\lambda}, [\![\boldsymbol{\varepsilon}]\!]$

and $[\![\boldsymbol{\sigma}]\!]$

$$\begin{aligned}
&\boldsymbol{\lambda}(n) = \overset{-1}{\boldsymbol{Q}}(\boldsymbol{n})[\boldsymbol{L}\Delta\boldsymbol{\Gamma}\boldsymbol{n}] \; ; \quad Q_{in} = L_{ijmn}n_j n_m \\
&-[\![\varepsilon]\!] = (\boldsymbol{\lambda}\otimes\boldsymbol{n} + \boldsymbol{n}\otimes\boldsymbol{\lambda})/2 = \boldsymbol{K}(\boldsymbol{n})\boldsymbol{L}\Delta\boldsymbol{\Gamma} \\
&K_{ijmn} = \{n_i \overset{-1}{Q}_{jm} n_n + n_i \overset{-1}{Q}_{jn} n_m + n_j \overset{-1}{Q}_{im} n_n + n_j \overset{-1}{Q}_{in} n_m\}/4 \\
&-[\![\boldsymbol{\sigma}]\!] = \boldsymbol{\sigma}_2 - \boldsymbol{\sigma}_1 = -\boldsymbol{G}^*(\boldsymbol{n})\Delta\boldsymbol{\Gamma} \; ; \; \boldsymbol{G}^* = \boldsymbol{L} - \boldsymbol{LKL}
\end{aligned} \tag{6}$$

Here $\boldsymbol{Q}$ is the usual "acoustic" tensor, and $\boldsymbol{G}^*$ is called the tensor of "plane moduli" of elasticity (because of the property $\boldsymbol{G}^*\boldsymbol{n} = 0$). Both $\boldsymbol{G}^*$ and $\boldsymbol{K}$ posses a number of interesting mathematical properties [4]. For example, $\boldsymbol{G}^*\boldsymbol{M}\boldsymbol{G}^* = \boldsymbol{G}^*$ , and $\boldsymbol{KL}\Delta\boldsymbol{\Gamma} = \Delta\boldsymbol{\Gamma}$ whenever $2\Delta\boldsymbol{\Gamma} = \boldsymbol{l}\otimes\boldsymbol{n}+\boldsymbol{n}\otimes\boldsymbol{l}$. By substituting (6) into (3) the driving forces $\Sigma^f$ can be expressed (in the considered situation) in the form

$$\begin{aligned}
&\Sigma^f = \pi_0^f(T) + \frac{1}{\rho}\boldsymbol{\sigma}_1\cdot\Delta\boldsymbol{\Gamma} - \frac{1}{\rho}e^{\infty}(\Delta\boldsymbol{\Gamma},\boldsymbol{n}) \; ; \quad e^{\infty} = \Delta\boldsymbol{\Gamma}\cdot\boldsymbol{G}^*\Delta\boldsymbol{\Gamma}/2 \\
&\pi_0^f(T) = \Delta u^* - T\Delta s^* \; ; \quad \Delta u^* = \overset{*}{u}{}_1^0 - \overset{*}{u}{}_2^0 \; ; \quad \Delta s^* = \overset{*}{s}{}_1^0 - \overset{*}{s}{}_2^0
\end{aligned} \tag{7}$$

where $\boldsymbol{\sigma}_1$ is the stress in front of the interface (stress in parent phase), $\rho$ may be identified (within infinitesimal theory) with the mass density in any reference state ($\rho \approx \rho_1 \approx \rho_0$), and $e^{\infty}$ is termed "interfacial coherency energy" (cf. Sec. 2). It vanishes whenever $\Delta\boldsymbol{\Gamma}$ satisfies the compatibility condition, i.e., whenever it can be expressed in terms of simple dyad containing vector $\boldsymbol{n}$ . The equation (7) plays the fundamental role in the general theory of martensitic structures developed by Roitburd [5].
**iii)** It is often postulated (on heuristic ground) that normal speed of the interface $c$ is a constitutive function of the force $\Sigma^f$ , such that inequality (3) is satisfied [3] [6]. If such function is continuous then $c = 0$ when $\Sigma^f = 0$ . The conditions $\Sigma^f = 0$ is, therefore, called the condition of interface equilibrium. If this condition is satisfied at every point of the interface then the shape of interface is stable. The relevant study may be found in [6]. Since $\Sigma^f$ depends on $\boldsymbol{n}$ one can expect that planes are stable interfaces (against arbitrary $\Delta\boldsymbol{\Gamma}$) provided that $\boldsymbol{\sigma}_1$ is uniform field.

In this paper we shall study another type of instability that is characterized by spontaneous growth of the product phase with stable interfaces, i.e., we shall study the situation when $c \neq 0$ even though $\Sigma^f = 0$ .This is typical situation when parent phase reaches the unstable equilibrium state. We shall restrict our attention to the thermally-induced $\mathcal{A} \to \mathcal{M}$ p.t. A free boundary condition at $\mathcal{A} \leftrightarrow \mathcal{M}$ interfaces in a thin wire were formulated in [7]. However, the approach presented in [7] is quite different from our. Moreover, the rod theory does not allow to display the role of the coherency energy.

## 2. Formulation of Free Boundary Conditions at the Interfaces

**i)** Consider an elastic infinite plate of thickness $2L$ . Its normal $\boldsymbol{n}(1,0,0)$ is collinear with $x_1$ direction, such that the plate occupies the region $0 \leq x_1 \leq 2L$ , $-\infty < x_i < +\infty (i = 2,3)$. Suppose that one-dimensional field of eigenstrains $\breve{\boldsymbol{\Gamma}}(x_1)$ is prescribed which is symmetric with respect to the mid-plane $x_1 = L$ ($\breve{\boldsymbol{\Gamma}}(L+\Delta x_1) = \breve{\boldsymbol{\Gamma}}(L-\Delta x)$). No surface tractions act at the boundaries of a plate. The problem is to find the distributions of residual stresses $\breve{\boldsymbol{\sigma}}(x)$ and total strains $\breve{\boldsymbol{\varepsilon}}(x)$ created by field $\breve{\boldsymbol{\Gamma}}$ when $\breve{\boldsymbol{\sigma}} = \boldsymbol{L}[\breve{\boldsymbol{\varepsilon}} - \breve{\boldsymbol{\Gamma}}]$, $\boldsymbol{L}$ =const. Its approximate solution satisfies in infinity the condition of zero resultant forces and moments. It is also symmetric with respect to the mid plane $x_1 = L$ , therefore, we shall restrict further considerations to the half plate $0 \leq x_1 \leq L$ . The solution is (cf. $(6)_{3,4}$) [8].

$$
\begin{aligned}
&\breve{\boldsymbol{\varepsilon}} = \boldsymbol{K}(\boldsymbol{n})\boldsymbol{L}[\breve{\boldsymbol{\Gamma}}(x_1) - \boldsymbol{\Gamma}^{\vee}] + \boldsymbol{\Gamma}^{\vee} \\
&\breve{\boldsymbol{\sigma}} = -\boldsymbol{G}^*(\boldsymbol{n})[\breve{\boldsymbol{\Gamma}}(x_1) - \boldsymbol{\Gamma}^{\vee}] \; ; \quad \boldsymbol{\Gamma}^{\vee} = \frac{1}{L}\int_0^L \breve{\boldsymbol{\Gamma}}(x_1)dx_1
\end{aligned}
\tag{8}
$$

Since we study here the martensitic p.t. it is expedient to neglect the thermal expansion and to identify $\breve{\boldsymbol{\Gamma}}(x_1)$-field with the field of transformation strain $\Delta\boldsymbol{\Gamma}$ i.e., $\breve{\boldsymbol{\Gamma}}(x_1) = \Delta\boldsymbol{\Gamma}$ =const. for $0 \leq x_1 < z^*(t)$ (region $R_2$ occupied by martensite ($\mathcal{M}$)) and $\breve{\boldsymbol{\Gamma}}(x_1) = 0$ for $z^* < x_1 \leq L$ (region $R_1$ occupied by austenity $\mathcal{A}$). In such situation from (8) it follows that the stress field is piece-wise constant

$$
\breve{\boldsymbol{\sigma}} \equiv \boldsymbol{\sigma}_2 = -[(1-z)\boldsymbol{G}^*\Delta\boldsymbol{\Gamma}] \text{ for } \mathcal{M} \, , \; \breve{\boldsymbol{\sigma}} \equiv \boldsymbol{\sigma}_1 = z(t)\boldsymbol{G}^*\Delta\boldsymbol{\Gamma} \text{ for } \mathcal{A}\text{--phase} \tag{9}
$$

Here $z(t) = z^*(t)/L$ is the volume fraction of $\mathcal{M}$, and $z^*(t)$ is the position of the singular interface at time $t$. Hence, one can conclude that $-\boldsymbol{G}^*\Delta\boldsymbol{\Gamma}$ is the interior (interfacial) stress generated in the thin layer ($z \approx 0$) of martensite just after its nucleation in an infinite crystal. Moreover, the mean elastic energy stored in such thin layer is $e^{\infty} = \Delta\boldsymbol{\Gamma}\cdot\boldsymbol{G}^*\Delta\boldsymbol{\Gamma}/2$ . The investigation of extremal conditions for $e^{\infty}$ gives information upon possible mutual orientations of $\boldsymbol{n}$, principal directions of $\Delta\boldsymbol{\Gamma}$ and principal directions of $\boldsymbol{\sigma}_2$ (cf. e.g. [9]).

**ii)** Substitute $(9)_2$ into $(7)_1$ and regard the condition $\Sigma^f = 0$ as the criterion that determines the dependence of $\mathcal{A} \to \mathcal{M}$ transformation temperature $T^{AM}$ upon the position of the interface.

$$\begin{aligned} T^{AM}(t) &= M_s^0 + 2z(t)(e^\infty/\rho\Delta s^*) \; ; \\ M_s^0 &= [\Delta u^* - (e^\infty/\rho)]/\Delta s^* = T^{eq} - (e^\infty/\rho\Delta s^*) \end{aligned} \tag{10}$$

Here $M_s^0$ is the theoretical martensite - start temperature i.e. the highest temperature at which the thin layer of $\mathcal{M}$ can be nucleated in infinite austenitic crystal $(z \approx 0)$, $T^{eq} = \Delta u^*/\Delta s^*$ is the equilibrium temperature in the ideal situation when $e^\infty = 0$. Applying the similar procedure to the reverse $\mathcal{A} \to \mathcal{M}$ p.t. one finds that the lowest temperature $A_s^0$ at which the thin layer of $\mathcal{A}$ nucleates in infinite martensitic crystal is $A_s^0 = [\Delta u^* + (e^\infty/\rho)]/\Delta s^*$ . Hence [5], one concludes that *interfacial coherency energy* $e^\infty$ *is responsible for the bifurcation of* $T^{eq}$ *into* $M_s^0$ *and* $A_s^0$ .

The balance of energy (2) when combined with Fourier law of heat conductivity written for the present situation becomes

$$-k\left[\!\left[\frac{\partial T(x_1,t)}{\partial x_1}\right]\!\right] = \rho\frac{dz^*(t)}{dt}T^{AM}(t)\Delta s^* \quad \text{for } x_1 = z^*(t) \tag{11}$$

where $k$ — is heat conductivity coefficient (assumed to be the same in both phases). Equations (10) and (11) are the basis of the following Stefan-like free boundary value problem. Find continuous temperature field $T(x_1,t)$ and the motion of the interface $z^*(t)$ such that: **a)** $T$ satisfies usual homogeneous heat conductivity equations in the region $R_2(0 < x_1 < z^*(t))$ occupied by $\mathcal{M}$-phase, and in the region $R_1(z^* < x_1 < L)$ occupied by $\mathcal{A}$-phase, and the initial-boundary conditions $T(x_1,0) = T_0 \geq M_s^0$ ; $T(0,t) = T_b \leq M_s^0$ , $\partial T(x_1,t)/\partial x_1 = 0$ for $x_1 = L$ , where $T_0$ and $T_b$ are constants. **b)** At the unknown moving boundary $x_1 = z^*(t)$ should be $T = T^{AM}(t)$ (cf. (10)), and the jump of the temperature gradient must satisfy the condition (11).

## 3. The Solution for the Half-space. Conclusion

**i)** It is expedient to consider first the analogous problem for the half-space (keeping all terms in Eq. (10)) since its solution is a good approximation for the solution of the underlying problem when time $t$ is sufficiently small. To solve the problem for the half-space we have first derived the integral equation for $z^*(t)$ using the method of Lightfoot (cf. [10]). The integral equation was solved by application of the perturbation method, regarding $e = e^\infty/(\rho\Delta s^* M_s^0) = (T^{eq}/M_s^0) - 1$ as a small parameter. The fundamental solution $(e = 0)$ is that of Neumann [10]. The initial motion of the interface was found to be

$$z^*(t) = 2\Lambda\sqrt{at} + \frac{a}{L}\chi\cdot e\cdot t + O(e^2) \; ; \; a \equiv k/\rho C_w \tag{12}$$

where $\Lambda$ and $\chi$ are the functions of dimensionless parameters $(\beta_M \leq \beta_0)$

$$\beta_M = l_c[1-(T_b/M_s^0)] \; ; \; \beta_0 = l_c[(T_0-T_b)/M_s^0] \; ; l_c = C_w/\Delta s^* \tag{13}$$

The dependence of $\Lambda$ on $\beta_0$ and $\beta_M$ can be found from the implicit algebraic equation (Neumann solution)

$$\beta_M = \mathrm{erf}(\Lambda)[\beta_0 + b(\Lambda)] \; ; \; b(\Lambda) \equiv \sqrt{\pi}\Lambda \cdot \exp(\Lambda^2) \cdot \mathrm{erfc}(\Lambda) \tag{14}$$

It is illustrated in Fig. 1. Having known $\Lambda = \Lambda(\beta_0, \beta_M)$ the quantity $\chi$ can be calculated from

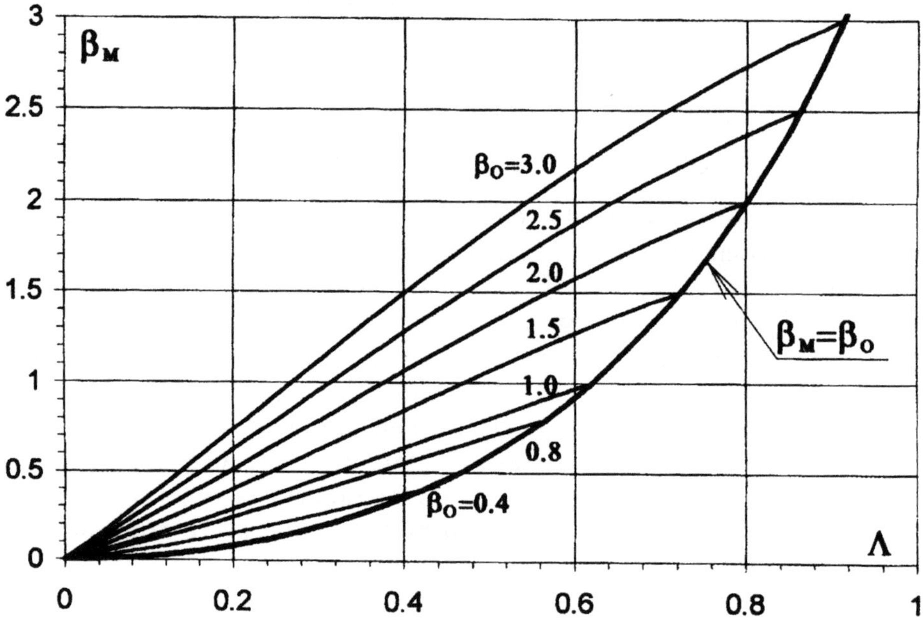


Figure 1

$$\chi = \frac{4\Lambda[\sqrt{\pi}l_c - 2\Lambda I(\Lambda^2)]}{\beta_0 \exp(-\Lambda^2) + \sqrt{\pi}\Lambda[\mathrm{erfc}(\Lambda) - b(\Lambda)] + 2(1+\Lambda^2)I(\Lambda^2)} \tag{15}$$

where $I(y)$ is the Ivantsov function [11]

$$I(y) = ye^y E_1(y) \; ; \; E_1(y) = \int_y^\infty (e^{-z}/z)dz \tag{16}$$

The relationships between $\chi, \beta_0$ and $\Lambda$ are shown in Fig. 2 and 3 for $l_c = 10$ and $l_c = 15$ , respectively

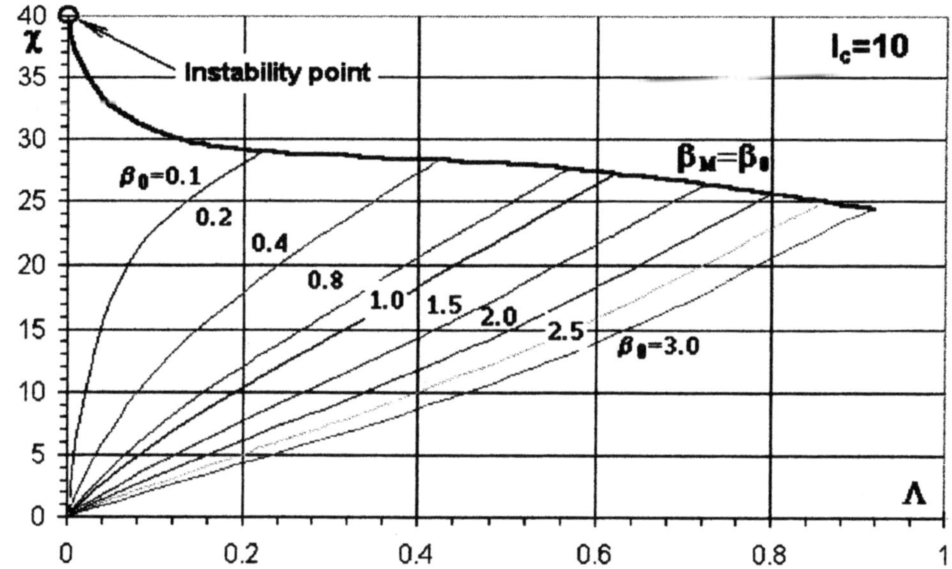


Figure 2

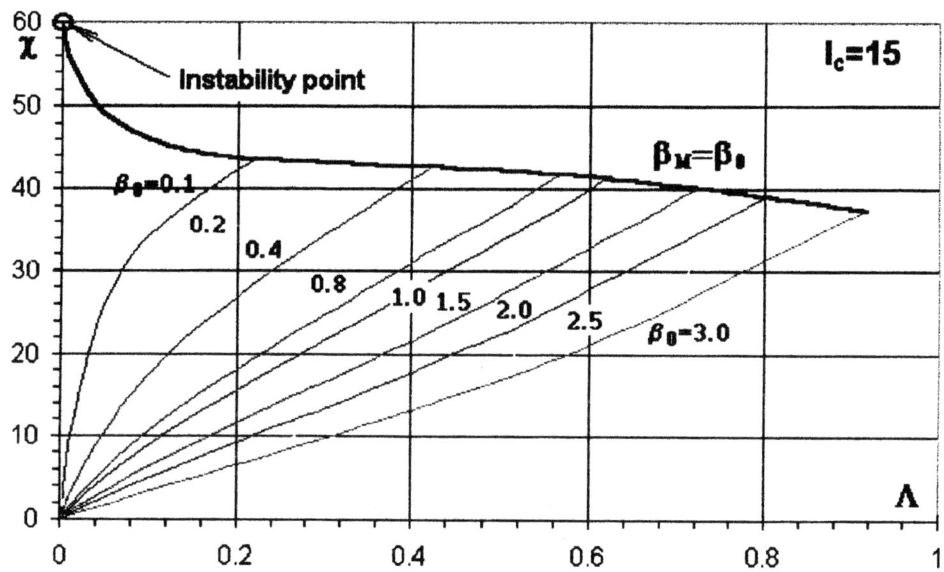


Figure 3

**ii)** Interesting is the special case when the initial temperature of the austenitic half-space is equal to $M_s^0, T_0 = M_s^0$ , so that $\beta_M = \beta_0 \equiv \beta^*$ . When the difference between $M_s^0$ and the temperature of the boundary

$x_1 = 0$ is small then $\beta^*$ is also small. In such case a good approximations for $\Lambda$ and $\chi$ are $\Lambda^2 = \beta^*/2$ and $\chi = 4\sqrt{\pi} l_c / \{\sqrt{\pi}(1+\beta^*) - \sqrt{2\beta^*}[0.57721566 + ln(\beta^*/2) + \pi/2]\}$ . In the limit when $\beta^* \to 0$ one gets $\Lambda = 0$ , $\chi = 4l_c$ and

$$\begin{aligned} c_M^0 &= 4ke/(\rho L \Delta s^*) = 4ke^\infty/[(\rho\Delta s^*)^2 L M_s^0] = \\ &= 4k[(T^{eq}/M_s^0) - 1]/(\rho L \Delta s^*) \neq 0 \end{aligned} \tag{17}$$

where $c_M^0 = \lim_{\beta^* \to 0} c = \lim_{\beta^* \to 0} (dz^*/dt)$ . *Since* $c_M^0 \neq 0$ *then the arbitrary small overcooling of the outer face* $(x_1 = 0)$ *below* $M_s^0$ *causes propagation of the interface.* This clearly indicates that the parent phase $\mathcal{A}$ is in the unstable state at temperature $M_s^0$ . Note the role of interfacial coherency energy $e^\infty$ . If $e^\infty = 0$ then $c_M^0 = 0$ and no instability is observed within first order approximation.

## Acknowledgements

One of us (B.R.) gratefully acknowledges the financial support of the Polish National Committee of Sciences (KBN project No 7T07A02608).

## References

1. Raniecki, B., Tanaka, K. (1994), *Int. J. Engng Sci.*, 32, 12, pp.1845-1859.
2. Heidug, W., Lehner, F.K. (1985), *PAGEOPH*, 132, pp. 91-98.
3. Abeyaratne, R., Knowles, J.K. (1990), *J.Mech.Phys.Solids*, 38, 3, pp. 345-360.
4. Hill, R. (1983), *J.Mech.Phys.Solids*, 31, 4, pp. 347-357.
5. Roitburd, A.L. (1974), *Uspiekhy Fizicheskikh Nauk (Progress in Physical Sciences - in Russian)*, 113, 1, pp.69-104.
6. Liu, I-Sih. (1992), *Continuum Mech.Thermodyn.* 4, pp. 177-186.
7. Bruno, O.P. (1995), *PHYSICAL REVIEW LETTERS*, 74, 5, pp. 746-749.
8. Orlov, S.S., Indenbom, W.L. (1969), *Krystallografia (Crystallography - in Russian)*, 14, 5, pp.780-783.
9. Kostlan, E., Morris Jr, J.W. (1987), *Acta metall.*, 35, 8, pp. 2167-2175.
10. Carslow, H., Jaeger, J. (1959), Conduction of heat in solids, 2nd edition, Chapter XI, Oxford University Press.
11. Kurz, W., Fisher, D.J. (1986), Fundamentals of Solidification, Appendix 7, Trans Tech. Publications, Switzerland-Germany-UK-USA.

# ERROR CONTROL FOR PHASE CHANGE PROBLEMS

R.H. NOCHETTO
*Department of Mathematics and Institute for Physical Science and Technology, University of Maryland*
*College Park, MD 20742, USA*

A. SCHMIDT
*Institut für Angewandte Mathematik, Universität Freiburg*
*7800 Freiburg, Germany*

AND

C. VERDI
*Dipartimento di Matematica, Università di Milano*
*20133 Milano, Italy*

## 1. Introduction

The presence of interfaces, and associated lack of regularity, is responsible for global numerical pollution effects for phase change problems. A cure consists of equidistributing discretization errors in adequate norms by means of highly graded meshes and varying time steps. Their construction relies on *a posteriori* error estimates, which are a fundamental component for the design of reliable and efficient adaptive algorithms for PDEs. These issues have been recently tackled in [6], [7], [8], and are briefly discussed here.

We consider for simplicity the classical two-phase Stefan problem

$$\partial_t u - \Delta\beta(u) = f \quad \text{in } Q = \Omega \times (0, T), \tag{1}$$

where $\beta(s) = \min(s, 0) + \max(s - 1, 0)$; this corresponds to an ideal material with constant thermal properties and unit latent heat. A discrete solution $U$ of (1) satisfies

$$\partial_t U - \Delta\beta(U) = f - \mathcal{R} \quad \text{in } Q, \tag{2}$$

where $\mathcal{R}$, a distribution with singular components and oscillatory behavior, is the so-called *parabolic residual*. Its size is to be determined in negative norms which entail averaging and thus quantify oscillations better.



*P. Argoul et al. (eds.),*
*IUTAM Symposium on Variations of Domains and Free-Boundary Problems in Solid Mechanics*, 53–61.

In §2 we show how to represent the error $e_{\beta(u)} = \beta(u) - \beta(U)$ in terms of $\mathcal{R}$, and in §3 we state an error estimate of the form

$$\|e_{\beta(u)}\|_{L^2(Q)} \leq \mathcal{E}(U, f, T, \Omega; h, \tau),$$

with computable right hand side; hereafter $h$ stands for the meshsize and $\tau$ for the time step. This formula is the basis of the adaptive algorithm of §4 and the simulations of §5. We finally show in §6 that $\mathcal{E}(U, f, T, \Omega; h, \tau) \to 0$ as $h, \tau \to 0$, and thereby that the algorithm converges.

## 2. Error Representation Formula

Upon subtracting (2) from (1) and integrating by parts against a smooth test function $\varphi$, the error $e_u = u - U$ satisfies the equation

$$\int_\Omega e_u(T)\varphi(T) = \int_\Omega e_u(0)\varphi(0) + \int_Q e_u(\partial_t\varphi + b\Delta\varphi) + \mathcal{R}(\varphi), \tag{3}$$

where $b(x,t) = \dfrac{\beta(u) - \beta(U)}{u - U}$ provided $u \neq U$ and $b(x,t) = 1$ otherwise, and

$$\mathcal{R}(\varphi) = \int_\Omega U(0)\varphi(0) - \int_\Omega U(T)\varphi(T) + \int_Q (f\varphi + U\partial_t\varphi + \beta(U)\Delta\varphi). \tag{4}$$

This motivates the study of the backward parabolic problem in *nondivergence* form [4], with vanishing diffusion coefficient $0 \leq b \leq 1$,

$$\partial_t\psi + (b + \delta)\Delta\psi = -b\chi \quad \text{in } Q, \quad \psi(T) = 0 \quad \text{in } \Omega, \tag{5}$$

where $\chi \in L^2(Q)$ and $\delta \downarrow 0$. The theory of nonlinear strictly parabolic problems [3] yields existence of a unique solution $\psi$, which satisfies [6]

$$\|\nabla\psi(t)\|_{L^2(\Omega)},\ \tfrac{1}{2}\|\partial_t\psi\|_{L^2(Q)},\ \delta^{1/2}\|\Delta\psi\|_{L^2(Q)} \leq \|\chi\|_{L^2(Q)}. \tag{6}$$

Equation (5) does not exhibit a regularizing effect in that $\|\Delta\psi\|_{L^2(Q)}$ is never bounded uniformly in $\delta$; compare with [2]. Using (6), together with $be_u = e_{\beta(u)}$ and $|e_u| \leq 1 + |e_{\beta(u)}|$, (3) with $\varphi = \psi$ yields

$$(1 - \delta^{1/2})\|e_{\beta(u)}\|_{L^2(Q)} \leq \|e_u(0)\|_{H^{-1}(\Omega)} + \sup_{\substack{\chi\in L^2(Q)\\ 0<t<T}} \frac{|\mathcal{R}(\psi)|}{\|\nabla\psi(t)\|_{L^2(\Omega)}} + |Q|^{1/2}\delta^{1/2}.$$

Taking $\delta \downarrow 0$ we deduce a representation formula based on evaluating $\mathcal{R}$ in the negative norm $L^1_t H^{-1}_x$, and valid for *any* numerical method:

$$\|e_{\beta(u)}\|_{L^2(Q)} \leq \|e_u(0)\|_{H^{-1}(\Omega)} + \sup_{\substack{\chi\in L^2(Q)\\ 0<t<T}} \frac{|\mathcal{R}(\psi)|}{\|\nabla\psi(t)\|_{L^2(\Omega)}}.$$

## 3. A Posteriori Error Estimators

We denote by $\tau_n$ the time step at the $n$-th step and set $t^n = \sum_{i=1}^n \tau_i$. Let $\mathcal{M}^n$ be a uniformly regular partition of $\Omega$ into simplices, with meshsize density $h_n$, and $\mathcal{B}^n$ be the collection of interior sides $e$ of $\mathcal{M}^n$ in $\Omega$; $h_S$ ($h_e$) stands for the size of $S \in \mathcal{M}^n$ ($e \in \mathcal{B}^n$). Let $\mathbf{V}^n \subset H_0^1(\Omega)$ be the piecewise linear finite element space over $\mathcal{M}^n$ and $I^n$ the Lagrange interpolation operator.

The discrete problem reads as follows. The discrete initial enthalpy $U^0 \in \mathbf{V}^0$ is an "interpolant" of $u_0$. Given $U^{n-1}, \Theta^{n-1} \in \mathbf{V}^{n-1}$, then $\mathcal{M}^{n-1}$ and $\tau_{n-1}$ are modified as described below to get $\mathcal{M}^n$ and $\tau_n$ and thereafter $U^n, \Theta^n \in \mathbf{V}^n$ computed according to $\Theta^n = I^n\beta(U^n)$ and

$$\frac{1}{\tau_n}\int_\Omega I^n((U^n - U^{n-1})\varphi) + \int_\Omega \nabla\Theta^n \cdot \nabla\varphi = \int_\Omega I^n(f(t^n)\varphi) \quad \forall\, \varphi \in \mathbf{V}^n. \quad (7)$$

Using mass lumping and enforcing $\Theta^n = I^n\beta(U^n)$ solely at the nodes introduces some consistency errors but amounts to having a monotone problem easy to implement and solved via an optimized nonlinear SOR [5].

If we set $U_t^n = (U^n - I^nU^{n-1})/\tau_n$ and $R^n = I^nf(t^n) - U_t^n$, integrate (4) by parts, and use Galerkin orthogonality (7), we easily arrive at

$$\begin{aligned}\mathcal{R}(\psi) = \mathcal{C} + \sum_n \int_{t^{n-1}}^{t^n}\int_\Omega \Big(R^n(\psi - \Psi) - \nabla\Theta^n \cdot \nabla(\psi - \Psi)\Big) \\ + \sum_n \int_\Omega (U^{n-1} - I^nU^{n-1})\psi^{n-1} + \sum_{n=1}^N \int_{t^{n-1}}^{t^n}\int_\Omega U_t^n(\psi - \psi^{n-1}),\end{aligned}$$

where $\Psi$ is any discrete approximation of $\psi$ and $\mathcal{C}$ stands for consistency terms. Integrating $\nabla\Theta^n$ by parts elementwise and using (6), we conclude

$$\mathcal{E}(U, f, T, \Omega; h, \tau) = C(\mathcal{E}_0 + \mathcal{E}_1 + \mathcal{E}_2 + \mathcal{E}_3 + \mathcal{E}_4) + \mathcal{E}_C.$$

Here $C$ depends on the minimum angle of $\mathcal{M}^n$, the error indicators $\mathcal{E}_i$ are

$$\begin{aligned}
\mathcal{E}_0 &= \|u_0 - U^0\|_{H^{-1}(\Omega)} && \text{initial error,}\\
\mathcal{E}_1 &= \sum_n \tau_n\Big(\sum_{e\in\mathcal{B}^n} h_e\|[\![\nabla\Theta^n]\!]_e \cdot \nu_e\|_{L^2(e)}^2\Big)^{1/2} && \text{jump residual,}\\
\mathcal{E}_2 &= \sum_n \tau_n\Big(\sum_{S\in\mathcal{M}^n} h_S^2\|R^n\|_{L^2(S)}^2\Big)^{1/2} && \text{interior residual,}\\
\mathcal{E}_3 &= \sum_n \|I^nU^{n-1} - U^{n-1}\|_{H^{-1}(\Omega)} && \text{coarsening,}\\
\mathcal{E}_4 &= \Big(\sum_n \tau_n\|U^n - I^nU^{n-1}\|_{L^2(\Omega)}^2\Big)^{1/2} && \text{time residual,}
\end{aligned}$$

and $\mathcal{E}_C$ stands for consistency errors which we do not specify and could in principle be removed at the expense of complicating the implementation of (7); see [6]. All indicators $\mathcal{E}_i$ can be evaluated in terms of the computed solution, initial datum, and source term, and are essential; they are also present for the heat equation [2]. The weights of $\mathcal{E}_1, \mathcal{E}_2$ correspond to $H^1_x$ regularity of $\psi$, as opposed to $H^2_x$ regularity for the heat equation, thereby reflecting the degenerate nature of (1). An alternative approach exploiting the additional, but nonuniform, regularity $\delta\|\Delta\psi\|_{L^2(Q)} \leq \|\chi\|_{L^2(Q)}$ is proposed in [6] and improves on this distinctive aspect of (1).

## 4. Adaptive Algorithm

The error estimators of §3 entail an $L^1$ or $L^2$ norm in time, which is impractical in that the entire history would be needed to control $\mathcal{E}(U, f, T, \Omega; h, \tau)$. We therefore resort to an $L^\infty$ norm in time, and the ensuing equidistribution strategy of minimizing the spatial degrees of freedom for a uniform error distribution in time. For each $S \in \mathcal{M}^n$, the resulting *local* spatial error indicators are denoted by $E_0(S)$, for initial error, and

$$
\begin{aligned}
E_1(S) &= \tfrac{1}{2}T^2\, h_S\, \|[\![\nabla\Theta^n]\!]_e \cdot \nu_e\|^2_{L^2(\partial S)} && \text{local jump residual,}\\
E_2(S) &= T^2\, h_S^2\, \|R^n\|^2_{L^2(S)} && \text{local interior residual,}\\
E_3(S) &= T^2\, \tau_n^{-2}\, \|I^n U^{n-1} - U^{n-1}\|^2_{L^2(S)} && \text{local coarsening.}
\end{aligned}
$$

Let $\varepsilon$ be a given error tolerance and $M^n = \operatorname{card}\mathcal{M}^n$. The objective is to select adaptively time steps and mesh densities in such a way that $E_i(S)$ have comparable size for all $S \in \mathcal{M}^n$ (equidistribution) and $\mathcal{E}(U, f, T, \Omega; h, \tau) \leq \varepsilon$. Elements $S$ are either refined or coarsened via ***bisection.*** This algorithm creates *compatible* consecutive meshes, extends naturally from 2D to 3D, and is handy for combined refinement/coarsening operations [1].

Given refinement and coarsening parameters satisfying $\Gamma_0+\Gamma_\tau+\Gamma_h \leq 1$, $\gamma_\tau < \Gamma_\tau$, $\gamma_h < \Gamma_h$, the initial mesh is obtained upon bisecting a coarse partition $\mathcal{M}^0$ until $E_0(S) \leq \Gamma_0^2\varepsilon^2/M^0$ for all $S \in \mathcal{M}^0$. Neglecting all terms $\mathcal{E}_C$ for simplicity, the meshsize and time step are modified as follows. We first check whether

$$\frac{\gamma_h^2}{M^n}\,\varepsilon^2 < \max(E_1(S), E_2(S), E_3(S)) < \frac{\Gamma_h^2}{M^n}\,\varepsilon^2 \quad \forall\, S \in \mathcal{M}^n,$$

or not. If the rightmost constraint is violated, then $S$ is bisected, whereas the leftmost constraint is used to flag $S$ for coarsening. The latter will be allowed only if the local error incurred is sufficiently small; this is a critical computational issue. We next verify whether

$$\gamma_\tau\,\varepsilon < T^{1/2}\|U^n - I^n U^{n-1}\|_{L^2(\Omega)} < \Gamma_\tau\,\varepsilon,$$

or not. Failure of the rightmost inequality forces $\tau_n$ to diminish, whereas that of the leftmost constraint causes a corresponding increase of next time step $\tau_{n+1}$ but acceptance of the current time step.

## 5. Numerical Experiments

Our approach is able to detect the presence of interfaces, and refine accordingly, and is insensitive to topological changes such as merging, extinction, and mush or singularity formation. The interface velocity need not be computed explicitly for mesh design, which is a major improvement with respect to [5]. Further simulations, comparisons with several other nonlinear solvers, and a detailed description of the adaptive algorithm (hierarchical data structures, mesh modification strategy, etc.) are examined in [8].

### 5.1. EXAMPLE 1: OSCILLATING CIRCLE

The interface is a circle of center $c(t)$ moving with velocity $V(t) = \dot{c}(t)$ and changing radius $R(t)$. The exact solution is given by

$$u(x,t) = \begin{cases} \alpha(r^2 - R^2) & \text{if } r \leq R \\ 1 + (2\alpha R - V \cdot \frac{x-c}{r} - \dot{R})(r - R) & \text{if } r > R, \end{cases}$$

where $r = |x - c|$ and $\alpha > 0$ is a constant, such that $u > 1$ when $r > R$. The numerical simulation was done in $\Omega = (-1,1)^2$ for $t \in (0,1)$ with

$$c(t) = (0.25, 0.4\sin(10\,t)), \quad R(t) = 0.35 + 0.2\sin(20\,t), \quad \alpha = 17.0\,.$$

Numerical parameters for the adaptive method were

$$\varepsilon = 175,\ \Gamma_0 = 0.1,\ \Gamma_\tau = 0.2,\ \Gamma_h = 0.6,\ \gamma_\tau = 0.155,\ \gamma_h = 0.268\,.$$

Figures 1 and 2 show discrete isothermal lines and corresponding meshes. The innermost isothermal $\Theta = 0$ is the free boundary. Values (and derivatives) of the solution, along with normal interface velocities, exhibit a large variation in time, depending on $R$ and $V$. This explains the various levels of mesh refinement and the variation in time of the element count.

### 5.2. EXAMPLE 2: OSCILLATING SOURCE

This is a phase change in a container $\Omega = (-1,1)^2$, $T = 20.0$, with initial temperature $\Theta(x,0) = 0.1\,x_2$, prescribed temperature at three walls $\Theta(x,t) = 0.1\,x_2$ for $x_2 > -1$, a fourth insulated wall $\partial_\nu\Theta(x,t) = 0$ for $x_2 = -1$, and two circular oscillating heat sources driving the evolution,

$$f(x,t) = \cos(0.2\,t)\max(0.0, 3.125 - 50|x - (-0.2,-0.5)|^2) \\ + \sin(0.2\,t)\max(0.0, 3.125 - 50|x - (-0.2,0.5)|^2).$$

The exact solution is unknown. The upper pictures of Figure 3 show a topology change consisting of two liquid phases merging. The lower pictures

*Figure 1.* Example 1: Isothermal lines $\Theta = 0, 2, 4, 6, \ldots$ at $t = 0.15, 0.20, 0.25, 0.30, 0.35, 0.40$

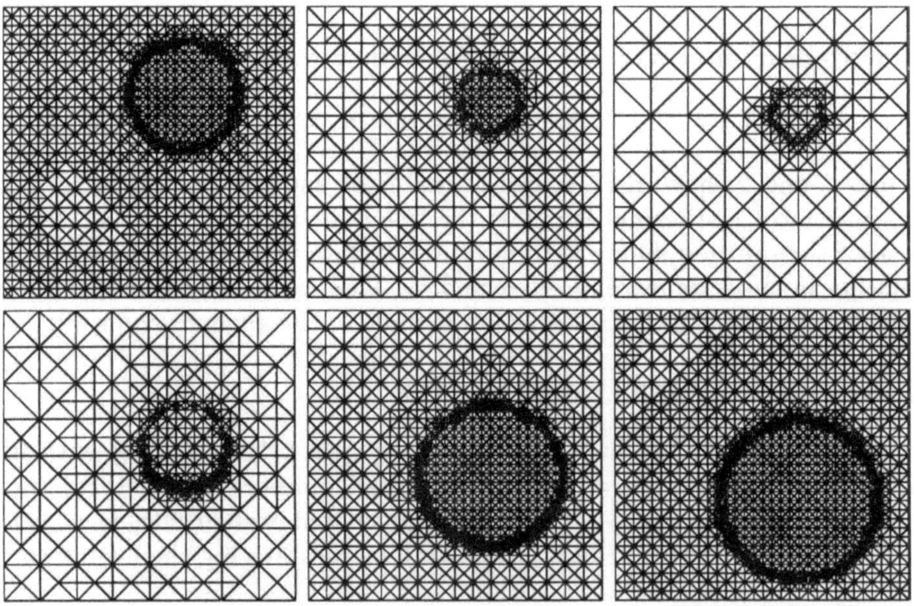

*Figure 2.* Example 1: Meshes at $t = 0.15, 0.20, 0.25, 0.30, 0.35, 0.40$; element count between 450 and 3600

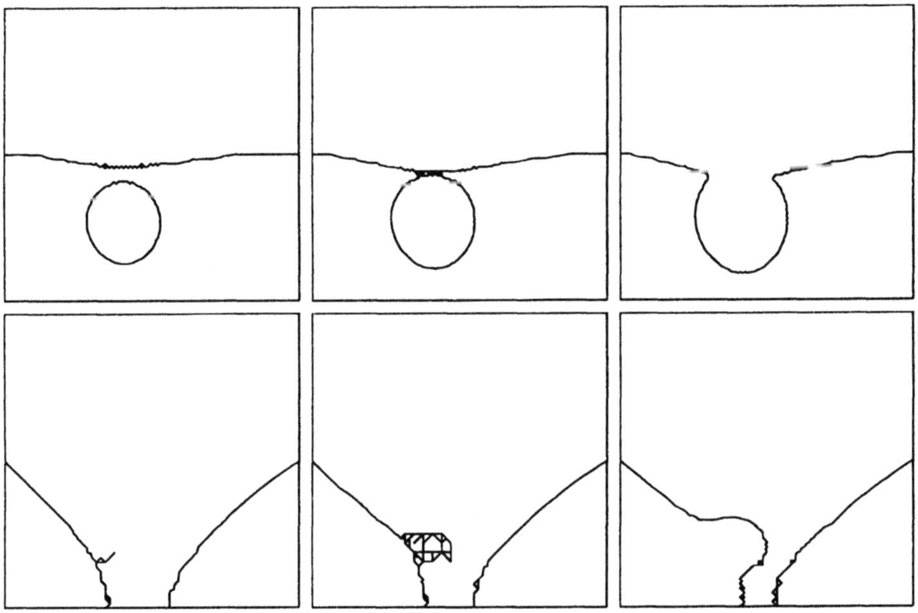

*Figure 3.* Example 2: Interfaces with topology changes and a mushy region at $t = 1.2, 1.6, 2.0, 8.6, 9.6, 10.6$

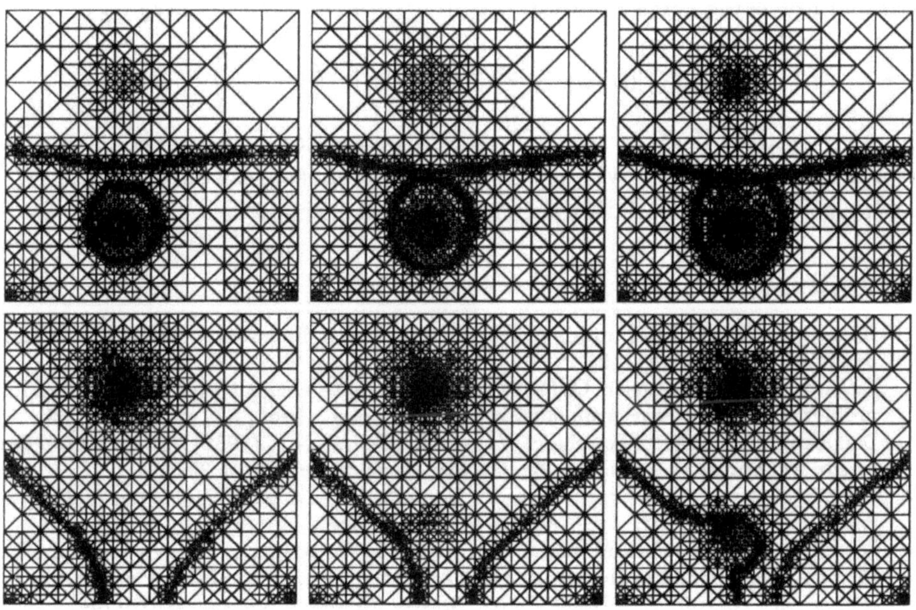

*Figure 4.* Example 2: Meshes at $t = 1.2, 1.6, 2.0, 8.6, 9.6, 10.6$

show a mushy region produced by a cooling source term. Figure 4 depicts the corresponding meshes with highly refined regions; they capture the interface location along with the presence of strong heat sources.

## 6. Convergence

It is not obvious that $\mathcal{E}(U, f, T, \Omega; h, \tau) \to 0$ as $h, \tau \to 0$ because $\mathcal{E}$ depends on discrete quantities that change with $h$ and $\tau$. To guarantee convergence further coarsening restrictions are enforced [7]. Assume that

$$\int_\Omega (I^n (U^{n-1})^2 - I^{n-1}(U^{n-1})^2) \le \tau_{n-1} \int_\Omega |\nabla \Theta^{n-1}|^2,$$

which accounts for the increase in time of $L^2_x$ energy due to mesh coarsening. Then (7) is *weakly* stable:

$$\max_n \|U^n\|^2_{L^2(\Omega)} + \sum_n \|U^n - I^n U^{n-1}\|^2_{L^2(\Omega)} + \sum_n \tau_n \|\nabla \Theta^n\|^2_{L^2(\Omega)} \le C. \quad (8)$$

If the additional, and rather restrictive, coarsening constraint is imposed

$$\|U^{n-1} - I^n U^{n-1}\|_{H^{-1}(\Omega)} \le C\tau_{n-1} \|h_{n-1} \nabla \Theta^{n-1}\|_{L^2(\Omega)}, \quad (9)$$

then the following error estimate results from (8) ($H_n = \max_\Omega h_n$)

$$\|e_{\beta(u)}\|_{L^2(Q)} \le CT \max_n \left(\tau_n + H_n + \tfrac{H_n^2}{\tau_n}\right).$$

This proves that if $H_n = o(\sqrt{\tau_n})$, which allows for highly graded meshes, the adaptive method converges. Condition (9) measures coarsening in $H_x^{-1}$, in which $I^n$ does not exhibit superconvergence properties.

## References


1. Bänsch, E. (1991) Local mesh refinement in 2 and 3 dimensions, *IMPACT Comput. Sci. Engrg.*, **3**, pp. 181–191
2. Eriksson, K. and Johnson C. (1995) Adaptive finite element methods for parabolic problems IV: nonlinear problems, *SIAM J. Numer. Anal.*, **32**, pp. 1750–1763
3. Ladyzenskaja, O.A., Solonnikov, V. and Ural'ceva, N.N (1968) *Linear and Quasilinear Equations of Parabolic Type*. AMS, Providence, TMM 23.
4. Nochetto, R.H. (1987) Error estimates for multidimensional singular parabolic problems, *Japan J. Indust. Appl. Math.*, **4**, pp. 111–138
5. Nochetto, R.H., Paolini, M. and Verdi, C. (1991) An adaptive finite elements method for two-phase Stefan problems in two space dimensions. Part II: implementation and numerical experiments, *SIAM J. Sci. Statist. Comput.*, **12**, pp. 1207–1244
6. Nochetto, R.H., Schmidt, A. and Verdi, C. (1997) A posteriori error estimation and adaptivity for degenerate parabolic problems, *Math. Comp.*, to appear.
7. Nochetto, R.H., Schmidt, A. and Verdi, C. (1997) Adapting meshes and time-steps for phase change problems, *Atti Accad. Naz. Lincei Cl. Sci. Fis. Mat. Natur. Rend. (9) Mat. Appl.*, **8**, pp. 273–292

8. Nochetto, R.H., Schmidt, A. and Verdi, C. (1997) Adaptive solution of phase change problems over unstructured tetrahedral meshes, in Grid Generation and Adaptive Algorithms (M. Luskin *et al.* eds.), Springer Verlag, New York, to appear.

# NUMERICAL SCHEMES FOR THE MEAN CURVATURE FLOW OF GRAPHS

G. DZIUK
*Institut für Angewandte Mathematik*
*Hermann–Herder–Straße 10*
*D–79104 Freiburg i. Br.*
*Germany*

## 1. Introduction

We study the numerical computation of the evolution of $n$–dimensional surfaces which are graphs in $\mathbb{R}^{n+1}$. The law of motion is given by

$$V = -H, \tag{1}$$

where $V$ represents the normal velocity and $H$ is the mean curvature of the interface

$$\Gamma(t) = \{(x, u(x,t) \mid x \in \Omega\}$$

above a bounded domain $\Omega \subset \mathbb{R}^n$. The normal to $\Gamma$ is chosen to be

$$\nu(u) = \frac{(\nabla u, -1)}{Q(u)}, \quad Q(u) = \sqrt{1 + |\nabla u|^2}.$$

and thus the normal velocity can be written as $V = -u_t/Q(u)$. Equation (1) then leads to the quasilinear parabolic problem

$$\begin{aligned} u_t - \sqrt{1+|\nabla u|^2}\, \nabla \cdot \frac{\nabla u}{\sqrt{1+|\nabla u|^2}} &= 0 \text{ in } \Omega \times (0,T) \\ u = u_0 \text{ on } \partial\Omega \times (0,T) &\cup \Omega \times \{0\}. \end{aligned} \tag{2}$$

From the formulation of the problem it is obvious that the Dirichlet boundary conditions are chosen to be independent of time. It is possible to treat more general boundary conditons. But we want to show the main numerical difficulties with this problem for the model case (2).



*P. Argoul et al. (eds.),*
*IUTAM Symposium on Variations of Domains and Free-Boundary Problems in Solid Mechanics*, 63–70.

Theoretical results on existence and uniqueness for (2) are included in [4] and are to the effect that for smooth convex boundary $\partial\Omega$ and smooth $u_0$ there exists a global smooth solution. For domains with boundaries which are not mean convex it is known that the gradient can blow up at the boundary after some time; see [5] for a detailed discussion.

A fairly complicated convergence proof for a semi–discrete Finite Element Method with piecewise linear elements for (2) is contained in [2]. There the authors used differentiation of the equation with respect to time and a deformation technique to prove optimal convergence rates for the case of two space dimensions ($n = 2$). In [3] the corresponding anisotropic problem is treated in arbitrary space dimensions. Theorem 4.1 there provides an adequate existence result for (2) exactly in the spaces which will be needed for the error estimate below.

The main difficulty with our problem (2) is that it is not in divergence form. Nevertheless (2) can be rewritten in a weak form which allows the application of a FEM. For any test function $\phi \in C_0^1(\Omega)$

$$\int_\Omega \frac{u_t\phi}{Q(u)} + \int_\Omega \frac{\nabla u \nabla \phi}{Q(u)} = 0. \tag{3}$$

Since we want to derive error estimates it is a good idea to have a look at the basic energy estimate for the problem. Inserting $\phi = u_t$ as a test function into (3) we get

$$\int_\Omega \frac{u_t^2}{Q(u)} + \frac{d}{dt}\int_\Omega Q(u) = \int_\Gamma V^2 + \frac{d}{dt}|\Gamma| = 0 \tag{4}$$

## 2. Discretisation in Space

Let $\mathcal{T}_h$ be an admissable nondegenerate triangulation of the domain $\Omega$ with mesh size bounded by $h$ and simplices $S$ and let $\Omega_h$ be the corresponding discrete domain. We assume that vertices on $\partial\Omega_h$ are contained in $\partial\Omega$. The discrete space is chosen to be

$$X_h = \left\{ v_h \in C^0(\Omega_h) \mid v_h \text{ is a linear polynomial on each } S \in \mathcal{T}_h \right\}.$$

The subspace containing functions with zero boundary values will be denoted by $X_{h0}$. By $I_h$ we denote the Clément interpolation operator modified in such a way that it respects zero boundary values, see [1] with some modifications.

The semi–discrete approximation of (3) then is to find $u_h(\cdot, t) \in X_h$ with $u_h(\cdot, t) - I_h u_0 \in X_{h0}$ and $u_h(\cdot, 0) = u_{h0} = I_h u_0$ such that

$$\int_{\Omega_h} \frac{u_{ht}\phi_h}{Q(u_h)} + \int_{\Omega_h} \frac{\nabla u_h \nabla \phi_h}{Q(u_h)} = 0 \tag{5}$$

for all $t \in (0,T)$ and all discrete test functions $\phi_h \in X_{h0}$.

Let us now prove the error estimate for the semi-discrete problem. The estimates will depend on the quantities

$$\sup_{(0,T)} \|u_t\|_{L^\infty(\Omega)}, \sup_{(0,T)} \|\nabla u\|_{L^\infty(\Omega)}, \int_0^T \|u_t\|^2_{H^2(\Omega)} \text{ and } \int_0^T \|\nabla u_t\|_{L^\infty(\Omega)} \quad (6)$$

which are finite under suitable assumptions on the data, see [3].

**Theorem 1** *Let $u$ be a solution of the continuous problem (2) Then there exists a unique solution $u_h \in C^1((0,T), X_h)$ of the semi-discrete problem (5) and*

$$\int_0^T \int_{\Omega\cap\Omega_h} \frac{(u_t - u_{ht})^2}{Q(u_h)} + \sup_{(0,T)} \int_{\Omega\cap\Omega_h} |\nu(u) - \nu(u_h)|^2 Q(u_h) \le ch^2$$

*The constant $c$ depends on the norms (6).*

*Proof:* We give the proof only for the case where the domain $\Omega$ is convex so that the discrete domain $\Omega_h$ is a subset of $\Omega$. The general case introduces some technical estimates only. The difference between weak continuous (3) and discrete equation (5) reads

$$\int_{\Omega_h} \frac{u_t - u_{ht}}{Q(u_h)} \varphi_h + \int_{\Omega_h} (\frac{\nabla u}{Q(u)} - \frac{\nabla u_h}{Q(u_h)}) \nabla \varphi_h$$
$$= \int_{\Omega_h} u_t (\frac{1}{Q(u_h)} - \frac{1}{Q(u)}) \varphi_h. \quad (7)$$

By assumption the boundary values do not depend on time and so $u_t \mid_{\partial\Omega} = 0$. We are allowed to choose $\varphi_h = I_h u_t - u_{ht}$ as test function in (7). This gives

$$\int_{\Omega_h} \frac{(u_t - u_{ht})^2}{Q(u_h)} + \int_{\Omega_h} \left(\frac{\nabla u}{Q(u)} - \frac{\nabla u_h}{Q(u_h)}\right) \nabla (u_t - u_{ht})$$
$$= \int_{\Omega_h} \frac{u_t - u_{ht}}{Q(u_h)} (u_t - I_h u_t) + \int_{\Omega_h} \left(\frac{\nabla u}{Q(u)} - \frac{\nabla u_h}{Q(u_h)}\right) \nabla (u_t - I_h u_t)$$
$$+ \int_{\Omega_h} u_t \left(\frac{1}{Q(u_h)} - \frac{1}{Q(u)}\right) (I_h u_t - u_{ht})$$
$$\le \left(\int_{\Omega_h} |\nu(u) - \nu(u_h)|^2 Q(u_h)\right)^{1/2} \{ \|u_t\|_{L^\infty(\Omega)} \left(\int_{\Omega_h} \frac{(u_t - u_{ht})^2}{Q(u_h)}\right)^{1/2}$$
$$+ch\|u_t\|_{H^2(\Omega)} \} + c\left(\int_{\Omega_h} \frac{(u_t - u_{ht})^2}{Q(u_h)}\right)^{1/2} h^2 \|u_t\|_{H^2(\Omega)}$$

where we used the interpolation estimates. With Young's inequality this can be continued as

$$\leq \frac{1}{2}\int_{\Omega_h} \frac{(u_t - u_{ht})^2}{Q(u_h)} + c\int_{\Omega_h} |\nu(u) - \nu(u_h)|^2 Q(u_h) + ch^2$$

where the constants depend on $\|u_t\|_{L^\infty(\Omega)}$ and $\|u_t\|_{H^2(\Omega)}$. Thus we have proved the inequality

$$\frac{1}{2}\int_{\Omega_h} \frac{(u_t - u_{ht})^2}{Q(u_h)} + \int_{\Omega_h} \left(\frac{\nabla u}{Q(u)} - \frac{\nabla u_h}{Q(u_h)}\right) \nabla(u_t - u_{ht}) \\ \leq c\int_{\Omega_h} |\nu(u) - \nu(u_h)|^2 Q(u_h) + ch^2. \tag{8}$$

The gradient term is written as a time derivative plus some error terms.

$$\int_{\Omega_h} \left(\frac{\nabla u}{Q(u)} - \frac{\nabla u_h}{Q(u_h)}\right) \nabla(u_t - u_{ht}) = \frac{d}{dt}\int_{\Omega_h} \frac{Q(u)Q(u_h) - (1 + \nabla u \cdot \nabla u_h)}{Q(u)} \\ + \int_{\Omega_h} \nabla u_t \cdot \left(\frac{\nabla u_h}{Q(u)} - \frac{\nabla u_h}{Q(u_h)} + \frac{\nabla u}{Q(u)} - \frac{1 + \nabla u \cdot \nabla u_h}{Q(u)^2}\frac{\nabla u}{Q(u)}\right).$$

The last term can be estimated by

$$c\|\nabla u_t\|_{L^\infty(\Omega)} \int_{\Omega_h} \frac{|\nabla(u - u_h)|^2}{Q(u_h)}.$$

Since

$$\frac{Q(u)Q(u_h) - (1 + \nabla u \cdot \nabla u_h)}{Q(u)} = \frac{1}{2}|\nu(u) - \nu(u_h)|^2 Q(u_h)$$

we can conclude that

$$\int_{\Omega_h} \frac{(u_t - u_{ht})^2}{Q(u_h)} + \frac{d}{dt}\int_{\Omega_h} |\nu(u) - \nu(u_h)|^2 Q(u_h) \\ \leq c\int_{\Omega_h} |\nu(u) - \nu(u_h)|^2 Q(u_h) + ch^2. \tag{9}$$

Note that here we used the boundedness of $\nabla u$ in the following way:

$$|\nu(u) - \nu(u_h)| \geq \left|\frac{\nabla u}{Q(u)} - \frac{\nabla u_h}{Q(u_h)}\right| \geq \frac{|\nabla(u - u_h)|}{Q(u_h)} - \frac{|\nabla u|}{Q(u)}\frac{|Q(u) - Q(u_h)|}{Q(u_h)} \\ \geq \left(1 - \frac{\|\nabla u\|_{L^\infty(\Omega)}}{\sqrt{1 + \|\nabla u\|^2_{L^\infty(\Omega)}}}\right)\frac{|\nabla(u - u_h)|}{Q(u_h)}.$$

A Gronwall argument completes the proof of the Theorem.

As an easy consequence of the Theorem we have the following estimate for the normal velocities $V = -u_t/Q(u)$ and $V_h := -u_{ht}/Q(u_h)$.

**Corollary 1** *Under the assumptions of Theorem 1 we have that*

$$\int_0^T \int_{\Omega\cap\Omega_h} (V - V_h)^2 Q(u_h) + \sup_{(0,T)} \int_{\Omega\cap\Omega_h} \frac{|\nabla(u-u_h)|^2}{Q(u_h)} \leq ch^2.$$

## 3. Fully Discrete Schemes

The derivation of fully discrete schemes for the mean curvature flow problem is an important task for both stability and efficiency of the algorithm. There are now several approaches to the time discretisation of curvature flow problems (see [3], [6] for example). In the following we discuss three different discretisations of (5) in time and prove stability estimates for them. We shall skip the index $h$ at the function $u$ and at the domain $\Omega$ because the estimates are based on the weak formulations only. We only use these for the proof so that the estimates remain valid both for continuous and spatially discrete solutions.

We use the notation $v_m(x) = v(x, m\tau)$ for $m = 0, \ldots, M$ with time step $\tau > 0$ and $M \leq [T/\tau]$.

The fully implicit scheme is given by

$$\frac{1}{\tau}\int_\Omega \frac{u_m - u_{m-1}}{Q(u_m)}\,\varphi + \int_\Omega \frac{\nabla u_m}{Q(u_m)}\,\nabla\varphi = 0 \quad (m = 1, \ldots, M). \tag{10}$$

So, in every time step one has to solve a highly nonlinear elliptic problem and thus has to use some solver for the nonlinear problem. The numerically most simple time discretisation is the explicit one,

$$\frac{1}{\tau}\int_\Omega \frac{u_m - u_{m-1}}{Q(u_{m-1})}\,\varphi + \int_\Omega \frac{\nabla u_{m-1}}{Q(u_{m-1})}\,\nabla\varphi = 0 \quad (m = 1, \ldots, M), \tag{11}$$

for which numerically one only has to invert the mass matrix (weighted with $Q(u_{m-1})$). If one uses mass lumping, every time step only requires one matrix multiplication. But as it is already for the linear heat equation the case, a restriction on the smallness of the time step size with respect to the grid size $h$ will be necessary. It then certainly is worth trying some semi–implicit scheme for time discretisation. We shall discuss the scheme

$$\frac{1}{\tau}\int_\Omega \frac{u_m - u_{m-1}}{Q(u_{m-1})}\,\varphi + \int_\Omega \frac{\nabla u_m}{Q(u_{m-1})}\,\nabla\varphi = 0 \quad (m = 1, \ldots, M). \tag{12}$$

Here in every time step one has to invert a Laplace–type equation with stiffness matrix weighted by $Q(u_{m-1})$.

As a stability criterion we use the basic energy norms introduced in (4).

**Theorem 2** *Let $u_m$, $m = 0, \ldots, M$ be a solution of the implicit scheme (10). Then*

$$\tau \sum_{m=1}^{M} \int_\Omega \left( \frac{u_m - u_{m-1}}{\tau} \right)^2 \frac{1}{Q(u_m)} + \int_\Omega Q(u_M)$$
$$+ \frac{1}{2} \tau \sum_{m=1}^{M} \int_\Omega \left| \frac{\nu(u_m) - \nu(u_{m-1})}{\sqrt{\tau}} \right|^2 Q(u_{m-1}) = \int_\Omega Q(u_0). \qquad (13)$$

*If $u_m$, $m = 0, \ldots, M$ is a solution of the explicit scheme (11) then*

$$\tau \sum_{m=1}^{M} \int_\Omega \left( \frac{u_m - u_{m-1}}{\tau} \right)^2 \frac{1}{Q(u_{m-1})} + \int_\Omega Q(u_M)$$
$$- \frac{1}{2} \tau \sum_{m=1}^{M} \int_\Omega \left| \frac{\nu(u_m) - \nu(u_{m-1})}{\sqrt{\tau}} \right|^2 Q(u_m) = \int_\Omega Q(u_0). \qquad (14)$$

*For a solution $u_m$, $m = 0, \ldots, M$ of the semi–implicit scheme (12) we have the stability estimate*

$$\tau \sum_{m=1}^{M} \int_\Omega \left( \frac{u_m - u_{m-1}}{\tau} \right)^2 \frac{1}{Q(u_{m-1})} + \int_\Omega Q(u_M)$$
$$+ \tau \sum_{m=1}^{M} \int_\Omega \left( \frac{Q(u_m) - Q(u_{m-1})}{\sqrt{\tau}} \right)^2 \frac{1}{Q(u_{m-1})}$$
$$+ \frac{1}{2} \tau \sum_{m=1}^{M} \int_\Omega \left| \frac{\nu(u_m) - \nu(u_{m-1})}{\sqrt{\tau}} \right|^2 Q(u_m) = \int_\Omega Q(u_0) \qquad (15)$$

It is worth to note that the semi–discrete scheme produces another positive term on the left hand side compared with the fully implicit scheme though it is weighted with $Q(u_{m-1})$ instead of $Q(u_m)$ in the first integral.

*Proof:* The very simple ingredients for the proof of the stability estimates are based on the observation that the gradient of $u_m - u_{m-1}$ can be decomposed in the following ways:

$$\begin{aligned} & |\nabla(u_m - u_{m-1})|^2 \\ = \; & |\nu(u_m) - \nu(u_{m-1})|^2 Q(u_m) Q(u_{m-1}) + (Q(u_m) - Q(u_{m-1}))^2 \\ = \; & Q(u_m)^2 + Q(u_{m-1})^2 - 2 - 2\, \nabla u_m \cdot \nabla u_{m-1} \end{aligned} \qquad (16)$$

In the implicit case we choose

$$\varphi = (u_m - u_{m-1})/\tau$$

as a test function in (10) and get

$$\int_\Omega \left(\frac{u_m - u_{m-1}}{\tau}\right)^2 \frac{1}{Q(u_m)} + \frac{1}{\tau}\int_\Omega Q(u_m) - \frac{1}{Q(u_m)} - \frac{\nabla u_{m-1}\cdot\nabla u_m}{Q(u_m)} = 0.$$

Together with (16) this then implies (13). The explicit case is analogue. For the semi–implicit scheme (12) we observe that

$$\int_\Omega \frac{\nabla u_m}{Q(u_{m-1})}\cdot\nabla(u_m - u_{m-1}) = \int_\Omega \frac{Q(u_m)^2 - 1}{Q(u_{m-1})} - \frac{\nabla u_m \cdot \nabla u_{m-1}}{Q(u_{m-1})}$$

$$= \int_\Omega \frac{Q(u_m)^2}{Q(u_{m-1})} - Q(u_m) + \frac{1}{2}\left|\nu(u_m) - \nu(u_{m-1})\right|^2 Q(u_m).$$

## 4. Example

On the next page we show computational results for a test problem for the equation (2) which illustrates the smoothing property of mean curvature flow. The domain $\Omega$ is the unit disk in $I\!R^2$, the initial data are given by the function $u_0(x_1, x_2) = \sin(4\pi x_1)\sin(4\pi x_2)$ and the Dirichlet data are $u_0|_{\partial\Omega}$. We used 16384 triangles and a time step $\tau = 0.001$. The graphics were produced using the *Grape* visualisation package (see e. g. [7]).

## References


1. Clément, P. (1975) Approximation by Finite Element Functions Using Local Regularization, *RAIRO Anal. Numér.*, **Vol. no. 2**, pp. 77–84
2. Deckelnick, K., Dziuk, G. (1995) Convergence of a Finite Element Method for Non-parametric Mean Curvature Flow, *Numer. Math.*, **Vol. no. 72**, pp. 197–222
3. Deckelnick, K., Dziuk, G. (1996) Discrete Anisotropic Curvature Flow of Graphs, *Preprint Mathematische Fakultät Freiburg i. Br.*, **Nr. 34/1996**
4. Lieberman, G. A. (1986) The First Initial–Boundary Value Problem for Quasilinear Second Order Parabolic Equations, *Ann. Sci. Norm. Sup. Pisa Ser. IV*, **8**, pp. 347–387
5. Oliker, V. I., Uraltseva, N. N. (1991) Evolution of Nonparametric Surfaces with Speed Depending on Curvature II, The Mean Curvature Case, *IMA Preprint Series*, **No. 832**
6. Walkington, N. J. (1996) Algorithms for Computing Motion by Mean Curvature, *SIAM J. Numer. Anal.*, **Vol. 33, no. 6**, pp. 2215–2218
7. Rumpf, M., Schmidt, A. et al (1990) GRAPE, Graphics Programming Environment, *Report SFB 256 Bonn*, **no. 8**

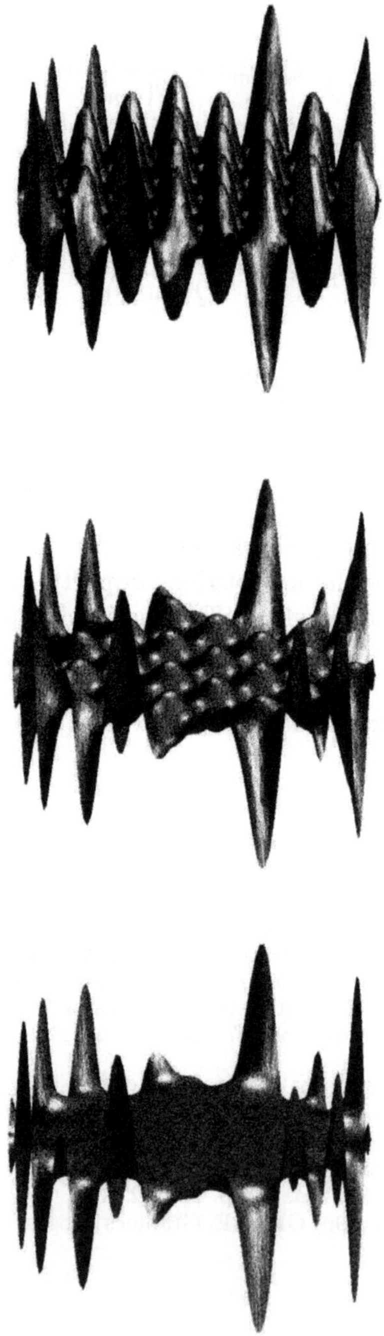

*Figure 1.* Solution at times 0.005, 0.02 and 3.4.

# PHASE DIAGRAMS – HEAT OF MIXING AND INTERFACIAL ENERGY

J. Ansorg and I. Müller
*TU Berlin, Thermodynamik, 10623 Berlin*

**Abstract.** The form of phase diagrams is affected by intermolecular energies and by interfacial energies. The former result in phenomena like azeotropy and miscibility gaps, while the latter deform the domains of the phase diagram, - in particular the miscibility gap.

## 1 Phase diagrams of liquid-vapour equilibrium in a binary mixture, when both phases are ideal mixtures, with and without interfacial penalty.

Under a high pressure a fluid mixture is in the liquid state, under a low pressure it is in the vapour state. The pressure p where the transition occurs depends on the mol fraction $X$ of one constituent of the mixture and it is plotted in the $(p, X)$-phase diagram. Actually boiling upon pressure decrease starts at a higher pressure than condensation upon pressure increase. Therefore the phase diagram contains two curves: the line $B$ of incipient boiling and the line $C$ of incipient condensation.

The phase diagram may be constructed by thermodynamic arguments: Indeed, the minimum of the Gibbs free energy $G$ determines the phase equilibrium. If both phases $'$ and $''$ are ideal mixtures, the free energies $G'$ and $G''$ - at fixed $p$ - are catenary-like curves, see Figure 1; and one might have expected that the transition $' \rightarrow ''$ occurs - upon an increase of $X$ - where the two curves intersect. This does not happen, however, because there is a range of mol fractions in which the phase *mixture* is energetically more favourable than single phases.

The free energy of the phase mixture is essentially the sum of the free energies of the two phases. But there may also be a penalty term due to the interfaces between the phases. If both phases are ideal mixtures we have

$$G = G'_{id} + G''_{id} + PX'X'',$$

with $X', X''$ as mol fractions of the phases, so that the penalty term is proportional to the expectation value of the number of interfaces. $P$ is the *interfacial penalty factor.*

If $P = 0$ holds, the free energy of the phase mixture lies on the common tangent of $G'$ and $G''$ and the points of incipient boiling and condensation in the $(p, X)$-diagram can be constructed as shown on the left-hand side of Figure 1 for one value of $p$, namely $p = p^*$. The states of liquid and vapour during the phase transition lie on the boiling and condensation lines; see the solid curves on the left-hand side of Figure 1.



*P. Argoul et al. (eds.),*
*IUTAM Symposium on Variations of Domains and Free-Boundary Problems in Solid Mechanics,* 71–76.

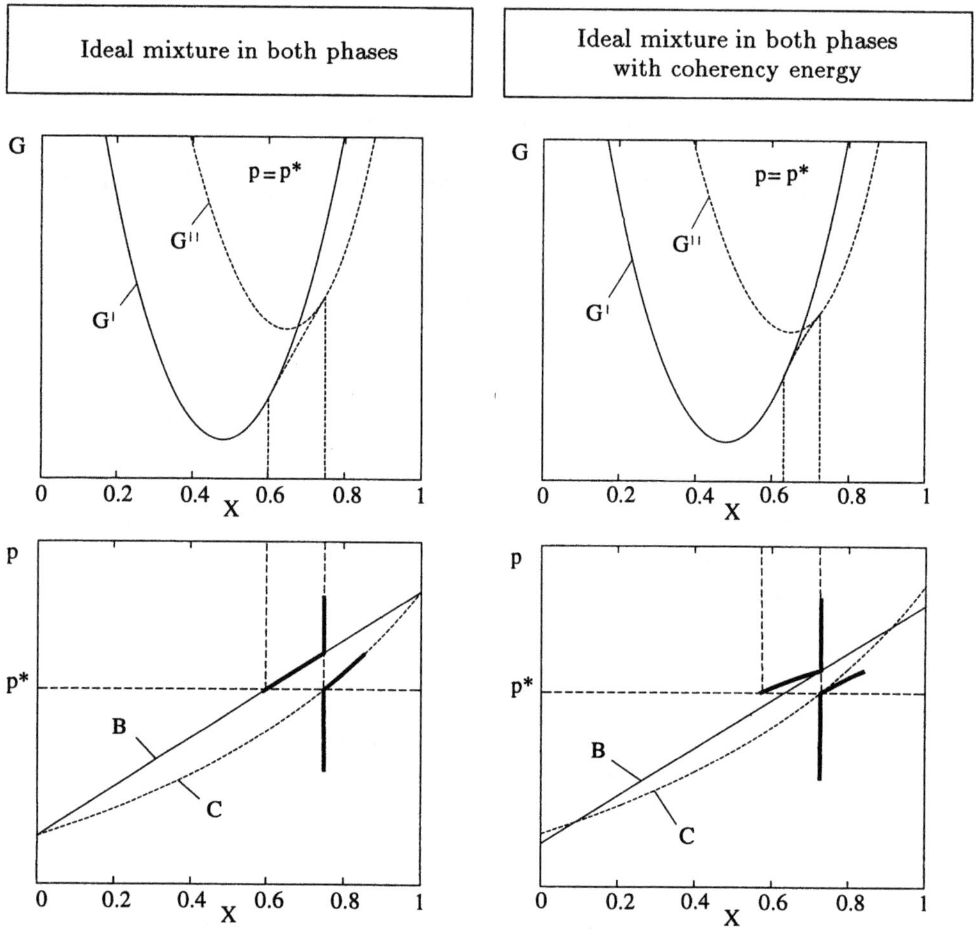


Figure 1 Gibbs free energies and phase diagrams without (left) and with (right) interfacial penalty.

If $P > 0$ holds, the free energy of the phase mixture lies on a concave curve squeezed smoothly into the corner where $G'$ and $G''$ intersect. It follows that the boiling line and the condensation line intersect and that the pure constituents exhibit a delay in the onset of boiling and condensation, - a phenomenon called supersaturation. Upon reflection it is understood that these phenomena are due to the necessity to furnish interfacial energy before the phase transition can start. Moreover, it follows upon reflection and calculation, e.g. see [1], [2], that the states of liquid and vapour during the phase transition do not lie on the boiling and condensation lines any more, but on the "spikes" that stick out of those lines, see the right-hand side of Figure 1.

## 2 Phase diagram of liquid-vapour equilibrium in a binary mixture without interfacial penalty, when the vapour is ideal but the liquid requires heat of mixing.

When two constituents are mixed, unequal molecules become next neighbours and often the formation of unequal pairs is energetically unfavourable. The mixture is then not ideal and a heat of mixing occurs. If only the liquid is non-ideal the free energy of the phase mixture reads

$$G = G'_{id} + G''_{id} + eX'_1X'_2,$$

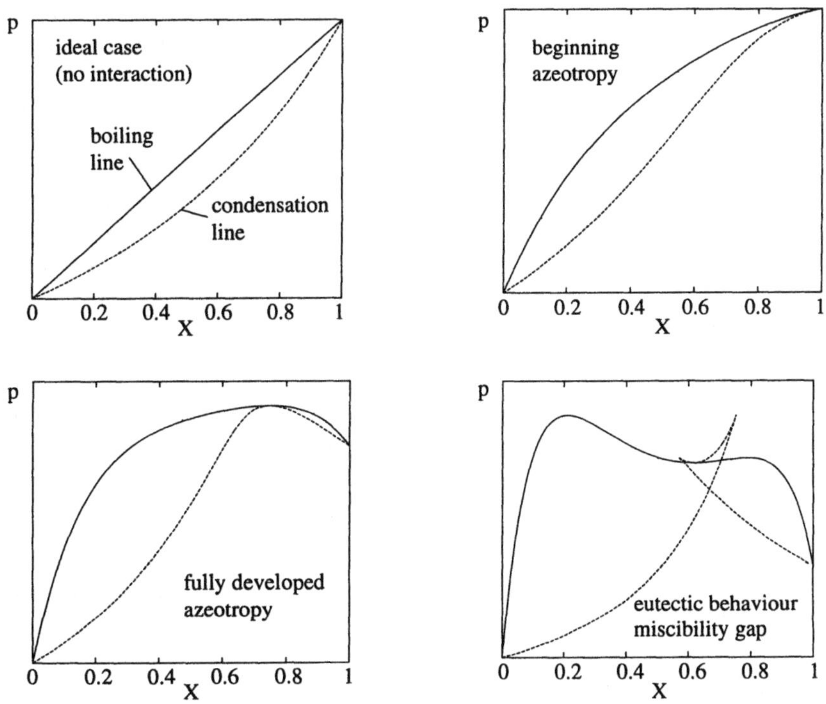


Figure 2 Phase diagrams for different values of molecular penalty.

where $X'_1, X'_2$ are the mol fractions of the constituents 1 and 2 in the liquid phase. $e$ is the *molecular penalty factor*, the energy required for the formation of unequal pairs of neighbouring molecules.

If $e > 0$ holds, the molecules of the liquid are less firmly attached to each other than without molecular penalty and therefore both boiling and condensation start at higher pressures: The boiling and condensation lines are deformed as shown in Figure 2 for increasing values of $e$, and the phenomena of azeotropy and immiscibility occur.

The fantastic complexity of the condensation line for the eutectic mixture is spurious in part, because all states above the intersection of the curve with itself are unstable. Let us consider this.

If $e$ is big enough, the penalty term in the liquid phase changes the ideal catenary of the free energy $G'_{id}$ by pushing a bump into it, see Figure 3. The figure illustrates the relative movement of the curves $G'_{id} + eX'_1X'_2$ and $G''_{id}$ and it shows a common tangent - or common tangents - when the curves intersect. The phase diagram can be constructed as before by projection of the common tangent(s) onto the appropriate line of constant pressure in a $(p, X)$-diagram. Thus two "ears" appear in that diagram. These are no longer relevant for $p > p_E$ where the common tangent of the convex parts of $G'$ lies lower than the common tangents of $G'$ and $G''$. From that pressure on upwards no stable equilibrium between liquid and vapour can occur. There is an equilibrium instead between two liquid phases with different values of $X$. In-between those values we have a miscibility gap.

The miscibility gap begins at the *eutectic line* for $p = p_E$ where equilibrium is possible between three phases: the vapour and the two immiscible liquids. The right-hand side of Figure 3 shows the phase diagram, where the relevant lines are the broad solid ones. As a rule the width of the miscibility gap grows with increasing pressure; that aspect is neglected in Figure 3, however.

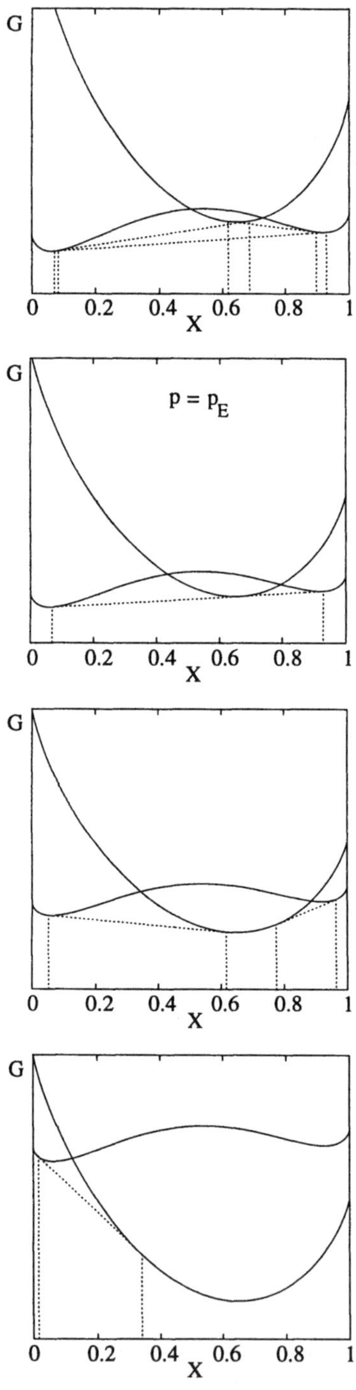


Figure 3

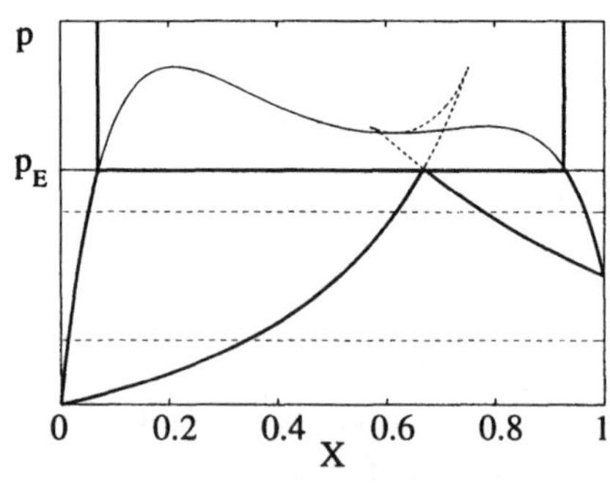


On the construction of a phase diagram with miscibility gap. (No interfacial penalty.)

## 3 Phase diagram of liquid-vapour equilibrium in a binary mixture with a miscibility gap in the liquid phase and interfacial penalty

When we have *both*, a molecular penalty in the liquid and an interfacial penalty between liquid and vapour the miscibility gap and the eutectic line will be affected as shown in Figure 4. The left-hand side of the figure has no interfacial energy but the right-hand side does.

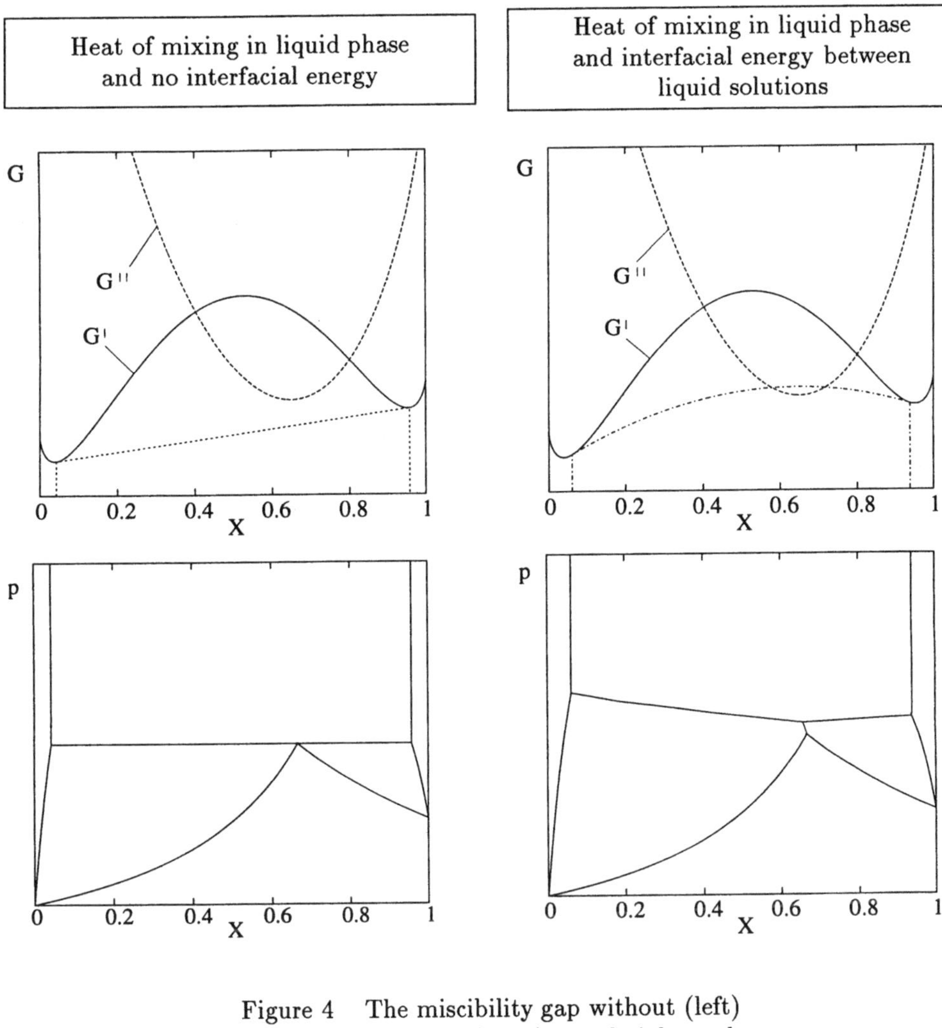


Figure 4 The miscibility gap without (left) and with (right) interfacial penalty.

The difference is due to the fact that the interfacial penalty converts the common tangent into a concave curve that fits smoothly into the lateral convex parts of $G'$. Thus $G'$ and $G''$ have stable common tangents below the concave equilibrium curve. These stable common tangents intersect and thus they draw out the eutectic point into a short line. Even when $G''$ lies entirely above the concave equilibrium curve, there is a short range of pressures where the common tangent - or tangents - of $G'$ and $G''$ lie below the concave curve. Therefore the formerly horizontal eutectic line is converted into two parts, both not straight. The details will depend on the exact shapes of $G'$ and $G''$, but there are two universal features:

- the miscibility gap is decreased
- the "ears" are enlarged.

Both are due to the fact, of course, that the fluid is reluctant to create interfaces between its phases, because interfaces cost energy.

## References

[1] Cahn, J.W. and Larché, F., 'A simple model for coherent equilibrium', *Acta Metall.* **32** (1984) pp. 1915-1923

[2] Müller, I., 'Boiling and condensation with interfacial energy', *Meccanica* **31** (1996) pp. 489-494

# PROPAGATION OF A FRONT BY KINK MOTION

*– From a Discrete Model to a Continuum Model*

ROHAN ABEYARATNE AND SRIKANTH VEDANTAM
*Massachusetts Institute of Technology*
*Cambridge, MA 02139, USA*

## 1. Introduction.

There are numerous problems in which an interface propagates through a body as it is loaded, where, when a material particle crosses this moving interface, it transforms from a metastable state to a more stable state. Examples of this include twin boundary and phase boundary motion in solids. Typically, such an interface propagates in a direction normal to itself at a speed, which in continuum theory, is determined by a *kinetic law*. The kinetic law is viewed as providing a continuum-scale description of the micromechanical processes underlying the front motion. In this paper we use a simple mechanical analog to illustrate how the kinetic law can be calculated by modeling the microscale mechanisms.

In many cases, the propagation of the interface in the apparently normal direction is in fact achieved by the rapid sideways propagation, at the micro-scale, of a "kink" (or step or ledge) along the front in a transverse direction (Figure 1), see for example Section 8.5 of [1]. A *mechanical analog* of this phenomenon is provided by the motion of an infinite chain of inter-connected particles on a corrugated surface. In some reference state, all of the particles lie within the same corrugation. The motion occurs because of forces that are applied on the particles in a direction normal to the corrugations. In the motion of interest, the chain of particles moves forward, from one corrugation to the next, but it does this one particle at a time. Thus, at some generic instant, all of the particles to the left of some particular one have advanced to the next corrugation whereas those to its right are still in the original corrugation. We are interested in calculating the velocity with which this pattern propagates along the chain (thereby determining the associated kinetic law).

After formulating the problem in Section 2, we show in Section 3 that the natural continuum approximation of this problem does *not* determine



*P. Argoul et al. (eds.),*
*IUTAM Symposium on Variations of Domains and Free-Boundary Problems in Solid Mechanics*, 77–84.

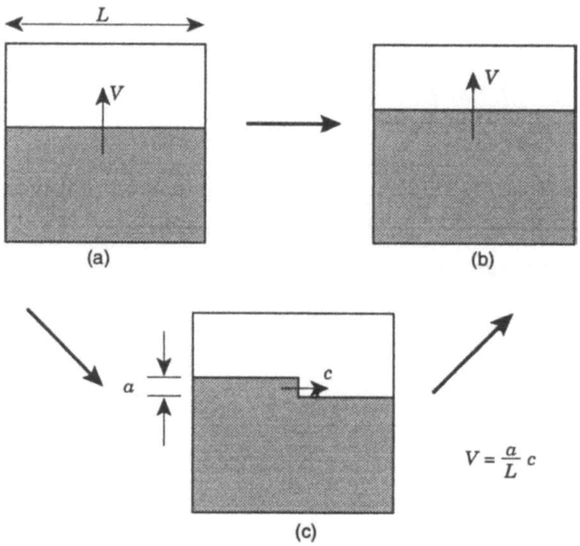


*Figure 1.* **The apparently forward motion of the interface from (a) to (b) is achieved by the sideways propagation of a kink as shown in (c).**

the propagation speed. In Section 4 we return to the discrete problem and determine the kinetic law by solving it. The result is then specialized to the continuum limit. In Section 5 we construct an improved continuum model which yields the same kinetic law as that obtained from the continuum limit of the discrete model. The results are described in Section 6. The present paper summarizes results that are described more fully in [2].

## 2. Formulation.

Consider a corrugated surface $z = \widehat{W}(x,y)$ which, on the average, is a plane that is parallel to the $x,y$-plane. Suppose that the corrugations extend uniformly in the $x$-direction, and that they are periodic with period $a$ in the $y$-direction:

$$\widehat{W}(x,y) = W(y) = W(y+a) \quad \text{for} \quad -\infty < x < \infty, \ -\infty < y < \infty. \qquad (1)$$

The local minima of $W$ are located at $y = ia$, $i = 0, \pm 1, \pm 2, \pm 3, \ldots$ The interval

$$ia - a/2 \ < \ y \ < \ ia + a/2, \qquad i = 0, \pm 1, \pm 2, \ldots, \qquad (2)$$

represents the *ith energy-well* of $W$ and $y = ia$ denotes the bottom of this energy-well.

Now consider an infinite chain of particles which lie in equilibrium at the bottom of the *zeroth* energy-well, i.e. along the $x$-axis. Each particle has mass $m$ and interacts with its two immediate neighbors through a central-force potential $\phi(r)$ where $r$ is the distance between a pair of particles. The associated force is $-\phi'(r)$. In this equilibrium configuration, the spacing between any pair of particles is $a$ and the corresponding force between them is $T = -\phi'(a)$. It is convenient to set $s = T/a$ so that $s$ has the dimensions of stiffness.

Suppose now that a constant force $f > 0$ is applied to each particle in the $y$-direction and assume that the particles are constrained so that they can only move in the $y$-direction. If $y = u_n(t)$ denotes the position of the $nth$ particle at time $t$, the y-component of the force applied to that particle by its two adjacent particles is $-\phi'(r_n)\ (u_{n+1} - u_n)/r_n\ + \phi'(r_{n-1})\ (u_n - u_{n-1})/r_{n-1}$ where $r_n^2 = a^2 + (u_{n+1} - u_n)^2$. If $u_{n+1} - u_n << a$, then on neglecting terms of $O((u_{n+1} - u_n)^3)$, this simplifies to $s(u_{n+1} - u_n) - s(u_n - u_{n-1})$. Therefore the equation of motion of the $nth$ particle is

$$m\ddot{u}_n = s(u_{n+1} - 2u_n + u_{n-1}) \ - \ W'(u_n) \ + f, \qquad n = 0, \pm 1, \pm 2, \dots . \quad (3)$$

The $nth$ particle is *not* of course restricted to being located in any particular energy-well and is free to move between energy-wells.

In this mechanical analog, the row of particles represents the "atoms" along a propagating front. The periodic potential $W$ represents the effect of the rest of the crystal on these particles. The particle spacing $a$ represents the interatomic distance; the forces $f$ represent the shear stress applied on the crystal, and $\mu = W''(0)$ represents the infinitesimal shear modulus of the material. Thus it is convenient to set

$$\sigma = f/a^2, \quad G = \mu/a, \quad \rho = m/a^3, \quad c_s = \sqrt{G/\rho}, \quad \text{and} \quad \Omega = \sqrt{\mu/s}, \quad (4)$$

so that $\sigma$ represents the shear stress, $G$ the shear modulus, $\rho$ the mass density and $c_s$ the shear wave velocity. Weiner and Sanders state that the ratio $\Omega^2$ between the two stiffnesses $\mu$ and $s$ represents the ratio of noncentral to central forces, see [3].

The applied forces $f (> 0)$ tend to cause the chain of particles to move forward from the *zeroth* energy-well to the $1^{st}$ energy-well. Suppose that this occurs by the steady motion of a *kink* to the right, i.e. suppose that each particle remains in the *zeroth* energy-well until the kink reaches it, at which time it moves over into the $1^{st}$ energy-well. Thus, the chain advances in a steady pattern "one particle at a time". More specifically, suppose that the $nth$ particle remains in the *zeroth* energy-well for all times less than a certain critical instant $n\tau$, and that at $t = n\tau$ it moves into the $1^{st}$ energy-well, and remains there for all subsequent times. This motion continues steadily with the $(n+1)th$ particle moving into the $1^{st}$ energy-well at time $t = (n+1)\tau$ and so on. If $c$ denotes the speed with which the kink propagates to the right, and $v$ is its non-dimensionalization with respect to the shear wave speed, then

$$c = a/\tau, \qquad v = c/c_s. \tag{5}$$

Note that at the instant $t = n\tau$, all of the particles to the left of the $nth$ one lie in the $1^{st}$ energy-well, while all of the particles to its right lie in the *zeroth* energy-well. The $nth$ particle itself is located at the boundary between the two energy-wells. Thus, for each $n$,

$$u_n(t) \begin{cases} \in \ (-\frac{a}{2}, \frac{a}{2}) & \text{for } -\infty < t < n\tau, \\ = \ \frac{a}{2} & \text{for } \ t = n\tau, \\ \in \ \left(\frac{a}{2}, \frac{3a}{2}\right) & \text{for } \ n\tau < t < \infty. \end{cases} \tag{6}$$

Since we are limiting attention to *steady* motions of the preceding type, the displacement of the $nth$ particle at time $t$ must equal the displacement of the $(n+m)th$ particle at time $t + m\tau$:

$$u_n(t) = u_{n+m}(t + m\tau) \tag{7}$$

for all integers $n$ and $m$.

In *summary* the steady motion of a kink is characterized by the solutions $u_n(t)$ of the set of differential equations (3) subject to the range requirement (6) and the periodicity requirement (7). Our interest here is primarily in determining the propagation speed $v$.

Frenkel-Kontorova models of this type have been studied extensively in the context of dislocation motion, see for example, [3-6]. In contrast to most of these studies, our interest here is in the continuum implications of the discrete model.

## 3. Analysis of kink motion: a continuum model.

In order to analyze the preceding discrete problem it is natural to consider replacing it by an appropriate continuum approximation. Let $u(x,t)$ be defined for $-\infty < x < \infty, -\infty < t < \infty$ and be such that

$$u(na, t) = u_n(t), \qquad n = 0, \pm 1, \pm 2, .... \tag{8}$$

Then it is natural to replace the system of ordinary differential equations (3) by the single partial differential equation

$$m\frac{\partial^2 u}{\partial t^2} = sa^2\frac{\partial^2 u}{\partial x^2} - W'(u) + f. \tag{9}$$

In order to study the steady motion of a kink, the periodicity requirement (7) suggests that we seek traveling wave type solutions of the form

$$u(x,t) = \hat{u}(\xi), \qquad \xi = x - ct. \tag{10}$$

The range condition (6) requires that

$$\hat{u}(\xi) \begin{cases} \in \left(-\frac{a}{2}, \frac{a}{2}\right) & \text{for } \xi > 0, \\ = \frac{a}{2}, & \text{for } \xi = 0, \\ \in \left(\frac{a}{2}, \frac{3a}{2}\right) & \text{for } \xi < 0. \end{cases} \tag{11}$$

The discrete problem (3), (6), (7) has now been replaced by the continuous problem (9)-(11). This latter problem can be analyzed using a phase plane analysis but we shall not do so here because a similar calculation cannot be carried out for the improved continuum model to be considered later.

In order to obtain an explicit solution, consider the *piecewise parabolic energy*

$$W(u) = (\mu/2)\ (u - na)^2 \quad \text{for } na - a/2 < u < na + a/2, \quad n = 0, \pm 1, \pm 2, \ldots \tag{12}$$

Because of (10)-(12), (4) and (5), the differential equation (9) can be written explicitly as

$$\frac{a}{\Omega^2}\ (v^2\Omega^2 - 1)\hat{u}'' + \frac{\hat{u}}{a} = \begin{cases} \frac{\sigma}{G} & \text{for } \ \xi > 0, \\ \frac{\sigma}{G} + 1 & \text{for } \ \xi < 0. \end{cases} \tag{13}$$

There are three cases to consider: *(i)* $1 - v^2\Omega^2 > 0$, *(ii)* $1 - v^2\Omega^2 < 0$ and *(iii)* $1 - v^2\Omega^2 = 0$. Note first that if a solution involves a shock, i.e. a discontinuity in $\hat{u}'$, the shock propagation speed must equal the wave speed associated with (9): $c = (sa^2/m)^{1/2}$ or equivalently $v = 1/\Omega$. Therefore in cases (i) and (ii) one only needs to consider $C^1$ solutions.

On solving (13) in case (i) and enforcing the range conditions (11) and the continuity of $\hat{u}$ and $\hat{u}'$ at $\xi = 0$, one finds that necessarily $\sigma = 0$ and that then $v$ can take any value less than $1/\Omega$.

In case (ii), the solution of (13) involves waves that propagate with a group velocity that is less than $c$. If we assume that such waves cannot be present ahead of the kink (whose speed is $c$), the terms which describe these waves must be omitted on the upstream side $\xi > 0$. Enforcing the requisite smoothness and the requirement $(11)_2$ leads to a function $\hat{u}$ which violates $(11)_3$. Thus there is no solution in this case.

In case (iii), (13) shows immediately that $\hat{u}$ itself is necessarily discontinuous.

Therefore we are led to the conclusion that the continuous problem (9)-(12) has a ($C^1$ or piecewise $C^1$) solution only if $\sigma = 0$ and that in this case the propagation speed $v(< 1/\Omega)$ is arbitrary.

## 4. Analysis of kink motion: the discrete model.

Since the preceding continuous problem is not very informative, we now return to the discrete problem. The periodicity condition (7) requires the displacement $u_n(t)$ to have the traveling wave form $u_n(t) = \bar{u}(\xi)$, $\xi = na - ct$. The discrete problem (3), (6), (7), (12) can be solved explicitly using Fourier transforms, see [2, 4], leading eventually to the following relation between the stress $\sigma$ and the propagation speed $v$:

$$\sigma = G\Omega^2 \sum_{j=1}^{M} \frac{1}{r_j |D'(r_j)|}, \tag{14}$$

$$D(k) = -v^2\Omega^2 a^2 k^2 + \Omega^2 + 4\sin^2(ak/2), \tag{15}$$

and the $2M$ real roots of the equation $D(k) = 0$ have been denoted by $k = \pm r_1, \pm r_2, \pm r_3, \ldots \pm r_M$ with $r_i > 0$.

Our interest is in the implications of this propagation law at the continuum level. The only waves which are "visible" at the continuum scale are those which have a long wavelength compared to the lattice dimension $a$, i.e. those with $ka << 1$. Thus the propagation law at the continuum scale is obtained from (14) by retaining only the roots $r_j$ relevant to the case $ka << 1$.

Thus we approximate $D(k)$, for small $ka$, by

$$D = \Omega^2 + (1 - v^2\Omega^2)(ka)^2 - \frac{1}{12}(ka)^4 + O((ka)^6). \tag{16}$$

If one formally replaces (15) with (16) in (14) one obtains the following explicit continuum-scale propagation law (kinetic law)

$$\sigma = \frac{G}{4}\left\{1 - \frac{1 - v^2\Omega^2}{\sqrt{(1 - v^2\Omega^2)^2 + \Omega^2/3}}\right\}, \tag{17}$$

which relates the applied stress $\sigma$ to the kink propagation speed $v$.

## 5. Analysis of kink motion: an improved continuum model.

The discrete model of the preceding section led to a kinetic law characterizing kink propagation whereas the previous continuum model did not. This points to the inadequacy of the continuum model rather than the absence of a kinetic law. On plotting a graph of the solution $\bar{u}(\xi)$ of the discrete problem one finds that $\bar{u}(\xi)$ undergoes rapid variations near $\xi = 0$, suggesting that the continuum approximation (9) may not be sufficiently accurate in the vicinity of the kink.

This suggests that we consider replacing the set of ordinary differential equations (3) by the partial differential equation

$$m\frac{\partial^2 u}{\partial t^2} = sa^2\frac{\partial^2 u}{\partial x^2} + \frac{sa^4}{12}\frac{\partial^4 u}{\partial x^4} - W'(u) + f, \tag{18}$$

where we have now kept a higher-order derivative when approximating $u_{n+1} - 2u_n + u_{n-1}$, cf. (9). It is expected that the term involving the coefficient $a^4$ will be very small except in regions where $u$ varies rapidly. The solution of (18), (10) for the piecewise parabolic energy (12) can be readily shown to be

$$\frac{\hat{u}(\xi)}{a} = \begin{cases} \dfrac{r^2}{2(r^2+b^2)}\, e^{-b\xi/a} + \dfrac{\sigma}{G} & \text{for } \xi > 0, \\[2ex] -\dfrac{r^2}{2(r^2+b^2)}\, e^{b\xi/a} - \dfrac{b^2}{r^2+b^2}\cos\dfrac{r\xi}{a} + \dfrac{\sigma}{G} + 1 & \text{for } \xi < 0, \end{cases} \tag{19}$$

where

$$\left.\begin{aligned} r &= \left[6(1-v^2\Omega^2) + \sqrt{36(1-v^2\Omega^2)^2 + 12\Omega^2}\right]^{1/2} && > 0, \\ b &= \left[-6(1-v^2\Omega^2) + \sqrt{36(1-v^2\Omega^2)^2 + 12\Omega^2}\right]^{1/2} && > 0. \end{aligned}\right\} \tag{20}$$

The range condition (11) requires that

$$\sigma = \frac{G}{4}\left\{1 - \frac{1 - v^2\Omega^2}{\sqrt{(1-v^2\Omega^2)^2 + \Omega^2/3}}\right\}, \tag{21}$$

and $\sigma < G/6$, the former being the same relationship between the applied stress $\sigma$ and the kink propagation speed $v$ that we found in Section 4 from the continuum limit of the discrete problem.

## 6. Summary.

Figure 2 shows a plot of the kinetic relation (21). Observe that $\sigma \to \sigma_{cr}$ as $v \to 0$ where

$$\sigma_{cr} = (G/4)\left\{1 - (1 + \Omega^2/3)^{-1/2}\right\}. \tag{22}$$

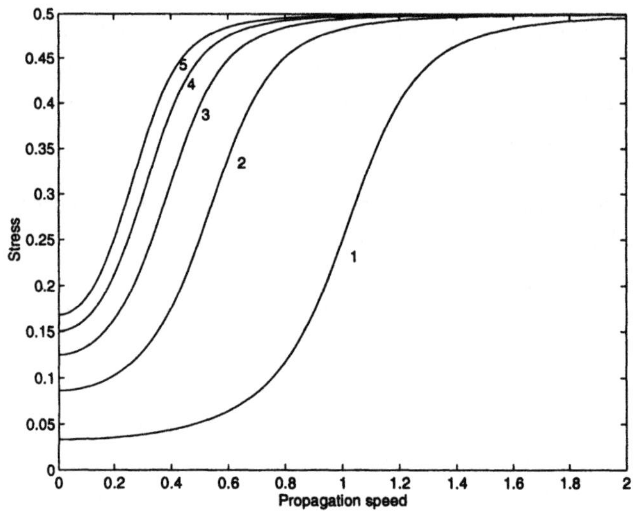


*Figure 2.* The kinetic relation. A graph of stress $\sigma/G$ versus propagation speed $v$ according to equation (21) for $\Omega = 1, 2, 3, 4$ and 5.

Consider the *equilibrium* of a chain of particles straddling two energy-wells, as characterized by the solution of the equilibrium version of (3) subject to $-a/2 < u_n < a/2$ for $n \geq 0$ and $a/2 < u_n < 3a/2$ for $n < 0$. This discrete problem can be readily solved to find the displacements $u_n$. One finds from the range conditions that an equilibrium solution exists only if the stress $\sigma$ does not exceed the critical value $\sigma_{st}$ given by

$$\sigma_{st} = (G/2)\ \Omega\ (\Omega^2 + 4)^{-1/2}. \tag{23}$$

If $\sigma > \sigma_{st}$ the kink must move. If $\sigma < \sigma_{cr}$ the kink cannot move. Note that $\sigma_{st} > \sigma_{cr}$.

**Acknowledgment.** The results reported here were obtained in the course of an investigation supported by the U.S. National Foundation.

## References

1. Hirth, J.P. and Lothe, J. (1982) *Theory of Dislocations*. Krieger, Florida.
2. Abeyaratne, R. and Vedantam, S. (1997) The Mechanics of Kink Motion and the Kinetics of Front Propagation, to appear.
3. Weiner J.H. and Sanders, W.T. (1964) Peierls Stress and Creep of a Linear Chain, *Physical Review*, **Vol. no. 134**, pp. A1007–1015.
4. Atkinson, W. and Cabrera, N. (1965) Motion of a Frenkel-Kontorowa Dislocation in a One-dimensional Crystal, *Physical Review*, **Vol. no. 138**, pp. A763–766.
5. Perchak, D. and Weiner, J.H. (1980) Local-mode Approximations in the Frenkel-Kontorova or Sine-Gordon Chain, *Physical Review B*, **Vol. no. 22**, pp. 2683–2692.
6. Weiner, J.H. (1964) Dislocation Velocities in a Linear Chain, *Physical Review*, **Vol. no. 136**, pp. A863–868.

# ON VARIATIONAL ASPECTS OF THE THEORY OF LIQUID CRYSTALS WITH VARIABLE DEGREE OF ORIENTATION

Z. NANIEWICZ
*Aristotle University of Thessaloniki, Department of Civil Engineering, GR-54006 Thessaloniki, Greece. On leave from Warsaw University, Institute of Applied Mathematics and Mechanics, Banacha 2, 02-097, Warsaw, Poland.*

AND

P.D. PANAGIOTOPOULOS
*Aristotle University of Thessaloniki, Department of Civil Engineering, GR-54006 Thessaloniki, Greece and Faculty of Mathematics and Physics, RWTH, Aachen, Germany.*

## 1. Introduction

Let us consider a nematic liquid crystal occupying in the three dimensional Euclidean space $\mathbf{R}^3$ a bounded region $\Omega$ with Lipschitz continuous boundary $\partial\Omega$. We put ourselves into the framework of Ericksen's theory of nematic liquid crystals with variable degree of orientation ([3], see also [4], p. 331) and we assume that the crystal is determined by the orientation vector $\mathbf{n} : \Omega \to \mathbf{R}^3$ with $|\mathbf{n}| = 1$ and the degree of orientation $s : \Omega \to [-\frac{1}{2}, 1]$. The Frank energy functional $\mathcal{F}_E$ in a setting of the one constant approximation is given by

$$\mathcal{F}_E(\mathbf{n}, s) = \int_\Omega \left\{ k_E(k|\nabla s|^2 + s^2|\nabla \mathbf{n}|^2) \right\} d\Omega + \int_\Omega \sigma_0(s) d\Omega,$$

where $\nabla = (\partial_j)$ stands for the gradient operator, $k_E$ and $k$ are positive module, $\sigma_0 : \Omega \to R$ is De Gennes potential of the form

$$\sigma_0(s) = \frac{1}{2}as^2 - \frac{1}{3}bs^3 + \frac{1}{4}cs^4 + d$$

with positive constants $a$, $b$, $c$ and $d$.



*P. Argoul et al. (eds.),*
*IUTAM Symposium on Variations of Domains and Free-Boundary Problems in Solid Mechanics*, 85–91.

The aim of the paper is to examine the optimization

**Problem (P)**: Find $(\mathbf{n}, s)$ such that

$$\mathcal{F}_E(\mathbf{n}, s) \longrightarrow \text{minimum};$$

under the assumption that both $\mathbf{n}$ and $s$ are not prescribed on any part of the boundary $\partial\Omega$ (the natural boundary conditions). Strong nonlinearity with respect to the higher order terms and nonconvexity of $\mathcal{F}_E$ make essential difficulties in the formulation of an adequate variational theory. However, after a suitable change in variables we shall derive a new type of variational problems as necessary conditions for the optimality of $\mathcal{F}_E$.

For another approach to the existence of solutions for (P) we refer to [1],[2] where also some regularity results for minimizers of $\mathcal{F}_E$ have been established.

## 2. Preliminary results

The following results are useful in our study and are given here without proofs.

**Proposition 1.** Let $\mathbf{u} \in H^1(\Omega; \mathbf{R}^3)$. Define $\mathbf{n} : \Omega \rightarrow \mathbf{S}^2 \cup \{\mathbf{0}\}$, $\mathbf{S}^2 = \{y \in \mathbf{R}^3 : |y| = 1\}$ being a unit sphere in $\mathbf{R}^3$, by the relations

$$\mathbf{n} = \begin{cases} \frac{\mathbf{u}}{|\mathbf{u}|} & \text{in } \Omega \setminus \mathcal{S}(\mathbf{u}) \\ \mathbf{0} & \text{in } \mathcal{S}(\mathbf{u}). \end{cases} \tag{1}$$

Then there exists a $\mathbf{N} \in L^2(\Omega; \mathbf{R}^{3\times 3})$ such that

$$\begin{array}{ll} \text{(i)} & \nabla \mathbf{u} = \mathbf{n} \otimes \nabla(|\mathbf{u}|) + \mathbf{N} \\ \text{(ii)} & \mathbf{N}^T \mathbf{n} = \mathbf{0} \\ \text{(iii)} & |\nabla \mathbf{u}|^2 = |\nabla(|\mathbf{u}|)|^2 + |\mathbf{N}|^2. \end{array} \tag{2}$$

The main idea of the approach presented here is to change variables $(\mathbf{n}, s)$ by $(\mathbf{u}, s)$ with $\mathbf{u} := s\mathbf{n}$ and to examine the existence of minimizers of $\mathcal{F}(\mathbf{u}, s) := \mathcal{F}_E(\mathbf{n}, s)$ on $\mathcal{A} \cap \mathcal{B}$, where

$$\mathcal{A} = \{(\mathbf{u}, s) \in H^1(\Omega; \mathbf{R}^3) \times H^1(\Omega) : |\mathbf{u}| = |s| \text{ a.e. in } \Omega\} \tag{3}$$

and

$$\mathcal{B} = \left\{(\mathbf{u}, s) \in H^1(\Omega; \mathbf{R}^3) \times H^1(\Omega) : s \in [-\frac{1}{2}, 1]\right\}. \tag{4}$$

The following properties of the admissible set can be pointed out.

**Proposition 2.** Let $\mathcal{A}$ and $\mathcal{B}$ be given by (1) and (2). Then

(i) $\mathcal{A} \cap \mathcal{B}$ is star-shaped with respect to 0
(ii) $\mathcal{A}$ is weakly closed in $H^1(\Omega; \mathbf{R}^3) \times H^1(\Omega; R)$
(iii) For each a pair $(\mathbf{u}, s) \in \mathcal{A}$ there exists $\mathbf{N} \in L^2(\Omega; R^{3\times 3})$ such that

$$\begin{aligned} &\nabla \mathbf{u} = \mathbf{n} \otimes \nabla s + \mathbf{N} \\ &\mathbf{N}^T \mathbf{n} = \mathbf{0} \\ &|\nabla \mathbf{u}|^2 = |\nabla s|^2 + |\mathbf{N}|^2 \end{aligned} \tag{5}$$

where $\mathbf{n} : \Omega \to \mathbf{S}^2 \cup \{\mathbf{0}\}$ is given by (1).
(iv) If $|s| \geq s_0 = \text{const} > 0$ a.e. in $\Omega$ then $\mathbf{n}$ in (1) admits distributional gradient $\nabla \mathbf{n} \in L^2(\Omega; \mathbf{R}^{3\times 3})$ and hence $\mathbf{n} \in H^1(\Omega; \mathbf{R}^3)$. Moreover

$$\mathbf{N} = s\nabla \mathbf{n}. \tag{6}$$

(v) If $\mathbf{u} \neq 0$ a.e. in $\Omega$ and $\frac{|\nabla \mathbf{u}|}{|\mathbf{u}|} \in L^1(\Omega)$ then $\mathbf{n}$ defined by (1) is an element of $W^{1,1}(\Omega; \mathbf{R}^3)$.

For any $(\mathbf{u}, s) \in \mathcal{A}$, due to a singular set $\mathcal{S}(\mathbf{u}) = \{x \in \Omega : \mathbf{u}(x) = \mathbf{0}\}$ the corresponding $\mathbf{n}$ given by (1) does need to lie in $H^1(\Omega; \mathbf{R}^3)$. Therefore we are not able to prevent coerciveness of $\mathcal{F}_E$ in the Sobolev space $H^1(\Omega; \mathbf{R}^3) \times H^1(\Omega)$. We shall follow the idea of [1],[2] consisting in changing variable and not to deal with $(\mathbf{n}, s)$-pairs, but to choose $(\mathbf{u}, s)$ as the main independent variables. It is justifiable by Proposition 2 which guarantees that each pair $(\mathbf{u}, s) \in \mathcal{A}$ generates uniquely the orientation vector $\mathbf{n}$ and tensor field $\mathbf{N}$ with the later gathering all the necessary properties attached to the term $s\nabla\mathbf{n}$.

We replace in the Frank energy functional $\mathcal{F}_E$ the term $s^2|\nabla \mathbf{n}|^2$ by $|\mathbf{N}|^2$ and use (iii) of Proposition 2 to define a new functional $\mathcal{F}$ of the variables $(\mathbf{u}, s) \in \mathcal{A} \cap \mathcal{B}$:

$$\begin{aligned} \mathcal{F}_E(\mathbf{n}, s) &= \int_\Omega \left\{ k_E(k|\nabla s|^2 + s^2|\nabla \mathbf{n}|^2) + \sigma_0(s) \right\} d\Omega \\ &= \int_\Omega \left\{ k_E(k|\nabla s|^2 + |\mathbf{N}|^2) + \sigma_0(s) \right\} d\Omega \\ &= \int_\Omega \left\{ k_E(|\nabla \mathbf{u}|^2 + (k-1)|\nabla s|^2) + \sigma_0(s) \right\} d\Omega \\ &=: \mathcal{F}(\mathbf{u}, s) \qquad (\mathbf{u}, s) \in \mathcal{A}. \end{aligned} \tag{7}$$

**Remark.** Note that the constant $(k-1)$ in the foregoing expression is negative for $0 < k < 1$. This makes $\mathcal{F}$ nonconvex with respect to its variable.

**Proposition 3.** Let $k > 0$. Then the functional

$$F(\mathbf{u}, s) := \int_{\Omega} \left\{ k_E(|\nabla \mathbf{u}|^2 + (k-1)|\nabla s|^2) \right\} d\Omega$$

is coercive on $\mathcal{A} \cap \mathcal{B}$ with respect to $H^1(\Omega; \mathbf{R}^3) \times H^1(\Omega; R)$-norm, and it is weakly lower semicontinuous on $\mathcal{A}$.

From Proposition 2 and Proposition 3 the existence result follows.

**Proposition 4.** Functional $\mathcal{F}$ defined by (7) possesses at least one minimizer on $\mathcal{A} \cap \mathcal{B}$, i.e. there exists $(\mathbf{u}, s) \in \mathcal{A} \cap \mathcal{B}$ such that

$$\mathcal{F}(\mathbf{u}, s) = \inf_{(\mathbf{v}, r) \in \mathcal{A} \cap \mathcal{B}} \mathcal{F}(\mathbf{v}, r). \tag{8}$$

The corresponding pair $(\mathbf{n}, s)$ determined by (1) will be referred to as the solution of the primal problem (P).

## 3. Necessary conditions and variational formulations

Throughout this Section $(\mathbf{u}, s) \in H^1(\Omega; \mathbf{R}^3) \times H^1(\Omega)$ is assumed to lie in $\mathcal{A} \cap \mathcal{B}$. Let us denote by $T_{\mathcal{A}}(\mathbf{u}, s)$ the contingent cone to $\mathcal{A}$ at $(\mathbf{u}, s) \in \mathcal{A}$, i.e.

$$\begin{aligned} T_{\mathcal{A}}(\mathbf{u}, s) = \{(\mathbf{v}, r) \in H^1(\Omega; \mathbf{R}^3) \times H^1(\Omega) : \exists \lambda_k \to 0_+, \exists (\mathbf{v}_k, r_k) \to (\mathbf{v}, r) \\ \text{in } H^1(\Omega; \mathbf{R}^3) \times H^1(\Omega) \ \text{ such that } \ (\mathbf{u}, s) + \lambda_k(\mathbf{v}_k, r_k) \in \mathcal{A}\}. \end{aligned} \tag{9}$$

**Proposition 5.** The necessary condition for $(\mathbf{v}, r)$ to belong to the contingent cone $T_{\mathcal{A}}(\mathbf{u}, s)$ is that

$$\mathbf{v} \cdot \mathbf{u} - rs = 0 \quad \text{a.e. in} \quad \Omega. \tag{10}$$

If the measure of the singular set $\mathcal{S}(\mathbf{u})$ is greater than 0 then the additional requirement is necessary, namely

$$|\mathbf{v}| = |r| \quad \text{a.e. in} \quad \mathcal{S}(\mathbf{u}). \tag{11}$$

**Proposition 6.** If $(\mathbf{v}, r) \in T_{\mathcal{A}}(\mathbf{u}, s)$ then

$$\begin{aligned}(\mathbf{v}-\mathbf{u}, r-s) &\in T_{\mathcal{A}}(\mathbf{u}, s)\\ (\mathbf{v}+\mathbf{u}, r+s) &\in T_{\mathcal{A}}(\mathbf{u}, s).\end{aligned} \tag{12}$$

**Proposition 7.** If $\varphi \in H^1(\Omega)$ is such that $(\varphi\mathbf{u}, \varphi s) \in H^1(\Omega; \mathbf{R}^3) \times H^1(\Omega)$ then

$$(\varphi\mathbf{u}, \varphi s) \in T_{\mathcal{A}}(\mathbf{u}, s). \tag{13}$$

**Lemma.** Let $(\overline{\mathbf{u}}, \overline{s}) \in C^\infty(\overline{\Omega}; \mathbf{R}^3) \times C^\infty(\overline{\Omega})$ be such that $\mathbf{u} \cdot \overline{\mathbf{u}} - s\overline{s} = 0$ a.e. in $\Omega$. Then $(s\overline{\mathbf{u}}, s\overline{s}) \in T_{\mathcal{A}}(\mathbf{u}, s)$ and, moreover, for any $\lambda_k \to 0_+$ there exists a sequence $(\mathbf{y}_k) \subset C^\infty(\overline{\Omega}; \mathbf{R}^3)$ with $\mathbf{y}_k \to \mathbf{0}$ in $H^1(\Omega; \mathbf{R}^3)$ as $k \to \infty$, such that

$$\Big(\mathbf{u} + \lambda_k(s\overline{\mathbf{u}} + \mathbf{y}_k), s + \lambda_k s\overline{s}\Big) \in \mathcal{A}. \tag{14}$$

**Remark.** If in the proof of Lemma one replaces $(\overline{\mathbf{u}}, \overline{s})$ by $(s\mathbf{w}, 0)$ with $\mathbf{w} \in H^1(\Omega; \mathbf{R}^3) \cap L^\infty(\Omega; \mathbf{R}^3)$ and $\mathbf{w} \cdot \mathbf{u} = 0$ a.e. in $\Omega$, then the corresponding solution $\mathbf{y}_k$ takes the form

$$\mathbf{y}_k = -\lambda_k \mathbf{u} \frac{|\mathbf{w}|^2}{\sqrt{1 - \lambda_k^2 |\mathbf{w}|^2} + 1},$$

which amounts to

$$\Big(\mathbf{u} + \lambda_k(s\mathbf{w} + \mathbf{y}_k), s\Big) \in \mathcal{A}. \tag{15}$$

Since $\mathbf{w} \in L^\infty(\Omega; \mathbf{R}^3)$, so for sufficiently large $k$, $\mathbf{y}_k \in H^1(\Omega; \mathbf{R}^3)$ and we are easily led to the conclusion that $\mathbf{y}_k \to \mathbf{0}$ in $H^1(\Omega; \mathbf{R}^3)$ as $\lambda_k \to 0$. Accordingly, $(s\mathbf{w}, 0) \in T_{\mathcal{A}}(\mathbf{u}, s)$.

From this Remark it follows that each element of the form $(s\mathbf{w}, 0)$ fulfilling the requirements $\mathbf{w} \in H^1(\Omega; \mathbf{R}^3) \cap L^\infty(\Omega; \mathbf{R}^3)$ and $\mathbf{w} \cdot \mathbf{u} = 0$ a.e. in $\Omega$ is contained in $T_{\mathcal{A}}(\mathbf{u}, s)$. Using the truncation procedure one can extend this results as follows.

**Proposition 8.** If $\mathbf{w} \in H^1(\Omega; \mathbf{R}^3)$ is such that $\mathbf{w} \cdot \mathbf{u} = 0$ a.e. in $\Omega$ and $s\mathbf{w} \in H^1(\Omega; \mathbf{R}^3)$ then

$$(s\mathbf{w}, 0) \in T_{\mathcal{A}}(\mathbf{u}, s). \tag{16}$$

The obtained results justify the introduction of the cones

$$\begin{aligned}T_{\mathcal{A}}^{(s)}(\mathbf{u}, s) &= \{(\mathbf{u}, s)H^1(\Omega)\} \cap H^1(\Omega; \mathbf{R}^3) \times H^1(\Omega)\\ T_{\mathcal{A}}^{(u)}(\mathbf{u}, s) &= \{(s \operatorname{Ker} \Phi, 0)\} \cap H^1(\Omega; \mathbf{R}^3) \times H^1(\Omega),\end{aligned} \tag{17}$$

responsible for variations of the variable degree of orientation $s$ and of the orientation vector $\mathbf{u}$, respectively. Here $\Phi : H^1(\Omega; \mathbf{R}^3) \to L^2(\Omega)$ is a linear continuous operator defined by

$$\Phi(\mathbf{v}) := \mathbf{v} \cdot \mathbf{u} \qquad \forall\, \mathbf{v} \in H^1(\Omega; \mathbf{R}^3). \tag{18}$$

The kernel of $\Phi$ will be denoted by the symbol "Ker $\Phi$", i.e. $\operatorname{Ker}\Phi = \{\mathbf{v} \in H^1(\Omega; \mathbf{R}^3) : \mathbf{v} \cdot \mathbf{u} = 0\}$. Note that the both cones just introduced are linear subspaces not necessarily closed, which by Propositions 6 and 7 are contained in the contingent cone $T_{\mathcal{A}}(\mathbf{u}, s)$, i.e.

$$T_{\mathcal{A}}^{(s)}(\mathbf{u}, s) \cup T_{\mathcal{A}}^{(u)}(\mathbf{u}, s) \subset T_{\mathcal{A}}(\mathbf{u}, s). \tag{19}$$

To formulate the next result let us denote by $I_{[-\frac{1}{2},1]}$ the indicator function of the interval $[-\frac{1}{2}, 1]$ and set

$$\mathcal{I}(r) = \int_\Omega I_{[-\frac{1}{2},1]}(r)d\Omega, \quad r \in H^1(\Omega).$$

**Proposition 9.** Let $(\mathbf{u}, s) \in \mathcal{A} \cap \mathcal{B}$ be a solution of the optimization problem (8). Then the following variational inequality holds

$$\int_\Omega \{2k_E[\nabla\mathbf{u} \cdot (\nabla\mathbf{v} - \nabla\mathbf{u}) + (k-1)\nabla s \cdot (\nabla r - \nabla s)]\}d\Omega +$$

$$\int_\Omega [\sigma_0'(s)(r-s) + I_{[-\frac{1}{2},1]}(r)]d\Omega \geq 0 \qquad \forall\, (\mathbf{v}, r) \in T_{\mathcal{A}}^{(s)}(\mathbf{u}, s) \cup T_{\mathcal{A}}^{(u)}(\mathbf{u}, s). \tag{20}$$

Now we are in a position to present our main result concerning the necessary conditions for the optimality of $\mathcal{F} + \mathcal{I}$ on $\mathcal{A}$. They will be expressed as a system of two coupled variational problems characterizing the equilibrium position of the liquid crystal under consideration.

**Theorem 1.** If $(\mathbf{u}, s) \in \mathcal{A} \cap \mathcal{B}$ is a solution of the optimization problem (8) then the following variational problems hold:

$$\int_\Omega 2k_E[|\nabla\mathbf{u}|^2(\varphi - 1) + s\nabla s \cdot (\nabla\varphi - \nabla 1) + (k-1)\nabla s \cdot (\nabla(\varphi s) - \nabla s)]d\Omega +$$

$$+ \int_\Omega \sigma_0'(s)((\varphi s) - s)d\Omega + \mathcal{I}(\varphi s) \geq 0 \qquad \forall\, \varphi \in H^1(\Omega) \tag{21}$$

and

$$\int_{\Omega} 2k_E \nabla \mathbf{u} \cdot \nabla(s\mathbf{w}) d\Omega = 0 \quad \forall\, s\mathbf{w} \in \operatorname{Ker} \Phi, \tag{22}$$

where only such $\varphi$'s and $\mathbf{w}$'s are allowed that $s\varphi \in H^1(\Omega)$ and $s\mathbf{w} \in H^1(\Omega; \mathbf{R}^3)$.

A new type of variational inequalities have been derived as necessary conditions for optimality of $\mathcal{F} + \mathcal{I}$ on $\mathcal{A}$. Namely, the variational inequality (21) involves the differences of the form "$(\nabla(\varphi s) - \nabla s)$" and "$(\varphi s - s)$" due to the fact that the bilateral constraints: $|\mathbf{u}| = |s|$ a.e. in $\Omega$, are singular in $\mathcal{S}(\mathbf{u})$. This is not the case of the theory of classical variational inequalities where such differences would be replaced by "$(\nabla\varphi - \nabla s)$" and "$(\varphi - s)$", respectively.

## References

Ambrosio, L. (1990) Existence of Minimal Energy Configurations of Nematic Liquid Crystals with Variable Degree of Orientation, *Manuscripta Math.* **68** 215-228.

Lin, F.H. (1991) On Nematic Liquid Crystals with Variable Degree of Orientation, *Comm. Pure Appl. Math.* **XLIV** 453-468.

Ericksen, J.L. (1991) Liquid Crystals with Variable Degree of Orientation, *Arch. Rational Mech. Anal.* **113** 97-120.

Virga, E.G. (1994) Variational Theories for Liquid Crystals, *Chapman & Hall London.*

# DIFFUSION-LIMITED DYNAMICS IN GROWTH EXPERIMENTS

F. ARGOUL , J. ELEZGARAY AND C. LÉGER
*Centre de Recherche Paul Pascal*
*C.N.R.S., Av. Schweitzer, 33600 Pessac, France*

Since the begining of this century, diffusion-limited interfacial processes have been the object of strong interest among the electrochemist and physicist communities because they can provide sufficiently simple transport equations to enable some theoretical modelling [1, 2, 3, 4]. In the early eighties, there was a revival of interest for electrodeposition of metals [5, 6, 7, 8] and this system was designated by physicists as a paradigm for producing Diffusion–Limited Aggregation (DLA)-like morphologies [9, 10]. Even if this enthusiasm on fractals suspended some fundamental questions concerning the actual transport and reaction mechanisms which are to be encountered in an electrodepositon experiment, it brought in this field new teams who progressively focused on these issues. In particular the origin of the inconsistency between different morphological phase diagrams [7, 8, 11, 12, 13] could be qualitatively elucidated in thin gap cells, in terms of either competitive adsorption and reaction processes [12, 13, 14] or the implication of convection such as buoyancy driven convection [15, 16, 17, 18, 19] or electroconvection [20]. Many early experimental evidences for DLA-like morphology in electrodeposition experiments were seriously questioned by these new experimental studies. In a recent work [21, 22, 23], we have emphasized electroless deposition as a good candidate for producing DLA like arborescent patterns. We have performed some structural characterization of both electroless deposits and diffusion limited aggregates, using the wavelet transform [21, 22, 24, 25, 26]. We have shown that both the experimental and the numerical clusters display five-fold symmetry and Fibonacci hierarchical ordering as common predominant statistical features [21]. In a recent publication [27], we have analyzed the actual occurrence of DLA like dynamics from direct computation of the growth probability distribution [28, 29] along the interface of an electroless aggregate.

In this paper, our goal is to bring the experimental demonstration of the actual existence of diffusion-limited **dynamics** in electrodeposition (ECD)



*P. Argoul et al. (eds.),*
*IUTAM Symposium on Variations of Domains and Free-Boundary Problems in Solid Mechanics*, 93–101.

experiments. This demonstration will be organized in two parts, in a first section we present the general model of ECD and we discuss the conditions where diffusion can be globally considered as limiting the kinetics of deposition, in a second section we discuss the experimental recovery of diffusion limited dynamics, based on the real-time interferometric recording of the concentration fields. The experimental demonstration is split into two complementary parts, we first show that the diffusive transport equation is fullfilled everywhere in the cell and on a second step we verify the linear dependence of the velocity of the growth with respect to the local concentration gradients.

## 1. Modelling electrodeposition processes

Before getting into the detailed equations [30, 31, 32] which describe an ECD we list rapidly the physical and chemical ingredients which are involved in an electrochemical process. Roughly speaking, two family of processes can be encountered. These are respectively **global transport processes** such as diffusion, migration and convection and **interfacial processes** such as adsorption, reaction, charge transfer, surface diffusion which are strictly confined to the interface between the electrolyte and the metal electrode. The ECD kinetics involves the reduction of cationic metal species and their inclusion in a metallic substrate $M^{z^+} + z^+e^- \to M$. The details of the mechanism may include a succession of adsorption, surface diffusion, charge transfer and desorption processes. In the context of this paper we will assume that we are far enough from the electrochemical equilibrium (strong current of cation discharge) so that these interfacial processes can be considered as fast with respect to the volume transport processes. In the volume of the reactor the three different transport processes can be described by flux equations.

It can be shown [30] that in the case of a single electrolyte $M_{\nu^+}^{z^+} A_{\nu^-}^{z^-}$, if electroneutrality is fulfilled in the bulk of the cell, the transport equations of the two species $M^{z^+}$ and $A^{z^-}$ can be simplified into a single diffuso-convective equation:

$$\partial_t C + \vec{\nabla}.(C\vec{v}_f) = D\nabla^2 C \,, \tag{1}$$

with $D = \dfrac{z^+u^+D^- - z^-u^-D^+}{z^+u^+ - z^-u^-}$ is the ambipolar diffusion coefficient[33], $C = C^+/\nu^+ = C^-/\nu^-$ and $\vec{v}_f$ the velocity of the fluid. It is important to remark that this equation is valid only if the diffusion coefficient can be assumed to be constant in the range of concentrations of the experiment. In the general case, one has to include the dependence of $D$ and $t^+$ [33]

versus $C$ [30]:

$$\partial_t C + \vec{\nabla}.(C\vec{v}_f) = \vec{\nabla}.(D(C)\nabla\vec{C}) - \frac{\vec{i}.\vec{\nabla}t^+(C)}{z^+\nu^+F} \,. \tag{2}$$

In this paper we will assume that we can neglect the convective term of eq.(1). In the experiments this can be achieved with very thin cells (gap $\leq$ 50 $\mu$m) [18, 19]; we will therefore work with a diffusion equation:

$$\partial_t C = \vec{\nabla}.(D(C)\nabla\vec{C}) - \frac{\vec{i}.\vec{\nabla}t^+(C)}{z^+\nu^+F} \,. \tag{3}$$

Beside this "bulk" transport equation, boundary conservation and kinetics equations need to be formulated. As far as the conservation of metal ions is concerned, it reads:

$$-M_M\vec{J}_M.\vec{n} = \rho_{agg}\vec{v}.\vec{n} \,, \tag{4}$$

$M_M$ is the molar mass of the metal $M$, $\vec{J}_M$ the flux density of species $M^{z+}$ on the interface, $\rho_{agg}$ the density of the aggregate and $\vec{v}$ the velocity of the interface. With the same type of assumptions as for the bulk diffusion equation, one can show that the flux density is proportional to the gradient of the species at the interface [30, 33]

$$\vec{J}_M.\vec{n} = -\nu^+ D(C)\vec{\nabla}C.\vec{n}/(1-t^+) \,. \tag{5}$$

The kinetic equation for the rate of charge transfer at the interface, must include surface tension and surface potential:

$$i = nF\vec{J}_M.\vec{n} = i_0\left(e^{\beta(\phi-\gamma\kappa')} - \frac{C_0^+}{C_{eq}^+}e^{-\alpha(\phi-\gamma\kappa')}\right) \tag{6}$$

$\phi = nF(E - E_{eq})/RT$, $E$ and $E_{eq}$ are electrode potentials, the latter corresponds to the equilibrium ($i = 0$), $i_0$ is the exchange current density [35], $\gamma$ is the surface tension of the metal/electrolyte interface and $\kappa' = M_M\kappa/(\rho_{agg}RT)$ where $\kappa$ stands for the curvature. If we can assume that the interfacial kinetics is fast i.e. $i_0/i >> 1$, the equation (6) reduces to an equilibrium equation:

$$RT\log\frac{C^+}{C_{eq}^+} = nF(E - E_{eq}) - \kappa\gamma M_M/\rho_{agg}$$

which is the extension of the Gibbs-Thompson [34] equation for electrochemical interfaces. For a smooth surface this equation is equivalent to the Nernst equation [35].

Starting from a smooth surface, under galvanostatic conditions, the solution of the one-dimensional diffusion equation $\partial C/\partial t = D\nabla^2 C$, with a constant diffusion coefficient reads:

$$C(x,t) = C_{t=0}\left[1 - \sqrt{\frac{t}{\tau_S}}\exp\left(\frac{-x^2}{4Dt}\right) + \frac{\sqrt{\pi}x}{2\sqrt{D\tau_S}}\operatorname{erfc}\left(\frac{x}{2\sqrt{Dt}}\right)\right], \quad (7)$$

with $\tau_S = \left(nF\sqrt{\pi D}C_{t=0}/2i(1-t^+)\right)^2$ is Sand's time corresponding to a null interfacial concentration of metal ions $C_0^+$. In the galvanostatic experiments, eq. (6) gives the dependence of the potential E of the electrode versus i and the interfacial concentration $C_0^+$.

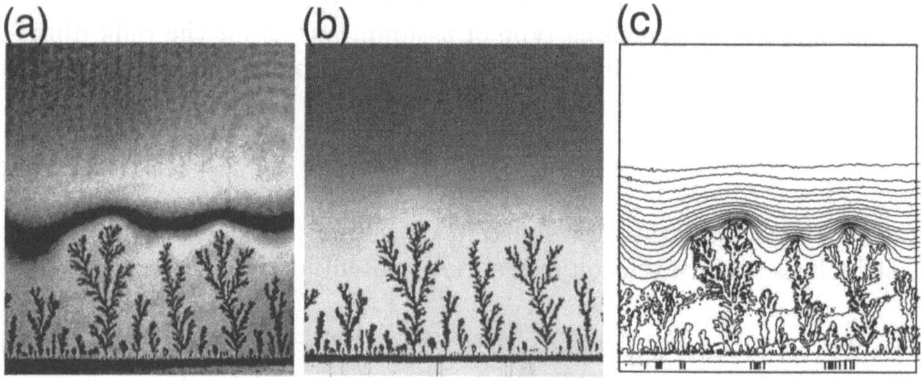


*Figure 1.* Interferometric analysis of the concentration field in a thin gap cell. The interferometric picture (a) is compared to the grey-level coded concentration (b) and level map (c) in two dimensions. Between each isoconcentration line $\Delta C \simeq 0.03$ mol.l$^{-1}$, i=40mA.cm$^{-2}$, [$CuCl_2$]=0.5mol.l$^{-1}$, cell geometry: 50$\mu$m$\times$5cm$\times$5cm.

## 2. Experimental observation of DLA-like dynamics in ECD

The next step of our discussion consists in the quantitative check of the occurrence of diffusion-limited dynamics in a real ECD experiment. After a brief presentation of the experimental set-up we analyze successively the relevance of the global transport equation (3) and the interfacial velocity relation (4).

*Experimental technique* Interferometric techniques have been applied since the early sixties for characterizing concentration fields in electrochemical cells, in particular the Mach-Zhender interferometer has proved its adequacy to revealing the time evolution of the depletion and dissolution layers in thick gap cells [36, 37]. We have recently improved this optical device by implementing a phase shift interferometer (PSI) [17, 38, 39, 40]. A description of the experimental set-up together with the experimental requirements for getting rid of convective effects is given in reference [40]. Globally, PSI enables the complete reconstruction of a concentration map from a set of successive interference pictures recorded for shifted values of the phase difference between the two wavefronts of the interferometer. Figure 1 presents an illustration of both the interference fringe pattern and of the reconstructed concentration maps.

*Bulk diffusion equation* By a straightforward computation, one can show that the last term of eq. (3) is negligible. The term $\vec{\nabla}.(D(C)\vec{\nabla}C)$ must be estimated from the experimental data and requires a previous filtering of $C(x,y,t)$ which becomes $\tilde{C}(x,y,t)$. The size of the Gaussian filter that we have used has been chosen to optimize the noise reduction and the preservation of the local properties of the concentration map. Moreover, we use a time averaged version of eq. (3):

$$\begin{aligned} \frac{\delta C(x,y,t)}{\delta t} &= \frac{C(x,y,t_f)-C(x,y,t_i)}{t_f-t_i} = \\ \overline{\psi(C)} &= \frac{1}{p}\sum_{j=1}^{p}\left\{D_j\Delta\tilde{C}_j + \left.\frac{\partial D}{\partial C}\right|_{C_j}|\vec{\nabla}\tilde{C}_j|^2\right\} \end{aligned}$$

where $C_j = C(x,y,t_i+(j-1)(t_f-t_i)/(p-1))$ and $D_j = D(C_j)$ are interpolated from the tabulated data of ref. [41] for zinc sulphate electrolytes at 25°C by a spline function. Fig. (2a) shows that eq. (3) is verified and confirms this non-linear diffusion equation as a pertinent model of the transport in thin gap electrolytic cells. Moreover, one can remark that this analysis does not depend on boundary conditions and that it could be applied to other experimental systems.

A more intuitive check consists in computing the evolution of the concentration profiles in front of the electrode. From our experimental data, we could perform this verification in lower concentration electrolytes, where one can assume that the diffusion coefficient is constant. Prior to the instability of the interface, there is a translational invariance along the electrode and we can approximate the interfacial process as a one-dimensional system, modelled by the evolution equation (7) with the boundary condition (5). By a rescaling of the concentration $C$ and the space coordinate $x$ into

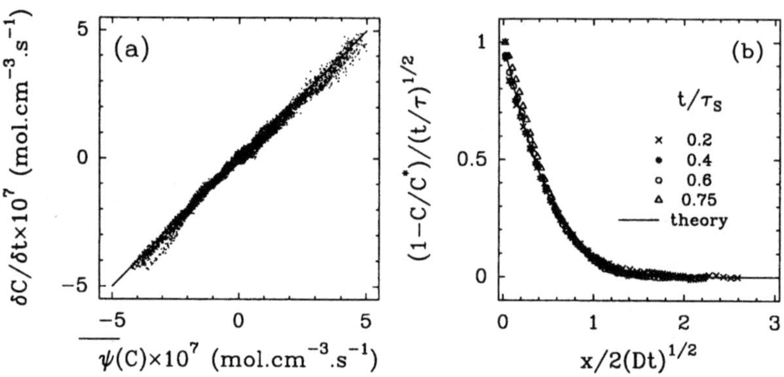


*Figure 2.* (a) Local analysis of the PSI concentration maps with equation (3). Lin-Lin plot of $\delta C/\delta t$ versus $\overline{\psi(C)}$ (see text). On the **right side** of the plot ($\delta C/\delta t > 0$) the points correspond to the neighborhood of the **anode** ([$ZnSO_4$]=0.1M, i=20mA.cm$^{-2}$, the sampling of the frame sets for PSI has been done between t=49s and t=417s at a 25 mHz rate). On the **left side** of the plot ($\delta C/\delta t < 0$) the points correspond to the neighborhood of the **cathode** ([$ZnSO_4$]=2M, i=20mA.cm$^{-2}$, the sampling of the frame sets for PSI has been done between t=128s and t=1080s at a 16 mHz rate). (b) Plot of the time evolution of interfacial concentration profiles in thin gap electrodeposition. ([$ZnSO_4$] 0.1 mol.l$^{-1}$, $i$ = 4mA.cm$^{-2}$). (theoretical fit with $D = 5.10^{-6}$cm$^2$.s$^{-1}$, $i(1-t^+) = 2.4$mA.cm$^{-2}$, $\tau_S(j,D) \simeq 251$ s). In (a) and (b) the cell geometry is 50$\mu$m×50mm×50mm.

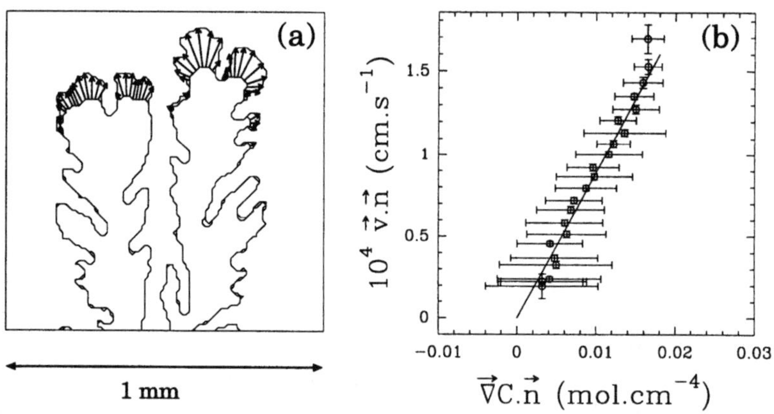


*Figure 3.* (a) Illustration of the contour velocity computation from two successive snapshots of electrodeposition (b) Plot of the normal velocity versus the projection of the concentration gradient as extracted from a statistics of 10 successive pictures by PSI. Parameter values: $Cu(NO_3)_2$=0.5M, i=40 mA.cm$^{-2}$, $\Delta t$ = 90s, gap of the cell: 50$\mu$m.

adimensional quantities [35], we show in Fig. (2b) that all the profiles at successive times superimpose with the profile of eq. (7). This is an additional confirmation of the diffusive character of the transport in the bulk.

*Interfacial velocity* In this section we want to push further the investigation of our experimental data by computing interfacial fluxes and comparing them with the velocity of the growth to check the relevance of eq. (4). From a set of successive pictures of the growth pattern, we extract the contours of the electrodeposit and we compute, for each pixel of the contour, the normal to this contour on which we project either the displacement of the interface or the gradient of concentration. Fig. (3a) illustrates this computation on two successive snapshots of the growth. The black arrows are the computed normal velocities. In Fig. (3b), by the plot of the normal velocities versus the normal gradients, obtained from a set of 10 pictures of the growth, we show that these two quantities are linearly related. These results are very encouraging in the sense that one can reasonably hope to extract from this type of measurements a dynamical estimation of the density of the aggregate. This work which requires a larger statistics than we got for this plot is under progress.

## 3. Conclusion

To pursue this study in thin gap cells, we plan to use PSI and normal velocity computations to track temporal morphological transitions from pure metal to metal oxides in copper system and to analyze the role of the counter ion (anion) on the interfacial kinetics. The application of this technique to thicker cells is also very challenging because in this case the buoyancy driven convection modifies the transport into convective rolls [18, 19] and as far as the boundary conditions are concerned, one can wonder if the presence of this convection, coupled with electroconvection processes could alter the no-slip boundary condition and modify the simple proportionality relation between the gradient of concentration and the velocity of the growth. We hope to elaborate on these questions in a forthcoming communication.

## Acknowledgements

We are very indebted to G. Gadret for his technical assistance for the optical setup. We thank A. Arneodo for stimulating discussions and fruitful interactions. We are very grateful to the Centre National d'Etudes Spatiales for supporting our work under Grant n°96/0233.

## References

1. T.R. Rosebrugh and W.L. Miller. Mathematical theory of the changes of concentration at the electrode, brought about by diffusion and by chemical reaction. *J. Phys. Chem.*, 14:816–884, 1910.

2. J.N. Agar. Diffusion and convection at electrodes. *Discussions of the Faraday Society*, 1:26–37, 1947.
3. N. Ibl, Y. Barrada, and G. Trumpler. Zur kenntis der naturlichen konvektion bei der electrolyse: Interferometrische untersuchungen der diffusionsschift i. *Helvetica Acta*, 37:583–597, 1954.
4. R.N. O'Brien and H.J. Axon. A study of concentration polarization at working electrodes. *Trans. Inst. Met. Finishing*, 34:41–52, 1957.
5. R.M. Brady and R.C. Ball. Fractal growth of copper electrodeposits. *Nature*, 309:225–229, 1984.
6. M. Matsushita, M. Sano, Y. Hayakawa, H. Honjo, and Y. Sawada. Fractal structures of zinc metal leaved grown by electrodeposition. *Phys. Rev. Lett.*, 53:286–289, 1984.
7. D.G. Grier, E. Ben Jacob, R. Clarke, and L.M. Sander. Morphology and microstructure in electrochemical deposition of zinc. *Phys. Rev. Lett.*, 56:1264–1267, 1986.
8. Y. Sawada, A. Dougherty, and J.P. Gollub. Dendritic and fractal patterns in electrolytic metal deposits. *Phys. Rev. Lett.*, 56:1260–1263, 1986.
9. T.A. Witten and L.M. Sander. Diffusion-limited aggregation, a critical phenomenon. *Phys. Rev. Lett.*, 47:1400–1430, 1981.
10. T.A. Witten and L.M. Sander. Diffusion-limited aggregation. *Phys. Rev. B*, 27:5686–5697, 1983.
11. P.P. Trigueros, J. Claret, F. Mas, and F. Sagues. Pattern morphologies in zinc electrodeposition. *J. Electroanal. Chem*, 219–235:165, 1991.
12. A. Kuhn and F. Argoul. Influence of chemical perturbations on the surface roughness of thin layer electrodeposits. *Fractals*, 1:451–459, 1993.
13. A. Kuhn and F. Argoul. Revisited experimental analysis of morphological changes in thin layer electrodeposition. *J. Electroanal. Chem*, 371:93, 1994.
14. P.P. Trigueros, F. Sagues, and J. Claret. Influence of an inert electrolyte on the morphology of quasi-two dimensional electrodeposits. *Phys. Rev. E*, 49:4328–4335, 1994.
15. V. Fleury, J.N. Chazalviel, and M. Rosso. *Phys. Rev. E*, 48:1279, 1993.
16. M. Rosso, J.-N. Chazalviel, V. Fleury, and E. Chassaing. *Electrochemica Acta*, 39:507, 1994.
17. D.P. Barkey, D. Watt, Z. Liu, and S. Raber. The role of induced convection in branched electrodeposit morphology selection. *J. Electrochem. Soc.*, 141:1206–1212, 1994.
18. J.M. Huth, H.L. Swinney, W.D. McCormick, A. Kuhn, and F. Argoul. Role of convection in thin-layer electrodeposition. *Phys. Rev. E*, 51:3444–3458, 1995.
19. J.-N. Chazalviel, M. Rosso, E. Chassaing, and V. Fleury. A quantitative study of gravity induced convection in two-dimensional parallel electrodeposition cells. *J. Electroanal. Chem*, 407:61–73, 1996.
20. V. Fleury, J.N. Chazalviel, and M. Rosso. Theory and experimental evidence of electroconvection around electrochemical deposits. *Phys. Rev. Lett.*, 68:2492–2495, 1992.
21. A. Kuhn, F. Argoul, J.F. Muzy, and A. Arneodo. Structural analysis of electroless deposits in the diffusion-limited regime. *Phys. Rev. Lett.*, 73:2998–3001, 1994.
22. A. Arneodo, F. Argoul, A. Kuhn, J.F. Muzy, and M. Tabard. Uncovering a multiplicative process in one-dimensional cuts of diffusion-limited aggregates. *Journal of Difference Equations and Applications*, 1:117–124, 1995.
23. A. Kuhn and F. Argoul. Diffusion-limited kinetics in thin-gap electroless deposition. *J. Electroanal. Chem*, 397:93–104, 1995.
24. F. Argoul, A. Arneodo, G. Grasseau, and R. Murenzi. Wavelet transform of fractal aggregates. *Phys. Lett. A*, 135:327, 1989.
25. A. Arneodo, E. Bacry, J.F. Muzy, and M. Tabard. Golden mean arithmetic in the fractal branching of diffusion-limited aggregates. *Phys. Rev. Lett.*, 68:3456–3459, 1992.
26. A. Arneodo, F. Argoul, J.F. Muzy, M. Tabard, and E. Bacry. Beyond classical mul-

tifractal analysis using wavelets: uncovering a multiplicative process hidden in the geometrical complexity of diffusion-limited aggregates. *Fractals*, 1:629–649, 1993. and references therein.

27. F. Argoul, A. Arneodo, J. Elezgaray, and A. Kuhn. Dynamical characterization of electroless deposition in the diffusion-limited regime. *Fractals*, 5, 1997. and references therein.
28. F. Mas and F. Sagues. Scaling properties of the growth probability distribution in electrochemical deposition. *Europhysics Letters*, 17:541–546, 1992.
29. J. Mach, F. Mas, and F. Sagues. Laplacian multifractality of the growth probability distribution in electrodeposition. *Europhysics Letters*, 25:271–276, 1994.
30. J.S. Newman. *Electrochemical Systems.* Prentice Hall, Englewood Cliff, New Jersey, 1991.
31. L. G. Sundstrom and F. H. Bark. On morphological instability during electrodeposition with a stagnant binary electrolyte. *Electrochemica Acta*, 40:599–614, 1995.
32. B.J. Bayly and M.Z. Bazant. Effective equations for electrodeposition in dilute binary aqueous solutions. preprint 1996.
33. $z^+$,$z^-$, $u^+$,$u^-$ and $D^+$,$D^-$ are respectively the charges of the cations and the anions, their mobility and their diffusion coefficient, $t^+ = 1/(1-(z^-u^-)/(z^+u^+))$ and $\nu^+$ the stoechiometric coefficient of the cation.
34. J. Souletie, J. Vannimenus, and R. Stora, editors. *Chance and Matter*, chapter 10, pages 629–711. Elsevier Science Publishers, Amsterdam, 1987.
35. A.J. Bard and L.R. Faulkner. *Electrochemical methods. Fundamentals and applications.* John Wiley and Sons, New York, 1980.
36. R.H. Muller. Double beam interferometry for electrochemical studies. In P. Delahay and C.W. Tobias, editors, *Advances in electrochemistry and electrochemical engineering*, volume 9, pages 281–368. Wiley Interscience, New York, 1972.
37. F. Argoul, E. Freysz, A. Kuhn, C. Léger, and L. Potin. Interferometric characterization of growth dynamics during dendritic electrodeposition of zinc. *Phys. Rev. E*, 53:1777–1788, 1996.
38. D.W Robinson and G.T. Reid, editors. *Interferogram analysis.* Institute of Physics Publishing, Techno House, Redcliffe Way, Bristol, BS1 6NX, England, 1993.
39. D. Malacara. *Optical Shop Testing.* John Wiley and Sons, Inc, New York, 1992.
40. J. Elezgaray C. Leger and F. Argoul. Experimental demonstration of diffusion--limited dynamics in electrodeposition experiments. *Phys. Rev. Lett.* , 78:5010, 1997.
41. J.G. Albright and D.G. Miller. Mutual diffusion coefficients of aqueous "$ZnSO_4$" at "25°c". *Journal of Solution Chemistry*, 4:809–816, 1975.

# MOLECULAR DYNAMICS SIMULATION OF MELTING/SOLIDIFICATION AND INDUCED STRESSES

TATSUO INOUE AND TAKUYA UEHARA
*Department of Energy Conversion Science,*
*Faculty of Energy Science, Kyoto University*
*Yoshida-hon-machi, Sakyo-ku, Kyoto, 606-8501, Japan*

**Abstract.** Molecular dynamics simulations are carried out to investigate the microscopic, or atomic scale mechanism of melting/solidification in relation to the macroscopic *materio-thermo-mechanics.* A simple model with 2-D array of atoms is treated, and uniform temperature distribution is considered. After evaluating the melting point and some other material properties for the model, mode of volume dilatation and generation of latent heat due to melting and hysteresis of temperature are simulated. The results represent that the feature of phase transformation depends on heating and cooling rates. Simulation of non-uniform heating process is followed where thermal stress is induced. The variation of temperature is compared with the solution of normal heat conduction equation, which shows good correspondence with MD simulation.

## 1. Introduction

Melting and solidification processes of metal or some other materials play quite important roll in manufacturing of machine parts and material processing, and the process is one of the interesting topics from the viewpoint of moving boundary problems. The authors have investigated such processes incorporating phase transformation including melting/solidification as well as in-solid transformation from macroscopic viewpoint, in the framework of continuum mechanics termed as *materio-thermo-mechanics*, and some successful results of simulation are reported[1-3].

Recent development of science and technology, especially in the electronic engineering, needs the adequate method to simulate sub-micron or-



*P. Argoul et al. (eds.),*
*IUTAM Symposium on Variations of Domains and Free-Boundary Problems in Solid Mechanics,* 103–111.

der process and microscopic consideration on melting and solidification. The molecular dynamics (MD) method is expected to be a useful tool for both sense of analytical and practical problems[4-5]. The continuum mechanics approach has, however, some difficulty to analyze the discontinuity on the interface of solid and liquid since the variation of volume fraction is a representative parameter to identify the phase change, and that no microscopic treatment is involved. This is why the authors try to approach the MD method to investigate microscopic feature of the phenomena[6-8], which is available to take account of the motion of all atoms or molecules composing the material.

## 2. Summary of Molecular Dynamics

The molecular dynamics is principally based on solving the Newton's equation of motion for each atom labeled by subscript "$I$",

$$\boldsymbol{F}_I = m_I \ddot{\boldsymbol{r}}_I, \qquad \textit{(No sum on I, and same hereafter)} \tag{1}$$

where $\boldsymbol{F}_I$ is force acting on the $I$-th atom, $m_I$ is mass of the atom and $\boldsymbol{r}_I$ is position vector.

In the present simulation, the force is assumed to be derived from Morse type potential $\phi$, which is one of two-body interatomic potential and represented by

$$\phi(r) = D[\exp\{2\alpha(r - r_0)\} - 2\exp\{-\alpha(r - r_0)\}] \tag{2}$$

Here, the material parameters take the values such that $D$=8.62×10$^{-20}$ (J), $\alpha$=1.12 (1/Å) and $r_0$=2.86 (Å) which are equivalent to aluminum.

In molecular dynamics, temperature $T$ and stress tensor $\boldsymbol{\sigma}$ are represented by statistical thermodynamics concept as follows;

$$T = \frac{1}{3Nk_B}\sum_I m_I v_I^2 \tag{3}$$

$$\boldsymbol{\sigma} = \frac{1}{V}\Big(\sum_I m_I \boldsymbol{v}_I \otimes \boldsymbol{v}_I - \frac{1}{2}\sum_I \sum_{J>I} \frac{d\phi}{dr}\Big|_{r=r_{IJ}} \frac{\boldsymbol{r}_{IJ}\otimes \boldsymbol{r}_{IJ}}{r_{IJ}}\Big) \tag{4}$$

where, $N$ is number of atoms treated, $k_B$ is Boltzmann constant, $v_I$ is velocity, and $\boldsymbol{r}_{IJ}$ means the position vector taken from the $I$-th atom to the $J$-th one, and $r_{IJ}$ is the distance between the two atoms.

## 3. Uniform Heating Process

### 3.1. MODEL AND CONDITIONS

Two dimensional simple model shown in fig.1 is treated as the first example in this section. It is composed of 400 atoms, and they are initially disposed

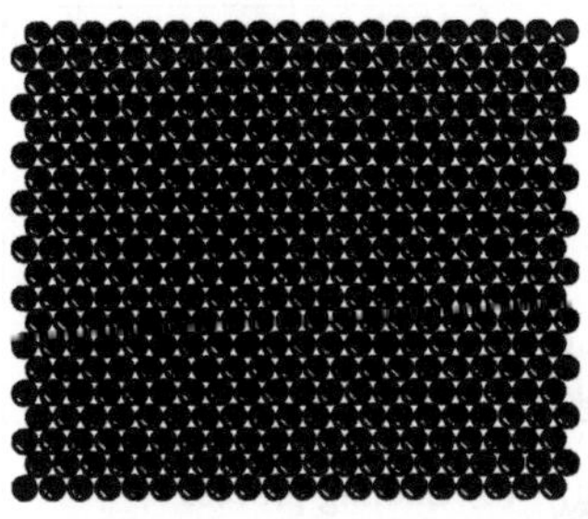

Fig. 1. Illustration of simulation model.

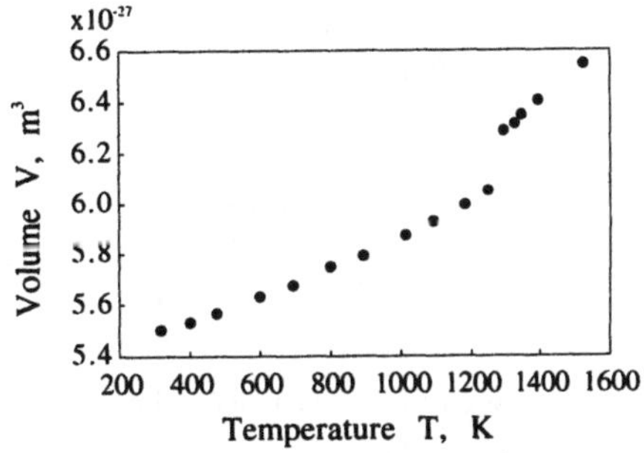


Fig. 2. Relation between volume and temperature.

at the lattice point in (111) plane of fcc structure. Periodic boundary condition is imposed to both directions of x and y to simulate bulk property. As a numerical integration method, Verlet's algorithm[9] is used, and the time interval is taken as $\Delta t{=}5{\times}10^{-15}$ (s).

Since the evaluation of the melting temperature for this model is necessary before simulating melting and solidification process, the following calculations are carried out. Temperature is kept steady throughout the calculation in the time step, and the temperature is reset at every 100 K in the range between 300 to 2000 K, while reset is made for every 20 K just above and below the melting point. The relation between temperature and density is represented in fig.2, which shows an apparent discontinuity approximately at $T{=}1300$ K. So, the melting temperature $T_M$ of this model is identified to be $T_M{=}1300$ K.

By use of the material data thus obtained, continuous heating and cooling operation is imposed over melting point. Here, two kinds of pattern are adopted for the manipulation of temperature: One is velocity-scaling scheme, or temperature is varied with time, and the other is energy-providing scheme, in which constant value of energy is successively provided to all atoms in every step.

### 3.2. VELOCITY-SCALING SCHEME

The model in fig.1 is heated and cooled by velocity-scaling scheme, and two kinds of the rate of temperature, rapid and moderate rate, are propounded. Figures 3 (a) and (b) show the calculated results representing the relation between density and temperature for two rates mentioned above. For moderate heating/cooling process, shown in fig.3(a), different shape of discontinuity in density variation is observed in heating and cooling processes, and a hysteresis loop forms in the temperature range between 1000 and 1500 K. Such hysteresis loop is found to be larger with increasing heating/cooling rate. When the cooling rate is rapid enough, the results are quite differ-

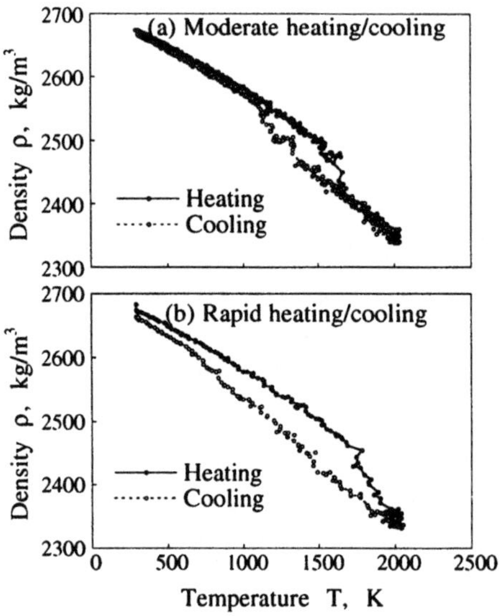


Fig. 3. Variation of density in heating and cooling process.

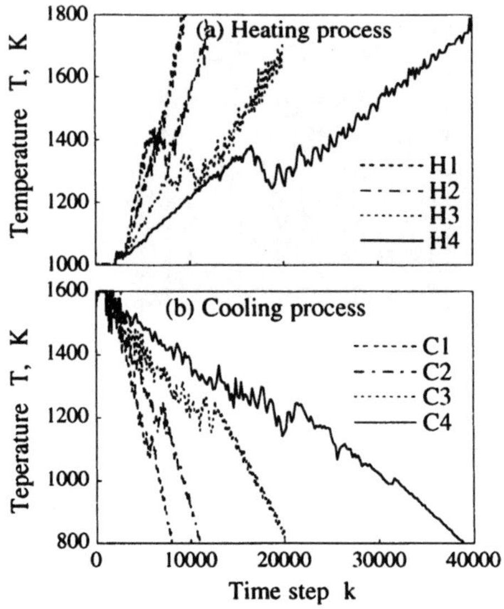


Fig. 4. Variation of temperature under 4 different rates.

ent as seen in fig.3(b), and structural disorder in liquid state is held until cooling is completed, and amorphous structure is obtained.

### 3.3. ENRGY-PROVIDING SCHENME

Following simulations are carried out by energy-providing scheme: Figures 4(a) and (b) respectively represent the results of variation of temperature during heating and cooling. Dependence of intensity of provided energy on the temperature variation is also depicted in the figure by denoting H1 - H4 and C1 - C4.

In such case of steady providing rate, temperature starts to increase almost linearly with time, as shown in fig.(a). The increasing of temperature is seen to be interrupted, and temperature decreases for a while, followed by reincreasing linearly. In cooling process referred in fig.4(b), similar but a little weaker tendency is also observed. The characteristic temperature for such interruption approximately corresponds to melting point. So it is clear that the variation of temperature is affected latent heat induced during melting/solidification.

## 4. Thermal Stresses at Steady State

### 4.1. MODEL AND CONDITIONS

From the macroscopic or mechanical viewpoint, thermal stresses are generated due to temperature gradient in material. Now, another model with

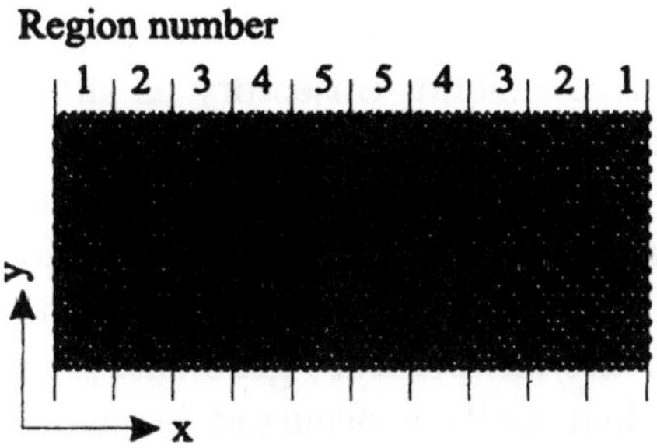


Fig. 5. Initial configuration and subregions.

(a) $T_s^*$=0.28

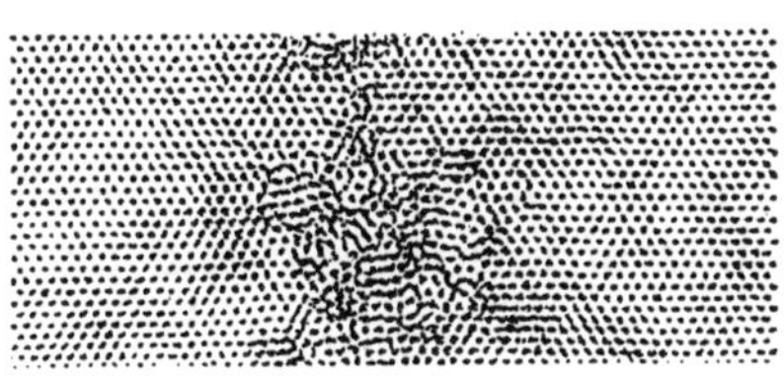

(b) $T_s^*$=0.32

Fig. 6. Trajectory of atoms.

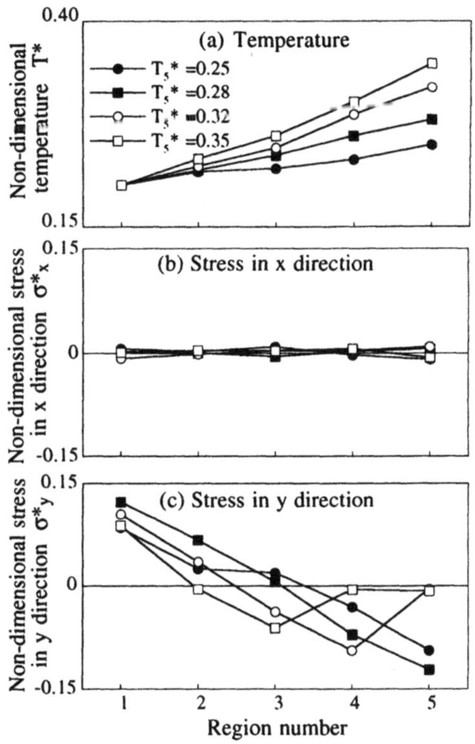


Fig. 7. Gradient of temperature and stresses.

rectangular shape consisting of 1800 atoms as shown in fig.5 is prepared. The whole region is divided into 10 subregions, which are symmetrically numbered because of the symmetry of the shape and conditions to be mentioned below, and temperature and stress in each subregion are evaluated. Those properties are expressed in non-dimensional form because the quantitative evaluation is not needed here.

Four different patterns of temperature gradient are applied as follows: Temperature of whole region is kept constant at $T_0^*$=0.20 (superscript "*" represents non-dimensional property) for 1000 steps, and then the central subregions (No.5 in fig.5) are heated up to $T^*=T_h^*$ (=0.25, 0.28, 0.32, 0.35 in each 4 case), and the outer subregions (No.1 in fig.5) are kept at $T^*=T_0^*$ (=0.20 for all cases). After relaxation process of some thousands of time steps, steady state is achieved.

## 4.2. RESULTS OF THERMAL STRESS SIMULATION

Trajectory of atoms at temperature for $T_h^*$=0.28 and 0.32 is illustrated in figs.6(a) and (b), respectively. When $T_h^*$ is below melting point, all atoms stay at the lattice point representing crystal state. In the case when $T_h^*$ increases over melting point, the motion of atoms in the region and periphery is violent to be molten state which gives random trajectory as shown in the figure, and solid/liquid interface expands progressively. Temperature and stress distributions are represented in fig.7. Almost linear temperature gradients are accomplished in every case, but stress in y-direction shows different mode for each conditions. When $T_h^*$ is below melting point, stress gradient is linear corresponding to the temperature gradient. However, stress in molten region vanishes when melting occurred in central regions (See hollow circles and squares), and stress in solid region gives non-linear pattern.

These results clearly satisfy the mechanical equilibrium such that stresses in high temperature part are in compression and those in low temperature part are tensile with no stress in molten state.

## 5. Heat Conduction Incorporated with Melting/Solidification

### 5.1. MODEL AND CONDITIONS

This section focuses the heat conduction mode for the similar model as fig.6(a), and comparison between the results of microscopic and macroscopic simulation are made. Temperature is controlled only in the central regions at $T^*=T_c^*$ and, after 1000 steps of relaxation, $T^*=T_0^*$. Two conditions are applied; $T_0^*=0.10$, $T_c^*=0.25$ (below melting point) in the first case and $T_0^*=0.20$, $T_c^*=0.35$ (above the melting point) in the other case.

### 5.2. RESULTS ON VARIATION OF TEMERATURE

Variations of temperature in central and outer regions ($T_c^*$ and $T_e^*$ respectively) simulated by MD method are shown in fig.8. When $T_c^*$ is below the melting point, $T_e^*$ increase smoothly by neglecting the fluctuations as shown in fig.8(a). In the case of $T_c^*$ is above the melting point, $T_e^*$ increases in the similar manner as fig.(a) in low temperature range. The average temperature does not exceed the melting point ($T^*=0.30$) in the beginning, and it takes some thousands time steps until it begins to increase again. This comes from the effect of latent heat during melting process stated in Sec.3.

Figure 9 represents the trajectory of atoms, which shows that solid/liquid interfaces moves outward and that liquidus area progresses. Variations of stresses and potential energy are shown in fig.10. Thermal stress is generated before melting, but the stress vanishes when melting starts in central region. Variation of potential energy represents in the way, corresponding to enlarging liquid area; potential energy increase at almost constant rate with increasing temperature, while it jumps up from approximately $\phi^*=-1.7$ to

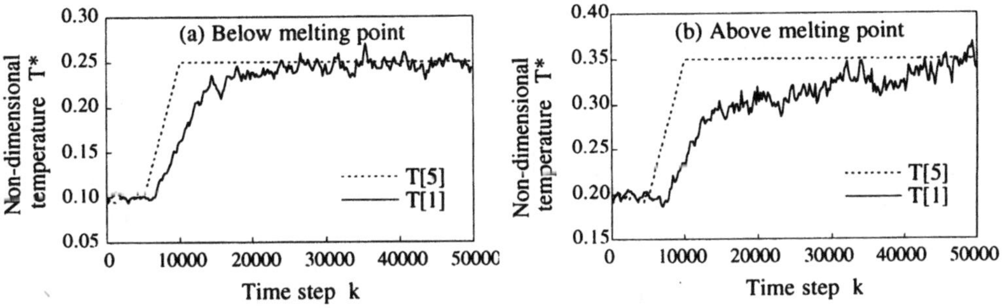


Fig. 8. Variation of temperature due to heat conduction.

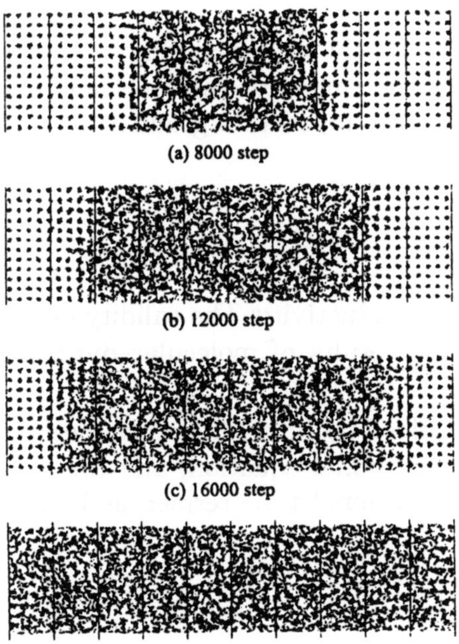


Fig. 9. Trajectory of atoms during melting.

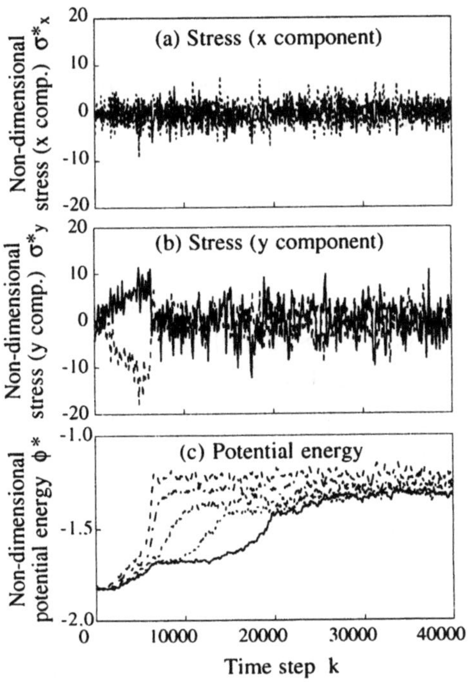


Fig. 10. Variation of stress and potential energy.

$-1.3$ in central region, and induces the increase in the neighbor region successively to complete melting in entire region.

## 6. Comparison between MD method and FDM

### 6.1. HEAT CONDUCTION EQUATION

Heat conduction equation is written as,

$$\rho c\dot{T} = -\frac{\partial h}{\partial x} \tag{5}$$

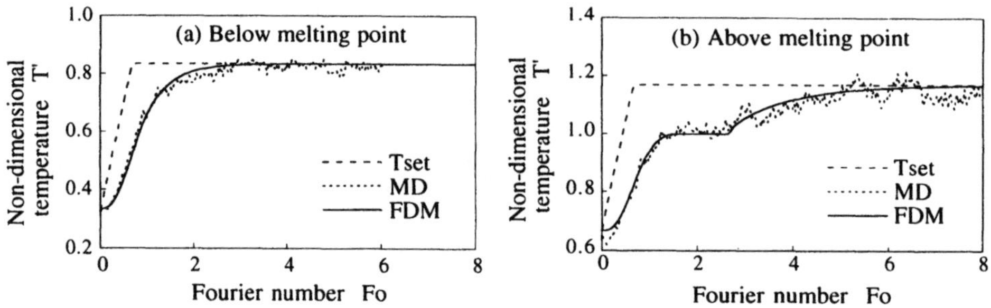


Fig. 11. Comparison between FDM and MD method.

where, $\rho$, $c$ and $\boldsymbol{h}$ are the density, specific heat and heat flux vector in the direction of $\boldsymbol{x}$, respectively. Following the empirical Fourier's law, $\boldsymbol{h}$ is written to be linear function with respect to temperature gradient as

$$\boldsymbol{h} = -k\frac{\partial T}{\partial \boldsymbol{x}} \tag{6}$$

where the proportional constant, $k$ is heat conductivity. The validity of this law is discussed bellow by comparing the results of molecular dynamics simulation stated in the previous section.

Now, eqs.(5) and (6 are numerically solved by finite difference method: The calculation is executed with non-dimensional equation, and the variation of temperature against the non-dimensional time termed as Fourier number,

$$F_o = \frac{k}{\rho c l^2} t \tag{7}$$

is plotted where $l$ is a characteristic length. To take into account of the effect of latent heat, equivalent specific heat method is adopted.

## 6.2. COMPARISON OF BOTH RESULTS

Simulated results by MD method and FDM are compared in fig.11, where, horizontal axis is non-dimensional time, Fourier number, and vertical axis is non-dimensional temperature by melting temperature $T'=T/T_M$. In the preparatory calculation, some kinds of thermal properties such as heat conductivity and specific heat are evaluated as in the same manner as in Section 2. Non-dimensional representation in fig.11 for both the results of temperature variation shows quite good correspondence with each other in spite of some fluctuations in MD method.

## 7. Concluding Remarks

Melting and solidification processes are simulated by molecular dynamics method, and the some leading parameters appearing in materio-thermo-mechanical effects, such as latent heat and thermal stress are simulated. Heat conduction mode incorporating melting are also simulated, and the results are compared with the solution of heat conduction equation.

## References

1. Inoue, T. and B. Raniecki, "Determination of Thermal-hardening Stresses in Steels by use of Thermoplasticity Theory", Journal of the Mechanics and Physics of Solids, Vol.**26**-3 (1978), pp. 187-212.
2. Inoue, T., Nagaki, S., Kishino, T. and Monkawa, M., "Description of Transformation Kinetics, Heat Conduction and Elastic-plastic Stresses in the Course of Quenching and Tempering of Some Steels", Ingenieur-Archiv., Vol.**50**-5 (1981), pp. 315-327.
3. Inoue, T. and Arimoto, K., "Development and Implementation of CAE System "HEARTS" for Heat Treatment Simulation based on Metallo-thermo-mechaniics", J. Materials Engineering Technology, ASM, Vol.**6**-1 (1997), pp. 51-60.
4. Broughton, J. Q., Gilmer, G. H. and Jackson, K. A., "Crystallization Rates of a Lennard-Jones Liquid", Phys. Rev. Lett., Vol.**49**-20 (1982), pp. 1496-1500.
5. Kotake, S. and Wakuri, S. "Molecular Dynamics Study of Heat Conduction in Solid Materials", JSME International Journal, Vol.**37**-1 B (1994), pp. 103-108.
6. Uehara, T. and Inoue, T., "Molecular Dynamics Simulation of Materio-Thermo-Mechanical Fields in Melting/Solidification Process", Journal of Japan Society of Mechanical Engineering, Vol.**63**-614 A (1997), pp. 2135-2141 (in Japanese).
7. Uehara, T. and Inoue, T., "Molecular Dynamics Simulation of Materio-Thermo-Mechanical Fields in Phase Transformation Process", Proc. of International Conferenc on Materials and Mechanics '97 (1997), pp. 431-432.
8. Uehara, T. and Inoue, T., "Molecular Dynamics Simulation of Heat Conduction and Thermal Stress in Relation with Continuum Mechanics", Materials Science Research International, Vol.**4**-1 (1998), pp. 45-52.
9. Verlet, L., "Computer "Experiments" on Classical Fluids. I. Thermodynamical Properties of Lennard-Jones Molecules", Phys. Rev., Vol.**159**-1 (1967), pp. 98-103.

# EXPLORING THE COMMONALITY BETWEEN FILLING: POROUS MEDIA AND PHASE CHANGE

V.R. VOLLER
*St . Anthony Falls Laboratory*
*University of Minnesota*
*Minneapolis MN 55414*

## 1. Introduction

Problems involving flow in porous media cover a wide spectrum of areas in engineering and science. An important class of porous media flow problems involves the infiltration of a fluid into an initially dry porous media and key element in such problems is the tracking of the free or moving interface between the filled and empty pores. Examples include, flow through a leaky dam (Crank, 1984), flow in the partially saturated region of a soil (Nieber, 1996) and injection of polymer in a mold (Bruyschke and Advani, 1990) (Lee and Sin, 1990) (Maier *et al.*, 1996) (Voller and Chen, 1996).

The area of interest in this paper will be the numerical treatment of problems related to tracking the polymer front during the injection stage of a polymer molding process. Interest will be focused on fixed space grid schemes in which the position of the moving polymer front is tracked via an auxiliary liquid fraction variable. The techniques and methods for generating and tracking the liquid fraction variable are directly borrowed from existing techniques for dealing with solid liquid phase change (Voller, 1996)

## 2. Problem of Interest

The immediate problem of interest is related to the Resin Transfer Molding Process (RTM), (Bruyschke and Advani, 1990) (Lee and Sin, 1990) (Maier *et al.*, 1996) (Voller and Chen, 1996). This process involves the injection of polymer into a mold that contains pre-existing fibers and the subsequent reaction and hardening of the polymer to form a lightweight and strong composite part. Note that the boundary conditions are fixed pressure at the injection gates and no flow (insulated) elsewhere.



*P. Argoul et al. (eds.),*
***IUTAM Symposium on Variations of Domains and Free-Boundary Problems in Solid Mechanics*, 113–119.**

The porous media formed by the fibers has two distinct scales. The pores between the fiber tows form the main flow paths. Flow in this region is governed by the Darcy law

$$q = -K\nabla P \tag{1}$$

Where due to the large pressures involved in the RTM process contributions from hydrostatic pressure (gravity terms) have been neglected.

The fiber tows, themselves, are made up of bundles of much finer fibers and as the polymer flows in through the pores it is absorbed into these bundles. In essence the RTM process can be though of as flow through an absorbing porous media. Assuming a simple first order relationship the amount absorbed (per unit volume) by the fiber tows can be modeled as

$$S = cP \tag{2}$$

where $c$ is a constant (think of it as a specific storage term). This model simply sates that the amount absorbed by a given fiber tow is directly proportional to the pressure in the adjacent pore. Clearly this model is ad-hoc in form, in general one would expect a complicated model to describe the absorbency of the fiber tows. Typically, however, the absorption by the fibers is small and in many RTM modeling studies (Bruyschke and Advani, 1990) (Lee and Sin, 1990) (Maier *et al.*, 1996) (Voller and Chen, 1996) it is ignored (i.e., $c$ is set to 0 ) with little consequence in the physical validity of the results. Further this choice of model will allow for a direct analogy between RTM and solid/liquid phase change processes; a step that opens up the door to a wide number of numerical solution strategies.

Under the assumptions that Eq. (1) and (2) hold the governing equation for the flow during the filling of an RTM mold is

$$\frac{\partial G}{\partial t} = \nabla(K\nabla P) \tag{3}$$

where

$$G = cP + F \tag{4}$$

is the total volume fraction filled (relative to the pore space, i.e., $G$ can take values larger than 1), $F$ is the volume fraction of pores filled ( a value that can not exceed 1) and (assuming no change with time; isothermal filling) $K$ is a lumped term accounting for spatial resistance to flow through the pores.

Note that the following relationships hold between the Pressure, $P$, pore filled fraction, $F$, and the total filled fraction $G$.

$$P = \begin{cases} 0 & \text{if } G < 1 \\ \frac{G-1}{c} & \text{otherwise} \end{cases} \tag{5}$$

and

$$F = \min(G, 1) \tag{6}$$

Essentially Eq. (5) states that at points in the domain which are empty or filling P is fixed at the gauge pressure of zero.

## 3. Commonality with Phase Change

A very basic model for solid/liquid phase change is the enthalpy model. In the case of a the melting of a pure material at the phase change temperature the governing enthalpy equation takes the form (Crank, 1984)

$$\frac{\partial H}{\partial t} = \nabla(K \nabla T) \tag{7}$$

where, in this case,

$$H = cT + FL \tag{8}$$

is enthalpy, $F$ is the liquid fraction, $L$ is the latent heat, $T$ is temperature and $K$ is the thermal conductivity.

The values $H$,$F$ and $T$ are related in the following manner

$$T = \begin{cases} 0 & \text{if } H < L \\ \frac{H-L}{c} & \text{otherwise} \end{cases} \tag{9}$$

$$F = \min(\frac{H}{L}, 1) \tag{10}$$

On setting the latent heat L = 1 and comparing Eqs. (3–6) with Eqs. (7–10) the analogy between modeling phase change and filling porous media is immediately obvious. This analogy allows us to borrow from well established phase change techniques in solving porous media problems (Voller, 1996).

## 4. A Simple Test Problem

Below two schemes borrowed from the phase change literature and applied to the filling of a porous media are presented; an explicit time integration scheme and an implicit time integration scheme. In order to present the

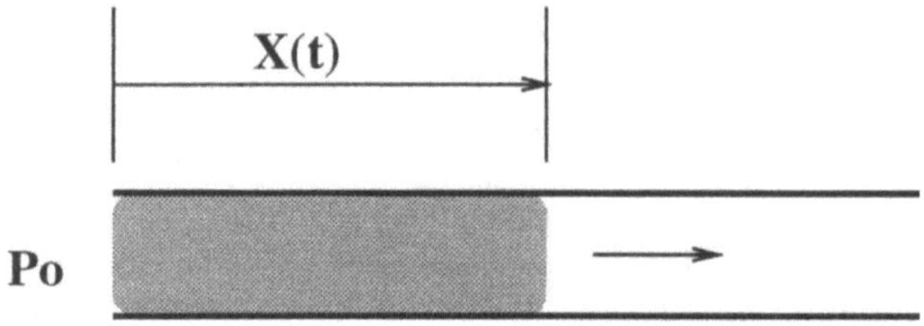


*Figure 1.* Filling a porous tube, $P_o = 1$

mechanisms involved in these solutions a simple one-dimensional problem of filling a porous tube (length 1 unit), with a unit properties and an applied Pressure of $P_o = 1$ at $x = 0$, will be considered, see Fig 1.

In the case where $c = 0$ (no absorbency by the fibers) this problem has a very straight forward analytical solution. The position of the front $s(t)$ with time is given by

$$X(t) = \sqrt{2t} \tag{11}$$

and the pressure field at any time( in the filled portion of the tube) is linear.

In solving this problem a very simple finite difference solution based on a grid of 51 uniformly spaced nodes (i=0, 1 ,2, ... 50) will be employed . It can not be expressed strongly enough, however, that at no time will the dimensionally of the problem be used to enhance the performance of a given numerical scheme. All of the methods presented can and have been used to solve problems in multi-dimensions using both Galerkin and Control Volume Finite Element Methods (e.g., (Voller and Chen, 1996))

## 5. An Explicit Solution

In the first place a simple explicit time integration solution is suggested for the filling of an RTM mold.

A discrete form of Eq. (3) –assuming $K = 1$ – is

$$G_{i,j} = G_{i,j-1} + \frac{\Delta t}{\Delta x^2} \left( P_{i-1,j-1} - 2P_{i,j-1} + P_{i+1,j-1} \right) \tag{12}$$

where $i$ is a space counter, $j$ is a time counter, $\Delta x = 0.02$ is the space step and $\Delta t$ is the time step. In any time step solution of Eq. (12) will generate a nodal field of current $G$ values. From this field a direct application of Eqs. (5) and (6) can be used to generate current nodal fields for $P$ and $F$. These fields can then be used in the next time step calculation and the $F$ field can be used to track the movement of the filling front.

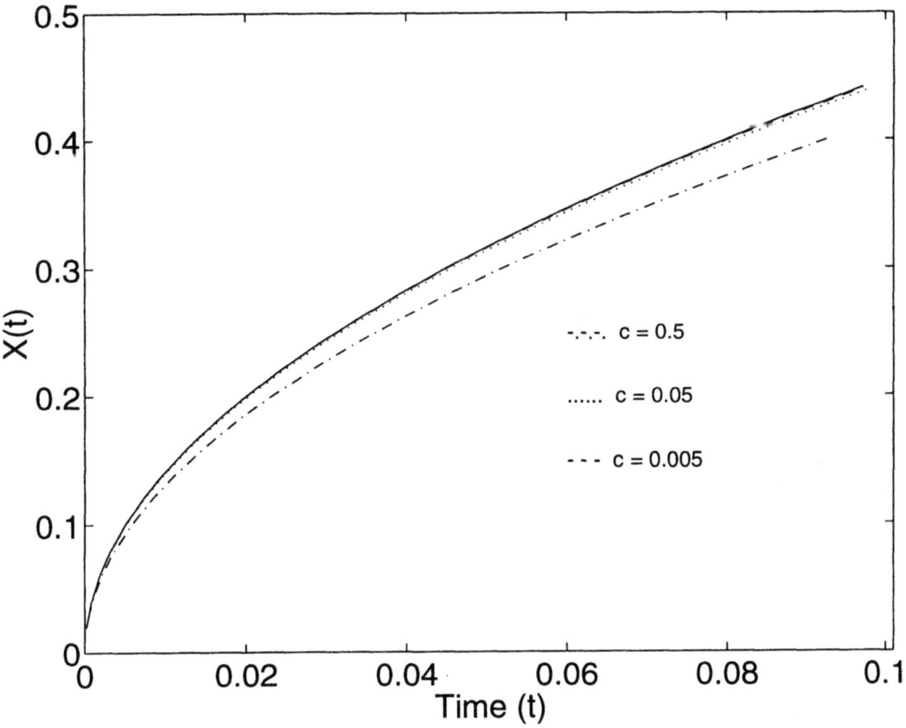


*Figure 2.* Explicit scheme predictions of front movement with various values of absorbency $c$

Predictions for front movement – with different values of absorbency ($c$) – are compared with the analytical solution in Fig. 2. This result indicates that with reasonably small values of $c$ the proposed explicit scheme can model the case of zero absorbency.

On the positive side the proposed explicit scheme is very easy to program and readily extendible to two-dimensions. The method is also very flexible and can accommodate complicated models. For example the modeling of weld lines where two filling surfaces meet. On the negative side a small value of absorbency $c$ – a reasonable physical value – places a severe stability restriction of the choice of time step $\Delta t$. One way around this problem is to use the implicit time integration scheme proposed below.

## 6. An Implicit Solution

For the case of a zero absorbency ($c = 0$) a one-dimensional implicit time integration of Eq. (3) is

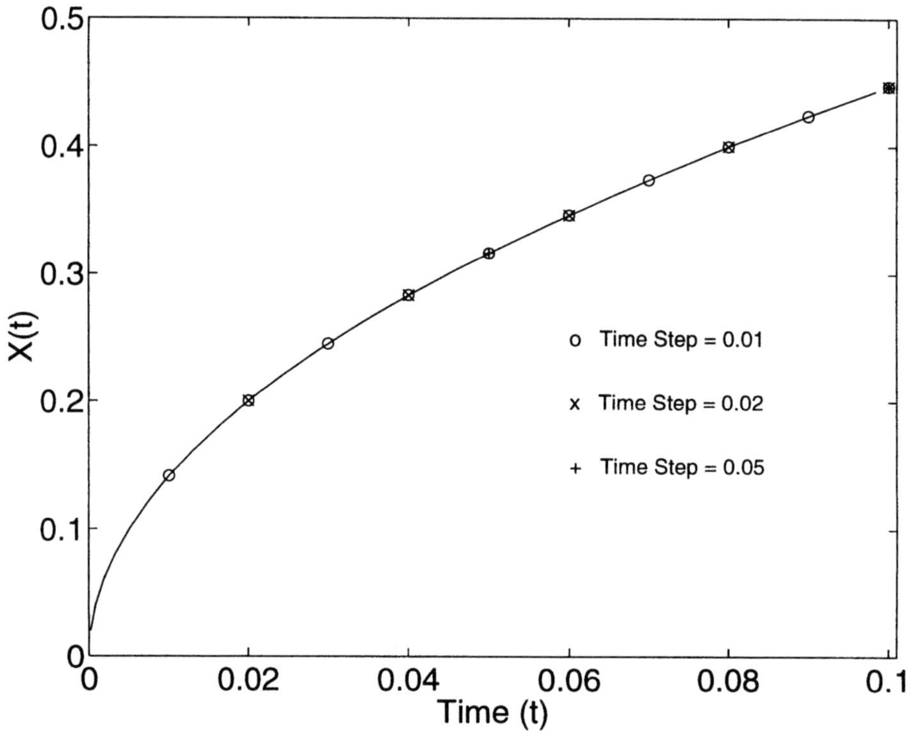


*Figure 3.* Implicit scheme predictions of front movement with various time steps

$$a_i P_i = P_{i-1} + P_{i+1} + \frac{\Delta x^2}{\Delta t}\left(F_{i,j-1} - F_{i,j}\right) \tag{13}$$

where in normal circumstance the coefficient takes the value $a_i = 2$.

In a time step an iterative solution of Eq. (13) subject to the condition that $P = 0$ when $F < 0$ needs to be devised. One efficient approach (Voller and Peng, 1995) is as follows:

1. With the most current liquid fraction (fill factor) field, $F^n$, solve Eq. (13) for the current pressure field (the superscript $n$ is an iteration counter) . In this solution at any node, $i$, where $F_i < 0$ the value of $a_i$ is set to $10^{25}$. This step will ensure that the solver will return nodal values of $P_i = 0$ at all the nodes which are empty or filling.
2. At every node in the field update the fill factor on rearrangement of Eq. (13) i.e,

$$F_{i,j} = F_{i,j-1} + \frac{\Delta t}{\Delta x^2}\left(P_{i-1} + P_{i+1}\right) \tag{14}$$

3. In using Eq. (14) it has been assumed that the cell around node $i$ is filling and as such $P_i = 0$. To account for filled cells – following use of Eq. (14)– nodal values of $F_i > 1$ are pegged back to a value of $F_i = 1$.
4. steps (1) to (3) above are repeated until the calculated $F$ field does not change between iterations.

The performance – prediction of filling front – of the implicit method – using a variety of time step – on the simple tube filling problem is illustrated in Fig. 3. The most striking part of these results is that the predictions are independent of time step. It has been rigorously shown that this result also hold in two-dimensions (Voller and Chen, 1996). The independent of time step result is also consistent with the analysis of the Hele-Shaw problem (Crank, 1984). A powerful follow on from the independence of time step is the development of algorithms to predict the last point to fill in a given mold (Voller and Chen, 1996).

## References

Bruyschke, M.V. and Advani, S.G. (1990) A Finite Element/Control Volume Approach to Mould Filling in Anisotropic Porous Media, *Polym. Compos.*,**11**, pp. 398–405

Lee, J. S. and Sin, H.C. (1990) Iterative Boundary Reflection Method for the Simulation of Injection Mould Filling, *Polym. Eng. Sci.*,**30**, pp. 1531-1542

Maier, R.S., Rohaly, T.F., Advani, S.G. and Fickie, K.D. (1996) A Fast Numerical Method for Isothermal Resin Transfer Mould Filling, *Int. J. Num. Meth. Eng.*,**39**, pp. 405-417

Crank, J. (1984) *Free and Moving Boundary Problems.* Clarendon Press, Oxford.

Voller, V.R. and Chen Y.F. (1996) Prediction of Filling Times of Porous Cavities, *Int. J. Num Meth. Fluids*,**23**, pp. 661-672

Nieber, J.L. (1996) Modeling Finger Development and Persistence in Initially Dry Porous Media, *Geoderma*,**70**, pp. 207-229

Voller, V.R. (1996) An Overview of Numerical Methods for Solving Phase Change Problems, *Adv. Num Heat Transfer*,**1**, pp. 341-380

Voller, V.R. and Peng, S. (1995) An Algorithm for Analysis of Filling in Polymer Molding, *Polym. Eng. Sci.*, **35**, pp. 1758-1765

# ABOUT NUCLEATION AND GROWTH

A. VISINTIN
*Dipartimento di Matematica,*
*via Sommarive 14, 38050 Povo di Trento, Italy.*

## 1. Free Energy and Gradient Flow

**Macroscopic and Mesoscopic Free Energies.** Let us consider a substance capable of attaining two phases (solid and liquid, say), which occupies a bounded region $\Omega \subset \mathbf{R}^3$. Let us assume that the relative temperature is a prescribed function $\theta \in L^\infty(\Omega)$, and set

$$\chi := \begin{cases} 1 & \text{in } \Omega^* \text{ (the liquid phase)}, \\ -1 & \text{in } \Omega \setminus \Omega^* \text{ (the solid phase)}. \end{cases}$$

Under several physical assumptions (incompressibility, homogeneity and anisotropy of the material, and so on) the *macroscopic* free energy reads

$$\tilde{F}_\theta(\chi) := -\frac{L}{2\tau_E} \int_\Omega \theta \chi \, dx \tag{1.1}$$

(plus an additive function of $\theta$). $L$ is the density of latent heat, $\tau_E$ is the absolute temperature of equilibrium for a solid-liquid mixture; both are positive constants. Obviously $\chi$ minimizes $\tilde{F}_\theta$ under the constraint "$|\chi| = 1$ a.e. in $\Omega$" iff

$$\theta \leq 0 \text{ in } \Omega \setminus \Omega^*, \quad \theta \geq 0 \text{ in } \Omega^*. \tag{1.2}$$

To account for *undercooling* prior to solid nucleation, at first we set

$$P(v) := \begin{cases} \dfrac{1}{2} \displaystyle\int_\Omega |\nabla v| & \text{if } |v| = 1 \text{ a.e.}, \\ +\infty & \text{otherwise}, \end{cases} \qquad \forall v \in L^1(\Omega).$$

Thus $P(v)$ represents the perimeter of the set $\{x \in \Omega : v(x) = 1\}$, that is, the area of the solid-liquid interface. We then introduce the *mesoscopic* free energy

$$F_\theta(\chi) := -\frac{L}{2\tau_E} \int_\Omega \theta \chi \, dx + \sigma P(\chi) \tag{1.3}$$



*P. Argoul et al. (eds.),*
*IUTAM Symposium on Variations of Domains and Free-Boundary Problems in Solid Mechanics,* 121–127.

(plus an additive function of $\theta$, and a boundary contribution which here we neglect). As we shall see, the *surface tension coefficient* $\sigma$ is very small in terms of the usual macroscopic system of units. This requires the use of a *mesoscopic* length-scale.

**Radial Setting.** We want to describe nucleation of a solid ball of radius $R$ in a liquid at a uniform temperature $\theta < 0$. The corresponding variation of $F_\theta$ only depends on $R$, and reads

$$\varphi(R) = \frac{4\pi L}{3\tau_E}\theta R^3 + 4\pi\sigma R^2 \qquad \forall R \geq 0 \text{ (cf. Fig. 1).} \tag{1.4}$$

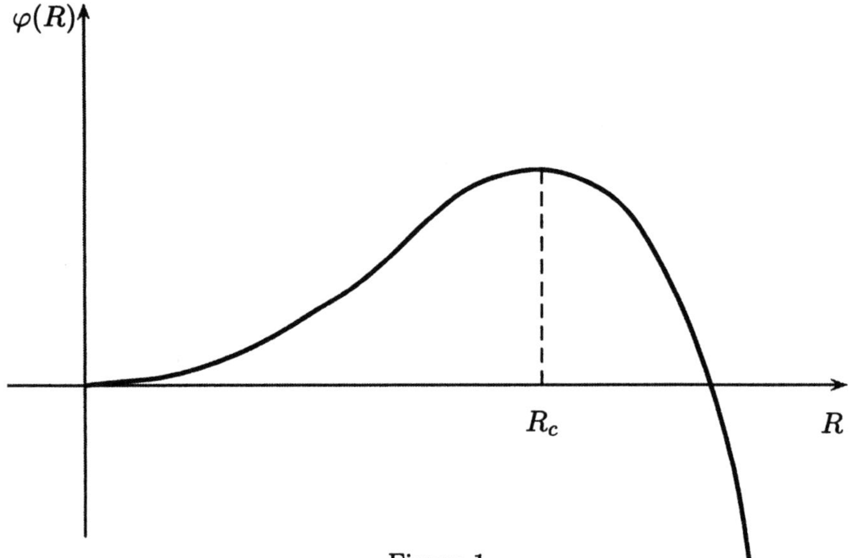


Figure 1.

Hence

$$\varphi'(R) = 0 \quad \Longleftrightarrow \quad R = -\frac{2\sigma\tau_E}{L\theta} \; (=: R_c > 0). \tag{1.5}$$

The dynamics is governed by descent along the potential $\varphi$. By a suitable choice of the relaxation coefficient $a$, the *gradient flow* for $\varphi$ yields the law of *motion by mean curvature:*

$$a\frac{dR}{dt} = -\frac{\varphi'(R)}{4\pi R^2} = -\frac{2\sigma}{R} - \frac{L\theta}{\tau_E}. \tag{1.6}$$

**General Setting: Gibbs-Thomson Law and Mean Curvature Flow.** Let us denote by $H(\cdot, t)$ the double of the mean curvature of the interface $\mathcal{S}_t := \partial\Omega^*(t)$ (with the convention that $H > 0$ for solid balls). Any stationary point of $F_\theta$ fulfils the classical *Gibbs-Thomson law*

$$\theta = -\frac{\sigma\tau_E}{L} H \qquad \text{on the interface } \mathcal{S}. \tag{1.7}$$

For instance, for water at atmospheric pressure at about $0^0$C we have $\sigma \simeq 1.8 \times 10^{-6}$ cal cm$^{-2}$, $L \simeq 80$ cal cm$^{-3}$, $\tau_E \simeq 273$ K. Hence $2\sigma\tau_E/L \simeq 1.2 \times 10^{-5}$ cm K. Hence nucleation at an undercooling of few degrees corresponds to a critical radius of the order of $10^{-5}$ cm.

Defining the *signed distance* from the solid-liquid interface

$$\rho(x,t) := \text{ ess-dist } (x, \Omega \setminus \Omega^*(t)) - \text{ ess-dist } (x, \Omega^*(t)) \qquad \forall (x,t),$$

$\partial\rho/\partial t$ equals the velocity of advancement of the liquid front. The *mean curvature flow* (with forcing term) then reads

$$a\frac{\partial \rho}{\partial t} = \sigma H + \frac{L}{\tau_E}\theta \qquad \text{on } \mathcal{S}. \tag{1.8}$$

**Remark.** For any bounded $\theta < 0$, the *whole liquid* state "$\chi \equiv 1$ in $\Omega$" is a relative (nonabsolute) minimizer of $F_\theta$ with respect to the metric of $L^1(\Omega)$. This is made possible by the nonconvexity of the potential, and by the presence of a (convex and coercive) higher order term, i.e., the surface tension term $\frac{\sigma}{2}\int_\Omega |\nabla\chi|$.

## 2. Nucleation and Growth

By (1.6) it is clear that phase nucleation cannot occur by mean curvature flow, but only as a fluctuation that takes $R$ from 0 to a value larger or equal to $R_c$. In a small time interval $\delta t$ this occurs with probability proportional to $\delta t \exp\left[-\varphi(R_c(\theta))/k\tau\right]$. So mean curvature flow is deterministic and continuous, whereas nucleation is stochastic and discontinuous.

By the classical theory of nucleation, for a metal and at the beginning of the process, the *nucleation rate* $I$ (:= number of nuclei formed per unit volume per unit time) varies from almost zero to a very large value in a very small range of temperatures, which can be identified with a single value $\theta_c (< 0)$, see Fig. 2. Therefore, for any prescribed time-scale, nucleation can be assumed to occur at a *critical temperature* $\theta_c$. Nuclei grow by mean curvature flow, until either they impinge on other nuclei, or undercooling is eliminated by release of latent heat.

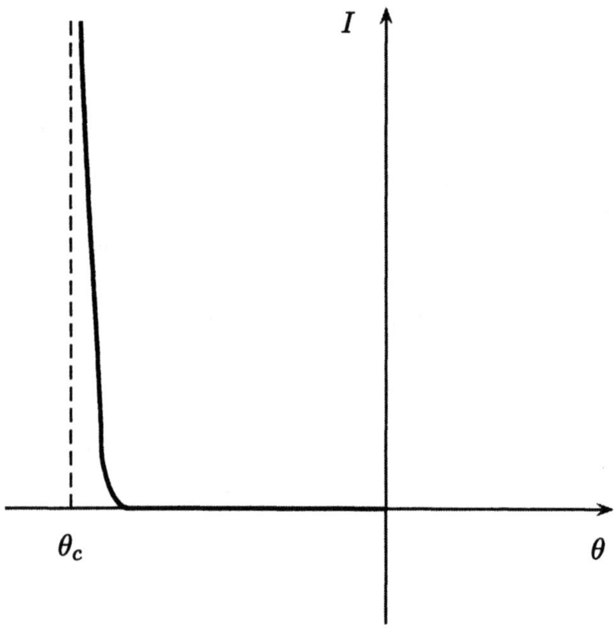


Figure 2.

**Nucleation and Hysteresis.** For the sake of simplicity here we assume that the thresholds for solid and liquid nucleation are symmetric. Neglecting the dependence on $x$, we set $\chi_{\pm}(t) := \chi(t \pm 0)$ for any $t$.

The undercooling and superheating required to trigger nucleation are schematically outlined in the Fig. 3, where $c$ is a positive parameter.

In analytic terms this behaviour can be written as

$$\begin{cases} \chi_+ \in \operatorname{sign}(\theta + c\chi_-) \\ |\chi| = 1 \end{cases} \qquad \forall t; \tag{2.1}$$

or also, setting $\psi(v) := 0$ if $|v| = 1$ and $\psi(v) := +\infty$ otherwise,

$$\psi(\chi_+) - \psi(v) \leq (\theta + c\chi_-)(\chi_+ - v) \qquad \forall v \in \mathbf{R}, \forall t. \tag{2.2}$$

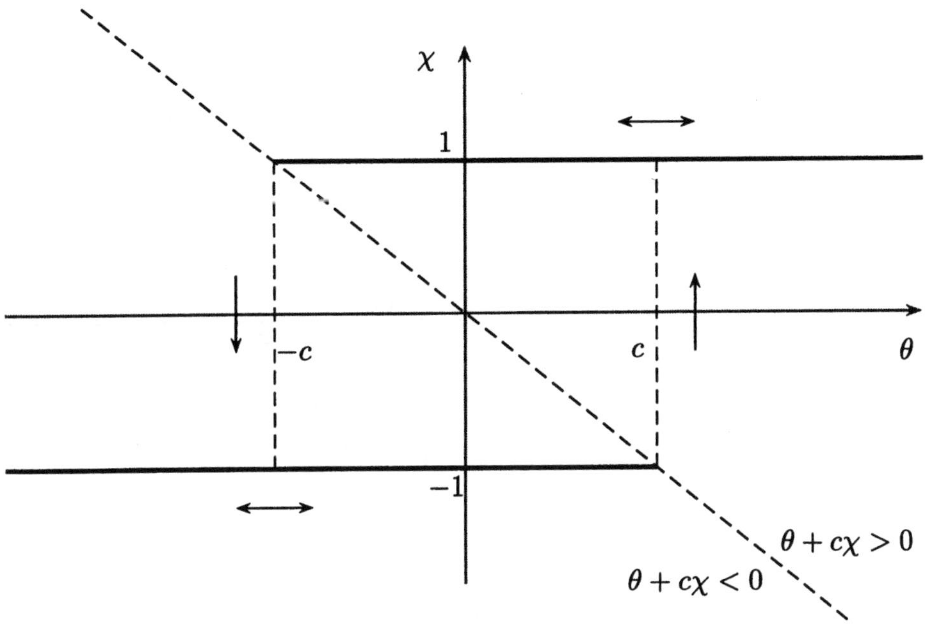


Figure 3.

Now we reinsert the $x$-dependence, assume that $\theta(\cdot, t) \in L^\infty(\Omega)$ for any $t$, and represent nucleation by the variational inequality

$$P(\chi_+) - P(v) \leq \frac{L}{2\tau_E} \int_\Omega (\theta + c\chi_-)\,(\chi_+ - v)dx \quad \forall v \in L^\infty(\Omega), \forall t. \tag{2.3}$$

**Nonlinear mean curvature flow.** At this point the exigence arises of matching nucleation and mean curvature flow. We then propose to modify (1.8) as follows:

$$\alpha\left(\frac{\partial \rho}{\partial t}\right) = \sigma H + \frac{L}{\tau_E}\theta \qquad \text{on } \mathcal{S}_t, \forall t \in I_c, \tag{2.4}$$

where $\alpha : \mathbf{R} \to \mathbf{R}$ is nonconstant, bounded and monotone. For instance, see Fig. 4,

$$\alpha(\xi) := \begin{cases} -M & \text{if } \xi < -M/a \\ a\xi & \text{if } -M/a \leq \xi \leq M/a \\ M & \text{if } \xi > M/a, \end{cases} \tag{2.5}$$

where $a, M$ are positive constants.

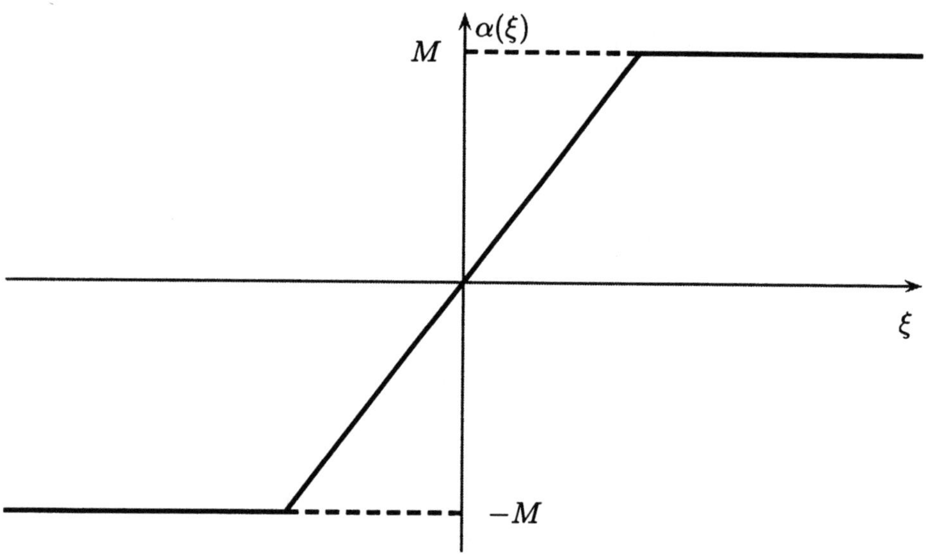


Figure 4.

In conclusion, we represent nucleation and curvature flow by coupling (2.3) and (2.4), which provide complementary and consistent information.

## 3. Weak Formulation

Let us assume that

$$\theta \in C^0(\bar{Q}), \qquad \chi^0 \in \mathrm{Dom}(P), \tag{3.1}$$

and set $L = \tau_E = a = M = 1$, for the sake of simplicity.

We then provide a weak formulation of our model of *nucleation and mean curvature flow* under a prescribed temperature field evolution.

**Problem P..** *To find $\chi \in BV(Q)$ such that (denoting by $\partial_e$ the "essential boundary")*

$$\mathcal{S}_t := \partial_e\{\chi(\cdot, t) = 1\} \text{ is a surface of class } C^1, \text{ for a.a. } t \in ]0, T[, \tag{3.2}$$

$$H(\cdot,t)\ (:= 2 \times \textit{mean curvature of } \mathcal{S}_t) \in L^2(\mathcal{S}), \textit{ for a.a. } t \in ]0,T[, \tag{3.3}$$

$$2\int_0^T [P(\chi) - P(v)]\, dt \le \iint_Q (\theta+\chi)(\chi - v)dxdt \quad \forall v \in L^1(Q), \tag{3.4}$$

$$\begin{aligned} &2P(\chi(\cdot,T)) - 2P(\chi^0) + \textstyle\int_0^T dt \int_{\mathcal{S}_t} \left[(H+\theta)^2 \quad v^2\right] dS \\ &+_{C^0(\bar{Q})}\langle v-\theta, \partial\chi/\partial t\rangle_{C^0(\bar{Q})'} \le 0 \qquad \forall v \in C^0(\bar{Q}), |v| \le 1. \end{aligned} \tag{3.5}$$

The system (3.4), (3.5) is *formally* equivalent to (2.3) and (2.4).

**Theorem ..** *(See [7]) If (3.1) holds and $\partial\theta/\partial t \in L^1(Q)$, then Problem P has a solution such that the perimeter function $P(\chi)$ is nonincreasing in $[0,T]$.*

In phase transitions the temperature field is coupled with phase evolution, and the energy balance equation (here also written with normalized coefficients)

$$\frac{\partial}{\partial t}(\theta+\chi) - \Delta\theta = f \qquad \text{in } \mathcal{D}'(Q) \tag{3.6}$$

must be coupled with Problem P. Existence of a weak solution can be proved for this problem, see [8].

[1,2,5,7] represent mean curvature flow by means of characteristic functions. [6,8] couple mean curvature flow and phase transitions. [3,4] are multi-author books on mean curvature flow. [9] is a recent monograph on phase transitions.

## References


1. Almgren, F. , Taylor, J.E. , Wang, L. (1993) Curvature-driven flows: a variational approach. *S.I.A.M. J. Control and Optimization* **31** 387–437
2. Almgren, F. , Wang, L. Mathematical existence of crystal growth with Gibbs-Thomson curvature effects. *Journal of Geometric Analysis (to appear)*
3. Buttazzo, G. , Visintin, A. (eds.) (1994) *Motion by Mean Curvature and Related Topics.* De Gruyter, Berlin.
4. Damlamian, A. , Spruck, J. , Visintin, A. (eds.) (1995) *Curvature Flow and Related Problems.* Gakkōtosho Scientific, Tokyo.
5. Luckhaus, S. , Sturzenhecker, T. (1995) Implicit time discretization for the mean curvature flow equation. *Calc. Var.*, **3** pp. 253–271
6. Soner, M. (1995) Convergence of the phase-field equation to the Mullins-Sekerka problem with kinetic undercooling. *Arch. Rational Mech. Anal.*, **131** pp. 139–197
7. Visintin, (1998) A. Nucleation and mean curvature flow. *Communications in P.D.E.s*, **23**, pp. 17-35
8. Visintin, A., Stefan problem with nucleation and mean curvature flow. (in preparation)
9. Visintin, A., (1996) *Models of Phase Transitions.* Birkhäuser, Boston.

# WAVELET METHODS FOR A STEFAN PROBLEM

J. ELEZGARAY AND P. PETIT
*Centre de Recherche Paul Pascal*
*C.N.R.S., Av. Schweitzer, 33600 Pessac, France*

One of the most interesting features of nonequilibrium systems is their ability to display pattern formation phenomena. A typical example is the formation of fingering or dendritic patterns when a crystal grows in a supercooled melt. In this work, we will consider numerical methods for the simulation of the two-dimensional Stefan problem arising in this system. More precisely, we focus on the one-sided diffusion limited model (considered, for instance, in [1]) in the limit of infinite diffusion length. Therefore, we compute the time evolution of a (finite length, closed) interface $\Gamma(\alpha, t), \alpha \in [0, 2\pi]$ separating two (time dependent) open domains $\Omega_1(t)$ and $\Omega_2(t)$ in which a dimensionless diffusion field (proportional either to the local temperature in the case of a pure liquid, or to the local chemical potential in the case of a liquid mixture) $u(\eta, t)$ satisfies:

$$u(\eta) = 0 \quad \text{if} \quad \eta \in \Omega_1(t), \tag{1}$$

$$\Delta u(\eta) = 0 \quad \text{if} \quad \eta \in \Omega_2(t), \tag{2}$$

$$[u]|_\Gamma = T\kappa, \tag{3}$$

where $\kappa$ is the local curvature and $T$ is the surface tension. The normal velocity of the interface is given by the continuity equation $v_n = -\nabla u \cdot n$ . This formulation also corresponds to the Hele-Shaw problem (a viscous fluid pushed by a less viscous fluid) in the Darcy approximation [2]. A similar system of equations can also be found in simple models of electrodeposition phenomena in thin gap cells.

A considerable amount of work has been devoted to the analysis of steady (self-similar) solutions of eqs (1-3), specially in the limit of small surface tension. This interest has been mainly sparked by the early discovery of exact solutions when $T = 0$ (the so called Saffman-Taylor fingers [2]). This is in constrast with the well established experimental observation that



*P. Argoul et al. (eds.),*
***IUTAM Symposium on Variations of Domains and Free-Boundary Problems in Solid Mechanics*, 129–136.**

only one of these solutions is observed when $T \sim 0$ (but positive), and the mechanism underlying this selection phenomenon has only been recently understood. Several families of exact unsteady solutions are known at $T = 0$. The selection mechanism of these solutions has been partially explained by S. Tanveer [3], using a perturbative approach. However, several aspects of the problem cannot be easily handled perturbatively, and the numerical simulation of eqs. (1-3) remains one of the basic tools for the study of this Stefan problem. In the second section of this paper, we will address the problem of numerically checking the results of S. Tanveer. In the first section, we discuss numerical methods to accelerate the solution of the integral equation determining the diffusion field $u(\eta, t)$.

## 1. The wavelet-multipole method

Let us consider the integral representation of the diffusion field

$$u(\eta,t) = \Re\left[Q(\eta) + \frac{1}{2i\pi}\int_0^{2\pi} \gamma(\alpha,t)\log(\eta - z(\alpha,t))d\alpha\right], \tag{4}$$

where $z(\alpha, t)$ is some parametrization of the interface $\Gamma(t)$, $\gamma(\alpha, t)$ is a (real) unknown fonction, $Q(\eta)$ a source term taking into account boundary conditions at infinity, and $\eta = x + iy$. Eq. (3) leads to an integral equation for $\gamma(\alpha, t)$ with a logarithmic kernel. We prefer instead to deal with its derivative, and rewrite eq. (3) as $\tau \cdot \nabla u = T\kappa_s$ ($\tau$ is the unit tangent to the interface and $s$ is the arclength). Taking the derivative of (4) with respect to $\eta$ and the limit $\eta \to z(\alpha)$, it is easily seen [4] that $\gamma(\alpha, t)$ satisfies the following integral equation (in the following, we drop the time index):

$$\frac{1}{2}\gamma(\alpha) = -\Re\left(z_\alpha \frac{1}{2i\pi} P\int_0^{2\pi} \frac{\gamma(\alpha')}{z(\alpha) - z(\alpha')} d\alpha'\right) - T\kappa_\alpha(\alpha) - \Re\left(Q'(z(\alpha))z_\alpha\right). \tag{5}$$

As in [4], we use the modified point vortex approximation to discretize eq. (5). This yields the linear system

$$\gamma_j = -2\Re\left(z'_j \frac{2h}{2i\pi} \sum_{k=0, j+k=odd}^{N-1} \frac{\gamma_k}{z_j - z_k}\right) + \theta_j, \tag{6}$$

where $\theta_j$ takes into account the second and third terms of the r.h.s. eq. (5), $z'_j$ is an approximation of $z_\alpha(jh)$ and $h = 2\pi/N$.

The linear system of eq. (6) is nonsymmetric and full. Consequently, the method of choice for its solution is an iterative method such as GMRES [5]. In terms of computational effort, the most time consuming part of any such algorithm is the matrix-vector product. For the particular type of kernel we consider here, the multipole method of Greengard and Rokhlin [6] provides a matrix-vector product with a $O(N)$ asymptotic complexity. We used the non adaptive version of this algorithm, and modified its structure in order to achieve a partial vectorization of the code (this modified code runs at least three times faster than the original *adaptive* code of ref. [6]).

Is there any way to improve the $O(N)$ estimate for the specific problem we consider here? The answer is affirmative, provided no high accuracy is required. The idea behind what we call the wavelet-multipole method is to use the fact that, in some cases, $\gamma(\alpha)$ is actually well approximated by the superposition of a (low) number $(N_s)$ of functions. Let us assume that this is the case: $\gamma(\alpha) = \sum_{l=1}^{N_s} \Psi_l(\alpha) d_l$, with $d_l = \int_0^{2\pi} \Psi_l(\alpha)\gamma(\alpha)d\alpha$. Using the multipole method, one can compute the $N$ sums $S_i = \sum_{j=1}^{N} \gamma_j/(z_i - z_j)$ in $O(N)$ operations. Using the assumption about $\gamma$, we can rewrite

$$S_i = \sum_{l=1}^{N_s} d_l \sum_{j=1}^{N} \frac{\Psi_l(j)}{z_i - z_j} \equiv \sum_{l=1}^{N_s} d_l \tilde{\Psi}_l(z_i), \tag{7}$$

where $\tilde{\Psi}_l(\eta) = \int_0^{2\pi} \Psi_l(\alpha)/(\eta - z(\alpha))d\alpha$. Let us assume that one can determine a subset of $N_s$ points $z_{i_0(i)}, i = 1, \ldots, N_s$ such that, from the knowledge of the $N_s$ sums $S_{i_0(i)}$ it is possible to recover the $N$ sums $S_i$ with a given accuracy (this step will be called the *reconstruction* step). Then, using the fact that $\tilde{\Psi}_l(\eta)$ is analytic with respect to $\eta$, we can rewrite the whole multipole method and evaluate the $S_{i_0(i)}, i = 1, \ldots, N_s$ in $O(N_s)$ operations, then compute the $S_i$'s with $O(N)$ operations (notice, however, that the multipole coefficients will depend upon the shape of the interface). The total cost of the method is $O(N_s) + O(N)$. Thus, this approach is faster than the standard multipole method as soon as $N_s << N$ and the $O(N_s)$ term is dominating. Furthermore, we need a procedure to implement the reconstruction step.

An important feature of the interface dynamics of eqs (1-3) is the fact that the interface develops a tree-like structure through a cascade of tip-splitting and side-branching instabilities. A typical example is shown in fig. 1(a). The corresponding $\gamma$ function is represented in fig. 1(b). It is apparent that the branched structure of the interface translates into a highly irregular $\gamma$ function, with very abrupt gradients corresponding to the successive 'fjords' characteristic of these growth patterns. Since the work of Meyer, Daubechies, Grossman and others [7], it is known that orthogonal wavelets are well suited to the 'compressed' description of functions with strong gradients. Here, we will use the multiresolution analysis introduced by S. Mallat [8]. In this approach, the wavelet basis is spanned by the periodic functions $\Psi_{j,k}(x) = \frac{2^{j/2}}{\sqrt{2\pi}} \sum_{q \in Z} \Psi(2^j(x/2\pi + q) - k)$, where $j = 0, 1, \ldots$ is the *scale* index, $k = 0, 1, \ldots, 2^j - 1$ is the *position* index and $\Psi(x)$ is in

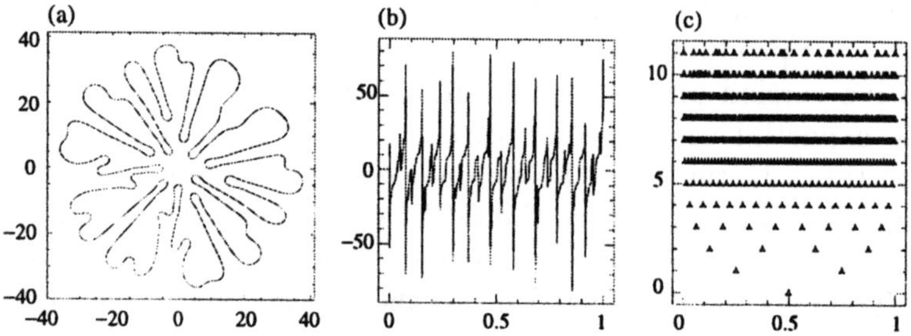


*Figure 1.* (a) A typical interface obtained from the evolution of a weak perturbation of the unit circle. (b) The function $\gamma(\alpha/2\pi)$ corresponding to the interface in (a). (c) Space (horizontal axis) - scale (vertical axis) representation of the wavelet coefficients $|d_{j,k}(\gamma)| \geq 10^{-5}$.

$L^2(R)$, and decreases (usually exponentially) at infinity. It is convenient to think of $\Psi_{j,k}(x)$ as a function localized around $2\pi(k+1/2)2^{-j}$, and of numerical support of $O(2^{-j})$. For particular choices of $\Psi(x)$, the set $\{\Psi_{j,k}(x)\}$ forms an orthogonal basis of $L^2_p([0,2\pi])$, and due to the good localization properties of $\Psi(x)$, the wavelet decomposition $\gamma(\alpha) = \sum_{j,k} d_{j,k}\Psi_{j,k}(\alpha)$ of a function like $\gamma(\alpha)$ in fig. 1(b) is mostly reduced to the coefficients $d_{j,k}$ with position $(k+1)2^{-j}$ and scale (characteristic width) $2^{-j}$ corresponding to those of the strongest gradients of the function. This is clearly seen in Fig. 1 (c), where a scale-position representation of the wavelet coefficients of $\gamma(\alpha)$ such that $|d_{j,k}| \geq 10^{-5}$ is given.

Wavelet decompositions are also characterized by the existence of fast algorithms for analysis (given $\gamma(k2^{-jmax}), k = 0, \ldots, 2^{jmax} - 1$, compute $d_{j,k} = \int_0^{2\pi} \gamma(x)\Psi_{j,k}(x)dx, j = 0, \ldots, jmax - 1$) and reconstruction (given $d_{j,k}$, compute $\gamma(k2^{-jmax})$ ) [9], both being of order $O(N) = O(2^{jmax})$, $jmax$ characterizing the finest scale for the sampling of $\gamma$. Correspondingly, it is possible [10, 11] to develop an approximate adaptive algorithm for the reconstruction step, provided the set $A_\epsilon(\gamma) = \{(j,k) \quad s.t. \quad |d_{j,k}| > \epsilon\}$ is known in advance. In order to implement the wavelet-multipole algorithm, we actually need an estimate of both $A_\epsilon(\gamma)$ and $A_\epsilon(S)$. As explained in [10, 11], several approximations for $A_\epsilon(S)$ are possible. One possibility is to add to each of the indices $(j,k)$ of $A_\epsilon(\gamma)$ the neighbour indices $(j, k+1)$ and $(j, k-1)$ as well as $(j-1, (k/2))$.

In figure 2(a), we show the CPU time required for the evaluation of the matrix-vector product (i.e., the $N$ sums $S_i$) as a function of $N_s$ for three different interfaces ($N = 8192$). It is clear that the complexity of this computation is $O(N_s)$,$N_s$ depending on the precision $\epsilon$ required. It turns out that for fairly 'complex' interfaces, resulting from the evolution of an

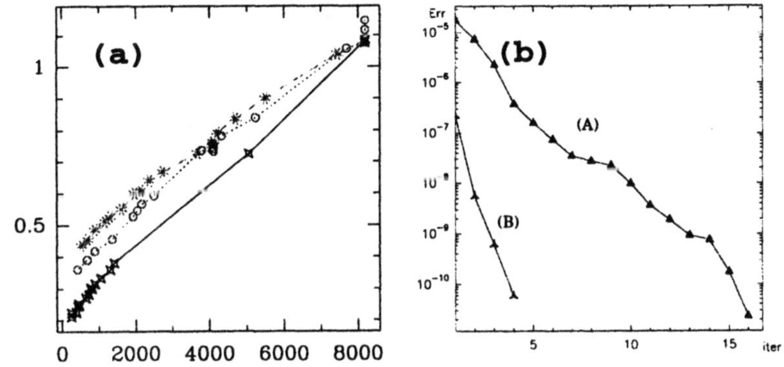


*Figure 2.* (a) CPU time for the computation of the $N$ sums $S_i$ versus the cardinal of $A_\epsilon(\gamma)$ for three different interfaces. (b) Comparison of the convergence of the GMRES method without (A) and with (B) preconditioning.

initially almost circular interface, $N_s \sim N/2$ if $\epsilon = 10^{-6}$, but $N_s \sim N/4$ if $\epsilon \sim 10^{-3}$ (these are only typical values), resulting in acceleration factors of two and four, respectively (the overhead induced by the decomposition-analysis step is negligible). It is worth emphasizing that, due to the high sensitivity to initial conditions of the eqs. (1-3), a faithful solution requires values of $\epsilon \sim 10^{-10}$ (or lower, for very low values of $T$), so that the matrix-vector product we have just described should only be used in the first steps of the iterative method.

An additional speed-up of the solution of the linear system (6) can be obtained by an adequate preconditioning. Although usual polynomial preconditioners work well, in the sense that the number of iterations required for convergence is significantly decreased, there is no actual gain in the effective computing time. We have rather used the fact that, as we consider a continuous evolution process, a reasonable approximation of the inverse of the matrix $M_n$ at time step $n$ is $M_{n-1}^{-1}$, or at least its approximation as given by the GMRES method. Using the same notations as in ref. [5], let $V_k^{(n-1)}$ the matrix whose colons are the Krylov vectors at step $(n-1)$, and $H_k^{(n-1)} \equiv V_k^{(n-1)T} M_{n-1} V_k^{(n-1)}$. Then, an efficient preconditioning is obtained with the approximation $M_n^{-1} \sim \tilde{M}_n^{-1} = V_k^{n-1} H_k^{n-1} V_k^{(n-1)T} + R_{n-1}$, $R_{n-1}$ being a $N \times N$ matrix, inversible on the subspace orthogonal to the Krylov subspace $K_k^{n-1}$. A possible choice for $R_{n-1}$ is $R_{n-1} = I - V_k^{(n-1)} V_k^{(n-1)T}$. It is easily seen that the complexity of the matrix-vector product $\tilde{M}_n^{-1} M_n \gamma$ is $O(kN)$, $k$ being the dimension of the Krylov subspace at time step $n$. Fig. 2(b) shows the convergence rate of the preconditioned and not preconditiones linear system, showing a decrease by a factor of four in the number of iterations, and three for the CPU time. It is important to note that $K_k^n$

is not easily determined when applying GMRES to the preconditioned matrix $\tilde{M}_n^{-1} M_n$. Our numerical experiments show that preconditioning $M_n$ by $\tilde{M}_{n-\delta n}^{-1}$, with $\delta n < 10$ still gives good results.

The algorithm resulting from the combination of the above improvements (partially vectorized multipoles, wavelet-based adaptive matrix-vector product, and preconditioning) is at least ten times faster than the standard (non preconditioned) matrix-vector product with adaptive multipoles. We address the reader to refs [10, 11] for a more detailed discussion of the influence of the choice of the wavelet basis. It seems likely that further improvements could be obtained using adapted wavelet decompositions, and particularly wavelet packets.

## 2. Singularity detection

As stated in the introduction, the theory concerning the selection mechanisms of the unsteady solutions of eqs. (1-3) has been mainly developed by S. Tanveer [3]. The basic problem is to quantify the perturbation induced by the surface tension term. This is done by comparing the time evolution of the same initial condition without and with (small) surface tension. The results of [3] are more easily phrased in terms of the singularities of the conformal mapping $f(\xi, t)$ such that, at each time $t$, the interior of the unit circle $|\xi| \leq 1$ is mapped to the exterior of $\Gamma(t)$. This mapping can always be written as $f(\xi, t) = A(t)/\xi + k(\xi, t)$, where $k(\xi, t)$ is *analytic* in $|\xi| \leq 1$. Let us assume that, at $t = 0$, $\partial_\xi f(\xi, 0)$ has an *isolated* singularity at $\xi_s(0)$, that is, locally $\partial_\xi f(\xi, 0)$ behaves as $(\xi - \xi_s(0))^\alpha$, and $\alpha < 0$. When $T = 0$, singularities move towards the unit circle, but their singularity exponent is constant in time. If zeros of $\partial_\xi f(\xi, 0)$ are present in the initial condition, they will also move towards the unit circle, and under some conditions they can hit it in finite time (in this case, the solution becomes ill-defined). When $T > 0$ and small, Tanveer has shown that the time evolution of a singularity depends strongly on the value of $\alpha$. If $\alpha \leq -4/3$, then the perturbation induced by the surface tension term is $O(T)$, and in particular the value of $\alpha$ is unchanged. However, if $-4/3 < \alpha < 0$, the singularity exponent becomes instantly equal to $-4/3$, and actually a *cluster* of $-4/3$ singularities forms around the $T = 0$ trajectory. If a zero of $\partial_\xi f(\xi, 0)$ exists at $t = 0$, it will also instantly generate a cluster of $-4/3$ singularities that eventually will separate from the $T = 0$ trajectory of the initial zero, and may induce a strong perturbation of the $T = 0$ solution in $O(1)$ time. Indirect confirmation of these results has been obtained recently [12], in particular concerning the size of the cluster of $-4/3$ singularities as a function of $T$. Here, we will focus on the numerical determination of the singularity exponent.

The conformal mapping $f(\xi, t)$ can be computed through the solution

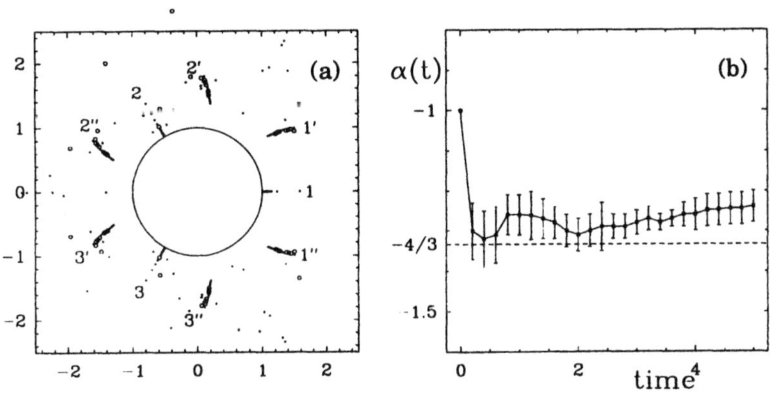


*Figure 3.* (a) Location of the zeros of $Q_M(\xi), M = 20$ for the initial interface $\partial_\xi f(\xi,0) = 1/(1-0.25\xi^3)/\xi^2$ and $T = 0.01$. Singularities with negative (resp. positive) exponent are represented by dots (resp. small circles). (b) Time evolution of the singularity strength along the trajectory marked 1 in (a).

of an integral equation (cf. [13]). Hence, for each $t$, we can determine the coefficients $a_n(t)$ such that $f(\xi,t) = A(t)/\xi + \sum_{n\geq 0} a_n(t)\xi^n$. In order to detect the singularities of $f(\xi,t)$, we need an analytic continuation outside $|\xi| = 1$. However, this is known to be an unstable procedure and we have to resort to an indirect method. A possibility is to determine an ordinary differential equation, satisfied *exactly* by the truncated series $k_M(\xi,t) = \sum_{n=1}^{M} a_n(t)\xi^n$ of $k(\xi,t)$. More precisely, it has been shown [14] that, for each value of $M$, there exist three polynomials $Q_M(\xi), R_M(\xi)$ and $S_M(\xi)$, such that

$$\xi Q_M(\xi)\frac{d^2k_M}{d\xi^2} + R_M(\xi)\frac{dk_M}{d\xi} + S_M(\xi)k_M = 0. \tag{8}$$

The behaviour of the solutions of these type of equations (called Fuchsian equations) is well known [15]. In particular, their singularities coincide with the zeros of $Q_M(\xi)$ (provided these are not also zeros of $R_M(\xi)$ or $S_M(\xi)$). Moreover, the singularity exponent at any zero $\xi_i$ of $Q_M(\xi)$ is given by

$$\alpha_i = 1 - \lim_{\xi\to\xi_i}(\xi-\xi_i)\frac{R_M(\xi)}{Q_M(\xi)}. \tag{9}$$

Thus, for each value of $M$, it is easy to compute the location and strength of the singularities of $k_M(\xi)$, and it remains to study their dependency with respect to $M$. We determine a lower bound for the values of $M$ by imposing that the relative error $\int_{|\xi|=1}|k(\xi,t)-k_M(\xi,t)|^2 d\xi/\int_{|\xi|=1}|k(\xi,t)|^2 d\xi$ is lower than some fixed accuracy. In addition, we only consider those of singularities with an amplitude $|A_i| > 10^{-3}$. Despite these precautions, the singularity distribution we obtain is rather noisy. We will term 'real' singularities those forming continuous trajectories in the $\xi$ plane, such as those shown in Fig. 3(a). These correspond to the evolution of an initial condition with three poles, distributed symmetrically around the origin. Each of them gives rise to two well defined trajectories, the singularity exponent along one of these trajectories being shown in Fig. 3(b). The error bars are computed from the values computed from eq. (9) for values of $M$ ranging from 30 to 50. Despite the strong fluctuations observed in the first steps, it is clear that these results [16] fully confirm the pertubative results of ref. [3].

To conclude, we would like to stress the fact that many aspects of the dynamics of the apparently simple Stefan problem given by eqs. (1-3) remain unclear. For instance, it is not known whether the branching process generating the tree-like structure observed in fig. 1(a) persists forever, or stops at some time (this may happen when $T = 0$). Our current research focuses on simple dynamical models for the creation of the $-4/3$ singularities.

## References

1. Saito, Y., Goldbeck-Wood, G. and Müller-Krumbhaar, H. (1988) Numerical simulation of dendritic growth, *Phys. Rev. A*, **38**, p. 2148.
2. Bensimon, D., Kadanoff, L.P., Liang, S., Shraiman, B.I. and Tang, C. (1986) Viscous flows in two dimensions, *Rev. Mod. Phys.*, **58**, p. 977.
3. Tanveer, S. (1993) Evolution of Hele-Shaw interface for small surface tension, *Phil. Trans. Roy. Soc. London A*, **343**, p. 155.
4. Dai, W. and Shelley, M.J. (1993) , A numerical study of the effect of surface tension and noise on an expanding Hele-Shaw bubble, *Phys. Fluids A*, **5**, p. 2131.
5. Saad, Y. and Schultz, M. (1986) GMRES: A generalized minimal residual algorithm for solving non-symmetric linear systems, *SIAM J. Sci. Stat. Comput.*, **7**, p. 856.
6. Carrier, J., Greengard, L. and Rokhlin, V. (1988) A fast adaptative multipole algorithm for particle simulations, *SIAM J. Sci. Stat. Comput.* , **9**, p. 669.
7. Meyer, Y., Ondelettes et opérateurs, Ed. Hermann (Paris) 1990.
8. Mallat, S. (1989) Multiresolution approximation and wavelet orthonormal bases of $L^2(R)$, *Trans. of AMS*, **315**, p. 69.
9. Rioul, O. and Duhamel, P. (1992) Fast algorithms for discrete and continuous wavelet transform, *IEEE Trans. Inform. Theory*, **38**, p. 569.
10. Petit, P. (1997) Thesis Univ. Bordeaux I.
11. Elezgaray, J. and Petit, P., in preparation.
12. Siegel, M., Tanveer, S. and Dai, W. (1996) *J. Fluid Mech.*, **323**, p. 201.
13. Henrici, P., Applied and computational complex analysis, vol. **3**, Ed. J. Whiley 1993.
14. Guttman, A.J. and Joyce, G.S. (1972) On a new method of series analysis in lattice statistics, *J. Phys. A*, **5**, p. 81.
15. Bender, C.M. and Orszag, S.A., Advanced mathematical methods for scientists and engineers, Ed. Mc Graw-Hill,1987.
16. Elezgaray, J., Petit, P. and Bonnier, B. (1997) Detecting complex singularities of a Hele-Shaw flow, *Europhysics Letters*, **37**, p. 263.

# Delamination of active layers in piezoelectric laminates

Paolo Bisegna and Franco Maceri
*Department of Civil Engineering, University of Rome "Tor Vergata", 00133 Rome, Italy.*

**Abstract.** A thin piezoelectric layer, bonded on a plane rigid support up to an initial circular defect of adhesion, is considered. The piezoelectric layer is actuated by an applied transversal difference of electric potential, which may induce a buckling of the debonded zone. The post-buckling behaviour is studied by using a suitable generalization of the von Kármán geometrically nonlinear plate theory. The quasi-static delamination process induced by the buckling phenomenon is investigated via the classical Griffith-Irwin energy criterion.

**Key words:** Piezoelectric laminates; delamination; Griffith-Irwin criterion; von Kármán plate theory; post-buckling behaviour.

## 1. Introduction

Adaptive structures can adjust their geometry, stiffness and damping, in order to achieve "optimal" performances under different environmental situations. The key-feature of an adaptive structure is the use of sensor/actuator devices, often embedded in, or patched on, the load-carrying members. Special examples of adaptive structures, which have been already vogue in aerospace industry, are given by piezoelectric laminates, obtained by bonding piezoelectric laminae to a base plate (which, in turn, is often a laminate).

Existing studies on piezoelectric laminates (see, e.g., [2] and the bibliography cited therein) assume a perfect bonding between the sensor/actuator laminae and the base plate. On the other hand, an imperfect bonding may take place, due to the manufacturing process or the high interlaminar stresses induced by the actuation of the piezoelectric laminae. As a consequence, the adaptive capabilities of the laminate may get significantly reduced.

In this work the delamination problem for a piezoelectric laminate is considered. The "thin-film" model introduced by Chai *et al.* [6] is adopted. In particular, the simple geometric configuration of a thin piezoelectric layer, bonded on a base plate up to an initial circular defect of adhesion, is considered. The base plate, which is usually much stiffer than the piezoelectric layer, is assumed to be rigid.

The piezoelectric layer is actuated by an applied transversal difference of electric potential, proportional to a monotonic evolution parameter, and is acted upon by normal surface forces. The applied electrical and mechanical loads may induce a buckling of the debonded part of the piezoelectric layer, which may separate from the base plate or partially remain in unilateral contact with it.

The buckling phenomenon, in turn, may induce an uniform-expansion growth of the circular debonded zone. In this paper, the simple and physically-oriented interface model based on the Griffith-Irwin energy criterion is employed. Hence,



*P. Argoul et al. (eds.),*
*IUTAM Symposium on Variations of Domains and Free-Boundary Problems in Solid Mechanics,* 137–150.

an accurate estimate of the energy-release rate associated with the delamination growth is needed at the post-buckling equilibrium configurations. In fact, the post-buckling behaviour is studied by using a suitable generalization of the von Kármán geometrically nonlinear plate theory.

The stability characteristic of the quasi-static delamination process is investigated via the relation between the evolution parameter and the radius of the debonded zone, at constant energy-release rate.

The present simple approach seems able to grasp the main features of the phenomenon. More detailed formulations could take into account the deformation of the base plate: on this subject important contributions were given, in the purely elastic case, by Kardomateas [12] and Bottega [4]. Moreover, a higher-order plate theory [16], a refined fracture criterion that includes mode mixity (ratio of Mode II to Mode I) [17], or other interface models [5, 13, 14, 15] could be adopted.

## 2. Von Kármán piezoelectric plate theory

In this section a von Kármán-type theory of piezoelectric plates is established. In particular, the theory of piezoelectric plates presented in [1] is modified, by taking into account also the von Kármán nonlinear terms in the expression of the strain tensor as a function of the displacement gradient.

Let $2h$ be the thickness of a piezoelectric layer, which is perfectly bonded to a rigid support, with the exception of a debonded zone $\Omega$. The present work is concerned with the analysis of the deformation experienced by the debonded part of the piezoelectric layer, whose undeformed configuration is the plate-like region $\Omega\times]-h,h[$.

A Cartesian frame $(O, x_1, x_2, x_3)$ is chosen with the origin $O$ on the middle plane of the piezoelectric layer and the $x_3$-axis oriented in the thickness direction away from the support. The structure of Cartesian product owned by the body suggests introducing a decomposition of the three-dimensional Euclidean vector space into the direct sum of the one-dimensional space $\mathbb{R}$ of vectors parallel to the $x_3$-axis and the two-dimensional space $\mathcal{V}$ of vectors parallel to the $x_1, x_2$-plane. Accordingly, the displacement vector is represented as the couple $(\mathbf{s}, s)$, where $\mathbf{s} = (s_1, s_2) \in \mathcal{V}$ is the in-plane displacement and $s \in \mathbb{R}$ is the transversal displacement. The space of symmetric second-order tensors over $\mathcal{V}$ is denoted by Sym. Hence, the strain tensor can be represented as the tern $(\mathbf{E}, \boldsymbol{\gamma}/2, \varepsilon)$, where

$$\mathbf{E} = \begin{pmatrix} E_{11} & E_{12} \\ E_{12} & E_{22} \end{pmatrix} \in \text{Sym}$$

is the in-plane strain, $\boldsymbol{\gamma} = (\gamma_1, \gamma_2) \in \mathcal{V}$ is the transversal shear strain and $\varepsilon \in \mathbb{R}$ is the transversal normal strain.

The piezoelectric layer is comprised by a homogeneous linearly piezoelectric transversely-isotropic material (Hermann-Maugin class $\infty mm$ [11]), with the axis of transverse isotropy oriented in the thickness direction. The constitutive behaviour of such a material is completely described by ten independent material constants: by adopting a standard notation [11], the "closed-circuit" elastic moduli are denoted by $c_{11}$, $c_{33}$, $c_{44}$, $c_{12}$ and $c_{13}$; the "clamped" permittivity constants are denoted by $\varepsilon_{11}$ and $\varepsilon_{33}$; and the "closed-circuit/clamped" piezoelectric constants are denoted by $e_{31}$, $e_{33}$ and $e_{15}$. For a later use, the following auxiliary material constants are introduced:

$$\begin{aligned}
&\bar{\varepsilon}_{33} = \varepsilon_{33} + e_{33}^2/c_{33}, && \bar{e}_{31} = e_{31} - c_{13}e_{33}/c_{33}, && \\
&\bar{c}_{11} = c_{11} - c_{13}^2/c_{33}, && \bar{c}_{12} = c_{12} - c_{13}^2/c_{33}, && c_{66} = (c_{11} - c_{12})/2, \\
&\hat{c}_{11} = \bar{c}_{11} + \bar{e}_{31}^2/\bar{\varepsilon}_{33}, && \hat{c}_{12} = \bar{c}_{12} + \bar{e}_{31}^2/\bar{\varepsilon}_{33}, && \\
&\nu = \bar{c}_{12}/\bar{c}_{11}, && \mu = \hat{c}_{12}/\hat{c}_{11}, && \eta = \sqrt{(\hat{c}_{11}/\bar{c}_{11})/[12(1-\nu^2)]}.
\end{aligned}$$

The piezoelectric layer is acted upon by a distributed normal surface force $p \in \mathbb{R}$ and is actuated by applied electric potentials $v^+ \in \mathbb{R}$ and $v^- = -v^+$ on the faces at $x_3 = +h$ and $x_3 = -h$, respectively, so that it is subjected to a transversal difference of electric potential $v = v^+ - v^-$.

The piezoelectric layer is modelled by adopting the theory of piezoelectric plates presented in [1]. As a matter of fact, the transversal normal strain, though taken into account in [1], is not of primary interest here and, for the sake of simplicity, in this work it is neglected. As a consequence, by recalling that $v^- = -v^+$, it turns out that the representation [1] of the displacement field reduces to the classical Kirchhoff-Love representation:

$$\mathbf{s} = \mathbf{u} - x_3 \nabla w \qquad \text{and} \qquad s = w,$$

where the unknown fields $\mathbf{u} = (u_1, u_2) \in \mathcal{V}$ and $w \in \mathbb{R}$ depend on the in-plane variable $\mathbf{x} = (x_1, x_2) \in \mathcal{V}$ only and $\nabla$ is the gradient operator with respect to $\mathbf{x}$. The debonded part of the piezoelectric layer is clamped to the remaining part: hence, the functions $\mathbf{u}$ and $w$ have to satisfy the boundary conditions:

$$\mathbf{u} = \mathbf{0} \quad \text{on} \quad \partial\Omega, \tag{1}$$

$$w = 0 \quad \text{and} \quad \nabla w \cdot \mathbf{n} = 0 \quad \text{on} \quad \partial\Omega, \tag{2}$$

where $\mathbf{n}$ is the outward unit normal field to $\partial\Omega$. Moreover, the debonded part of the piezoelectric layer may separate from the support, or partially remain in unilateral contact with it. Hence, $w$ has to satisfy also the unilateral contact condition:

$$w \geq 0 \quad \text{in} \quad \Omega. \tag{3}$$

In this work, the contact is assumed to be frictionless.

As is well known [20], the nonlinear von Kármán plate theory is based on the following representation of the in-plane strain:

$$\mathbf{E} = \operatorname{sym} \nabla \mathbf{u} + \nabla w \otimes \nabla w/2, \tag{4}$$

where sym denotes the symmetric part and $\otimes$ is the tensor product. The in-plane strain $\mathbf{E}$ and the curvature tensor $\mathbf{K} \in \mathrm{Sym}$, defined by:

$$\mathbf{K} = -\nabla\nabla w, \tag{5}$$

are the kinematical descriptors of the deformation of the plate. On the other hand, the stress state of the plate is summarized by the dual quantities of $\mathbf{E}$ and $\mathbf{K}$, i.e., the in-plane force $\mathbf{N} \in \mathrm{Sym}$ and the bending moment $\mathbf{M} \in \mathrm{Sym}$, which are respectively related to $\mathbf{E}$ and $\mathbf{K}$ by the constitutive equations [1]:

$$\begin{aligned} \mathbf{N} &= H[(1-\nu)\mathbf{E} + \nu\mathbf{I}\,\mathrm{tr}\,\mathbf{E}] + \mathbf{I}\,\bar{e}_{31}v, \\ \mathbf{M} &= D[(1-\mu)\mathbf{K} + \mu\mathbf{I}\,\mathrm{tr}\,\mathbf{K}], \end{aligned}$$

where $\mathbf{I} \in \mathrm{Sym}$ is the identity tensor, tr denotes the trace operator and the membranal stiffness $H$ and flexural stiffness $D$ of the plate are defined, respectively, as follows:

$$H = 2h\,\bar{c}_{11} \qquad \text{and} \qquad D = (2h^3/3)\hat{c}_{11}.$$

In particular, the quantity $\mathbf{I}\,\bar{e}_{31}v$ is the induced in-plane force, due to the piezoelectric effect, when the transversal difference of electric potential $v$ is applied at constrained in-plane deformation.

A weak formulation of the equilibrium equations can be obtained via the virtual displacement principle [7]:

$$\int_\Omega \mathbf{N}\cdot\delta\mathbf{E}\,da + \int_\Omega \mathbf{M}\cdot\delta\mathbf{K}\,da - \int_\Omega p\,\delta w\,da \geq 0, \tag{6}$$

to be satisfied for all the variations $\delta w$, $\delta\mathbf{K}$ and $\delta\mathbf{E}$ of the fields $w$, $\mathbf{K}$ and $\mathbf{E}$, respectively, verifying the constraints (1)-(5).

The weak formulation (6) can be derived as infimum condition of the functional

$$\begin{aligned} \mathcal{E} = &\int_\Omega (H/2)[(1-\nu)||\mathbf{E}||^2 + \nu(\mathrm{tr}\,\mathbf{E})^2]\,da + \int_\Omega \bar{e}_{31}v\,\mathrm{tr}\,\mathbf{E}\,da \\ &+ \int_\Omega (D/2)[(1-\mu)||\mathbf{K}||^2 + \mu(\mathrm{tr}\,\mathbf{K})^2]\,da - \int_\Omega p\,w\,da, \end{aligned}$$

over the constraint manifold (1)-(5). The symbol $||\cdot||$ denotes the Euclidean norm of the argument.

In order to deduce von Kármán-type equations for piezoelectric plates, the functional $\mathcal{E}$ is transformed into a mixed functional. In particular, the unilateral contact condition (3) and the compatibility equation (4), which are essential constraints for $\mathcal{E}$, are transformed into natural constraints. This task is accomplished by taking the Kuhn-Tucker multipliers of (3) and (4) as independent unknowns. They are, respectively, the reaction $q$ of the unilateral constraint and the in-plane force $\mathbf{N}$. Of course, the unknown field $q$ has to satisfy the unilateral contact condition

$$q \geq 0. \tag{7}$$

Moreover, $\mathbf{u}$ and $\mathbf{E}$ are eliminated from the set of unknowns, by enforcing a-priori the infimum conditions with respect to them. Hence, the following functional is built up:

$$\mathcal{K} = \inf_{\mathbf{u}} \inf_{\mathbf{E}} \left\{ \mathcal{E} - \int_\Omega q\, w\, da - \int_\Omega \mathbf{N} \cdot (\mathbf{E} - \operatorname{sym} \nabla \mathbf{u} - \nabla w \otimes \nabla w/2)\, da \right\}, \tag{8}$$

where the infimum with respect to $\mathbf{u}$ is taken over the constraint manifold (1). The supremum condition of $\mathcal{K}$ with respect to $q$, under the constraint (7), yields equation (3) and the complementarity condition $q\, w = 0$. The unknown field $\mathbf{N}$ has to satisfy the equilibrium equation $-\operatorname{div} \mathbf{N} = \mathbf{0}$: hence, it can be derived by the Airy stress function $\psi$, according to the equation:

$$\mathbf{N} = \operatorname{Curl} \operatorname{Curl} \psi,$$

where $\operatorname{Curl} \psi = (-\partial\psi/\partial x_2, \partial\psi/\partial x_1)$. Thus, the following von Kármán functional is obtained:

$$\begin{aligned} \mathcal{K} = & -\int_\Omega [(1+\nu)||\operatorname{Curl} \operatorname{Curl} \psi||^2 - \nu(\Delta\psi)^2]/(2H(1-\nu^2))\, da \\ & + \int_\Omega \bar{e}_{31} v(\Delta\psi - \bar{e}_{31} v)/(H(1+\nu))\, da + \int_\Omega \operatorname{Curl} \operatorname{Curl} \psi \cdot \nabla w \otimes \nabla w/2\, da \\ & + \int_\Omega (D/2)[(1-\mu)||\nabla\nabla w||^2 + \mu(\Delta w)^2]\, da - \int_\Omega (p+q) w\, da, \end{aligned} \tag{9}$$

defined over the constraint manifold (2) and (7). The von Kármán equations are obtained as infimum condition with respect to $w$ and supremum condition with respect to $\psi$ of the functional $\mathcal{K}$. They are, respectively:

$$D\Delta\Delta w - \operatorname{Curl} \operatorname{Curl} \psi \cdot \nabla\nabla w = p + q, \tag{10}$$

$$-\Delta\Delta\psi/(H(1-\nu^2)) - \operatorname{Curl} \operatorname{Curl} w \cdot \nabla\nabla w/2 = -\bar{e}_{31} \Delta v/(H(1+\nu)). \tag{11}$$

The constraints (1) and (3) are obtained as *natural* boundary conditions. In particular, (1) becomes:

$$\mathbf{R} \cdot \mathbf{t} \otimes \mathbf{t} = 0 \quad \text{and} \quad [\operatorname{curl} \mathbf{R} - \nabla(\mathbf{R} \cdot \operatorname{sym}(\mathbf{n} \otimes \mathbf{t}))] \cdot \mathbf{t} = 0 \quad \text{on } \partial\Omega, \tag{12}$$

where

$$\mathbf{R} = [(1+\nu)\mathbf{N} - \nu \mathbf{I} \operatorname{tr} \mathbf{N}]/(H(1-\nu^2)) - \nabla w \otimes \nabla w/2 - \bar{e}_{31} v \mathbf{I}/(H(1+\nu))$$

can be regarded as the symmetric part of $\nabla \mathbf{u}$ and the curl operator acts on a typical vector field $\mathbf{v}$ as follows: $\operatorname{curl} \mathbf{v} = \partial v_2/\partial x_1 - \partial v_1/\partial x_2$.

### 3. Energy-release rate

In this section the energy-release rate due to a quasi-static growth of the debonded zone is computed, through an overall energy-rate balance.

First, it is observed that the functional $\mathcal{E}$ is the increment of the total potential energy of the debonded plate with respect to the undeformed (but not unstressed) configuration. Then, it is pointed out that, *when the solution is reached*, $\mathcal{K}$ is equal to $\mathcal{E}$, as a consequence of (8).

It is supposed that a quasi-static delamination process takes place, at constant loads. Let $\dot{\mathbf{x}}$ be the pseudo-velocity of a point $\mathbf{x}$ belonging to the delamination front $\partial\Omega$. Here a dot denotes the partial differentiation with respect to a pseudo-time $t$. In fact, only the normal component $v_n = \dot{\mathbf{x}} \cdot \mathbf{n}$ is uniquely defined.

If $g$ denotes the energy-release rate per unit area, an overall energy-rate balance gives:

$$\int_{\partial\Omega} g v_n \, dl = -\frac{d\mathcal{E}}{dt} = -\frac{d\mathcal{K}}{dt}, \tag{13}$$

where $d\mathcal{E}/dt$ and $d\mathcal{K}/dt$ are the rates of the *optimal values* of the functionals $\mathcal{E}$ and $\mathcal{K}$, respectively. In order to compute $d\mathcal{K}/dt$, let $\mathcal{K} = \int_\Omega f \, da$, where $f$ is obviously defined by a comparison with equation (9). Then, it results:

$$\frac{d\mathcal{K}}{dt} = \int_\Omega \dot{f} \, da + \int_{\partial\Omega} f v_n \, dl. \tag{14}$$

The first integral at the right-hand side turns out to be:

$$\begin{aligned} \int_\Omega \dot{f} \, da &= -\int_\Omega \mathbf{R} \cdot \operatorname{Curl}\operatorname{Curl} \dot{\psi} \, da + \int_\Omega \operatorname{Curl}\operatorname{Curl} \psi[\nabla w] \cdot \nabla \dot{w} \, da \\ &\quad + \int_\Omega \mathbf{M} \cdot \nabla\nabla \dot{w} \, da - \int_\Omega (p+q) \, \dot{w} \, da - \int_\Omega w \, \dot{q} \, da. \end{aligned}$$

After an integration by parts, using (2), (10), (11) and (12), and observing that $\dot{q} = 0$ if $w > 0$, it follows that:

$$\begin{aligned} \int_\Omega \dot{f} \, da &= -\int_{\partial\Omega} [\operatorname{div} \mathbf{M} \cdot \mathbf{n} + \nabla(\mathbf{M} \cdot \operatorname{sym}(\mathbf{n} \otimes \mathbf{t})) \cdot \mathbf{t}] \, \dot{w} \, dl \\ &\quad + \int_{\partial\Omega} (\mathbf{M} \cdot \mathbf{n} \otimes \mathbf{n})(\nabla \dot{w} \cdot \mathbf{n}) \, dl. \end{aligned}$$

Since the total pseudo-time derivatives of $w$ and $\nabla w \cdot \mathbf{n}$ vanish at the delamination front, it results: $\dot{w} + v_n \nabla w \cdot \mathbf{n} = 0$ and $\nabla \dot{w} \cdot \mathbf{n} + v_n \nabla\nabla w \cdot \mathbf{n} \otimes \mathbf{n} = 0$ on $\partial\Omega$. Hence, from (2) it turns out that:

$$\int_\Omega \dot{f} \, da = -\int_{\partial\Omega} (\mathbf{M} \cdot \mathbf{n} \otimes \mathbf{n})(\nabla\nabla w \cdot \mathbf{n} \otimes \mathbf{n}) \, v_n \, dl. \tag{15}$$

By comparing (13), (14) and (15), using (2) and observing that $\nabla\nabla w[\mathbf{t}] = \mathbf{0}$ on $\partial\Omega$, it easily follows that:

$$\begin{aligned} g &= [(1+\nu)||\mathrm{Curl\,Curl}\,\psi||^2 - \nu(\Delta\psi)^2]/(2H(1-\nu^2)) \\ &\quad -\bar{e}_{31}v(\Delta\psi - \bar{e}_{31}v)/(H(1+\nu)) + (D/2)[(1-\mu)||\nabla\nabla w||^2 + \mu(\Delta w)^2]. \end{aligned}$$

## 4. Dimensionless equations

For the sake of simplicity, dimensionless quantities are introduced. To this end, a characteristic in-plane length $R$ is chosen (e.g., half the diameter of $\Omega$) and the slenderness $\lambda$ is defined as follows:

$$\lambda = R/(2h).$$

Moreover, it is posed:

$$\begin{array}{lll} \mathbf{X} = \mathbf{x}/R, & \overline{\Omega} = \{\mathbf{X} : \mathbf{x} \in \Omega\}, & dA = da/R^2, \\ W = w/(2h\eta), & \Psi = \psi/D, & P\,[\text{or } Q] = (\lambda/\eta)(R^3/D)\,p\,[\text{or } q], \\ \overline{\mathcal{K}} = (\lambda/\eta)^2\,\mathcal{K}/D, & G = R^2(\lambda/\eta)^2 g/D, & V = -\bar{e}_{31}\,(\lambda/\eta)^2\,v/(H(1+\nu)). \end{array}$$

Then, the functional $\mathcal{K}$ becomes:

$$\begin{aligned} \overline{\mathcal{K}} &= -\int_{\overline{\Omega}} [(1+\nu)||\mathrm{Curl\,Curl}\,\Psi||^2 - \nu(\Delta\Psi)^2]/2\,dA \\ &\quad + \int_{\overline{\Omega}} [(1-\mu)||\nabla\nabla W||^2 + \mu(\Delta W)^2]/2\,dA - \int_{\overline{\Omega}} V\,(\Delta\Psi + V/(1-\nu))\,dA \\ &\quad + \int_{\overline{\Omega}} \mathrm{Curl\,Curl}\,\Psi \cdot \nabla W \otimes \nabla W/2\,dA - \int_{\overline{\Omega}} (P+Q)\,W\,dA. \end{aligned}$$

Its optimality conditions supply the unilateral contact condition $W \geq 0$ and the dimensionless equations

$$\begin{aligned} \Delta\Delta W - \mathrm{Curl\,Curl}\,\Psi \cdot \nabla\nabla W &= P + Q \\ -\Delta\Delta\Psi - \mathrm{Curl\,Curl}\,W \cdot \nabla\nabla W/2 &= \Delta V, \end{aligned}$$

where the differential operators are now intended with respect to the dimensionless in-plane coordinate $\mathbf{X}$. The dimensionless energy-release rate $G$ becomes:

$$\begin{aligned} G &= [(1+\nu)||\mathrm{Curl\,Curl}\,\Psi||^2 - \nu(\Delta\Psi)^2]/2 + V(\Delta\Psi + V/(1-\nu)) \\ &\quad +[(1-\mu)||\nabla\nabla W||^2 + \mu(\Delta W)^2]/2. \end{aligned}$$

## 5. Axisymmetric problem

In this work the simple geometric configuration of an initial *circular* defect of adhesion with radius $R$ is considered. Hence, the domain $\overline{\Omega}$ is the two-dimensional unit ball and polar coordinates $(\rho, \theta)$ are suitable in the analysis.

The applied loads $v$ and $p$ are assumed to be axisymmetric (of course, they meet this assumption if they are uniform). Moreover, only axisymmetric post-buckling configurations are considered. As a consequence, all the functions involved in the present study are independent of $\theta$.

As is usual in the study of the post-buckling behaviour of circular plates [8, 9, 3], the following new unknown functions are introduced:

$$\begin{aligned} K &= \nabla\nabla W \cdot \mathbf{e}_\theta \otimes \mathbf{e}_\theta = W'/\rho \\ L &= \operatorname{Curl}\operatorname{Curl}\Psi \cdot \mathbf{e}_\rho \otimes \mathbf{e}_\rho = \Psi'/\rho, \end{aligned}$$

where $(\mathbf{e}_\rho, \mathbf{e}_\theta)$ are respectively the radial and circumferential unit vectors and a prime denotes the differentiation with respect to $\rho$. It is pointed out that $K$ is the dimensionless circumferential curvature and $L$ is the dimensionless radial membrane stress. The dimensionless deflexion $W$ is then recovered by integration, by using the boundary condition $W(1) = 0$:

$$W(\rho) = -\int_\rho^1 rK(r)\,dr.$$

In terms of the new unknowns, the functional $\overline{\mathcal{K}}$ becomes:

$$\begin{aligned} \overline{\mathcal{K}}/(2\pi) = &-\int_0^1 \rho[(\rho L')^2 + 2(1-\nu)(L^2 + \rho L L')]/2\,d\rho \\ &+ \int_0^1 \rho[(\rho K')^2 + 2(1+\mu)(K^2 + \rho K K')]/2\,d\rho + \int_0^1 \rho^3 L K^2/2\,d\rho \\ &- \int_0^1 \rho V\,(2L + \rho L' + V/(1-\nu))\,d\rho + \int_0^1 \rho K\,d\rho \int_0^\rho r(P(r) + Q(r))\,dr. \end{aligned}$$

Its optimality conditions supply the unilateral contact condition $W \geq 0$ and the final equations:

$$\Gamma K - KL = \rho^{-2}\int_0^\rho r(P(r) + Q(r))\,dr, \tag{16}$$

$$\Gamma L + K^2/2 = -V'/\rho, \tag{17}$$

where the differential operator $\Gamma$ acts on any scalar field $A$ as follows:

$$\Gamma A = \rho^{-3}(\rho^3 A')' = \rho^{-1}(\rho^{-1}(\rho^2 A)')' = A'' + 3A'/\rho.$$

| $c_{11}$ [GPa] | $c_{33}$ [GPa] | $c_{12}$ [GPa] | $c_{13}$ [GPa] | $c_{44}$ [GPa] | $e_{31}$ [C/m$^2$] | $e_{33}$ [C/m$^2$] | $e_{15}$ [C/m$^2$] | $\varepsilon_{11}$ [nF/m] | $\varepsilon_{33}$ [nF/m] |
|---|---|---|---|---|---|---|---|---|---|
| 126. | 117. | 79.5 | 84.1 | 23.0 | −6.5 | 23.3 | 17.0 | 15.0 | 13.0 |

TABLE I
Elastic, dielectric and piezoelectric constants of the piezoelectric ceramic PZT-5H.

The boundary conditions accompanying equations (16)-(17) are:

$$K'(0) = 0 \quad \text{and} \quad L'(0) = 0, \tag{18}$$

$$K(1) = 0 \quad \text{and} \quad (1-\nu)L(1) + L'(1) = -V(1). \tag{19}$$

In particular, equations (18) are regularity conditions, the first of (19) follows from the second of (2) and the second of (19) is a natural boundary condition.

The dimensionless energy-release rate $G$, in terms of the new unknowns, is:

$$G = K'(1)^2/2 + [(1+\nu)/(1-\nu)]L'(1)^2/2. \tag{20}$$

## 6. Bifurcation problem

In technical applications, the applied difference of electric potential is usually uniform ($V' = 0$) and no external surface force is imposed ($P = 0$). In this situation, the problem defined by equations (16)-(19) admits the "fundamental" solution: $K = 0$, $Q = 0$ and $L = -V/(1-\nu)$. Hence, the debonded part of the piezoelectric layer remains in passive contact with the base plate and is subjected to a uniform in-plane stress. This is the case until the applied difference of electric potential reaches a bifurcation value $V_b$, at which a post-buckling equilibrium path intersects the fundamental path. Such a bifurcation value is the lowest eigenvalue of the problem:

$$\Gamma K + V/(1-\nu)\, K = 0 \quad \text{in } (0,1)$$
$$K'(0) = 0, \quad K(1) = 0,$$

obtained by a linearization of equation (16) around the "fundamental" solution. It is easily found that [19, 18]:

$$V_b = (1-\nu)z_0^2,$$

where $z_0 = 3.832$ is the first zero of the first-order Bessel function of first kind $J_1$. For the piezoelectric ceramic PZT-5H, whose material properties are reported in table I, $\nu = 0.2906$ and hence $V_b = 10.415$. The physical value $v_b$ of the applied difference of electric potential, in the typical situation of a piezoelectric layer with

| $K(0)$ | $L(0)$ | $L(1)$ | $K'(1)$ | $L'(1)$ | $W(0)$ | $G$ | $V$ |
|---|---|---|---|---|---|---|---|
| -.10 | -14.68 | -14.68 | .08 | .00 | .02 | .00 | 10.42 |
| -.50 | -14.68 | -14.68 | .40 | .00 | .10 | .08 | 10.42 |
| -1.00 | -14.67 | -14.69 | .81 | -.01 | .19 | .32 | 10.43 |
| -2.00 | -14.63 | -14.72 | 1.62 | -.04 | .38 | 1.31 | 10.49 |
| -4.00 | -14.47 | -14.84 | 3.25 | -.18 | .77 | 5.33 | 10.71 |
| -6.00 | -14.21 | -15.05 | 4.95 | -.41 | 1.16 | 12.38 | 11.08 |
| -8.00 | -13.83 | -15.34 | 6.72 | -.74 | 1.56 | 23.05 | 11.62 |
| -10.00 | -13.33 | -15.73 | 8.60 | -1.19 | 1.97 | 38.27 | 12.35 |
| -12.00 | -12.71 | -16.23 | 10.64 | -1.78 | 2.40 | 59.51 | 13.30 |
| -14.00 | -11.94 | -16.86 | 12.90 | -2.53 | 2.86 | 89.00 | 14.49 |
| -16.00 | -11.01 | -17.64 | 15.44 | -3.49 | 3.34 | 130.35 | 16.01 |
| -18.00 | -9.90 | -18.61 | 18.39 | -4.73 | 3.87 | 189.46 | 17.93 |
| -20.00 | -8.55 | -19.83 | 21.92 | -6.34 | 4.46 | 276.74 | 20.41 |
| -22.00 | -6.91 | -21.39 | 26.31 | -8.52 | 5.12 | 412.12 | 23.70 |
| -24.00 | -4.84 | -23.49 | 32.14 | -11.62 | 5.92 | 639.26 | 28.28 |
| -26.00 | -2.06 | -26.55 | 40.76 | -16.52 | 6.95 | 1078.70 | 35.35 |
| -28.00 | 2.44 | -32.27 | 57.58 | -26.77 | 8.62 | 2309.56 | 49.67 |
| -28.50 | 4.50 | -35.25 | 66.81 | -32.68 | 9.40 | 3203.59 | 57.69 |
| -28.80 | 6.86 | -39.00 | 78.94 | -40.64 | 10.31 | 4618.02 | 68.31 |
| -28.50 | 12.43 | -49.55 | 115.93 | -65.99 | 12.61 | 10680.91 | 101.14 |
| -28.00 | 14.80 | -54.88 | 136.20 | -80.38 | 13.66 | 15152.47 | 119.31 |
| -27.00 | 18.06 | -63.23 | 169.99 | -104.91 | 15.18 | 24458.61 | 149.76 |

TABLE II

Post-buckling equilibrium path in the case $P = 0$. The columns from 1 to 6 are independent of the material. The columns 7 and 8 are obtained for $\nu = 0.2906$, which corresponds to PZT-5H.

thickness $2h = 500\ \mu$m and radius $R = 0.1$ m, turns out to be 63.3 Volts, a value which is usually reached in applications.

The post-buckling equilibrium path was obtained by a numerical integration of equations (16)-(19) [21]. The results are reported in table II. Both the fundamental path and the post-buckling path are plotted in figures 1-2: the bifurcation point clearly appears at $V = 10.415$ on the solid line, which corresponds to the case $P = 0$.

If a uniform external surface force $P \neq 0$ is applied, the buckling phenomenon disappears. In particular, when the surface force is tensile ($P > 0$), the fundamental and the post-buckling paths join together smoothly, as shown by the dotted lines in figures 1-2. On the other hand, the effect of a compressive surface force ($P < 0$) is more complex: indeed, partial or full unilateral contact between the base plate and the debonded layer may take place [10]. At low electric potential levels, the debonded plate remains completely in contact with the base plate. When the

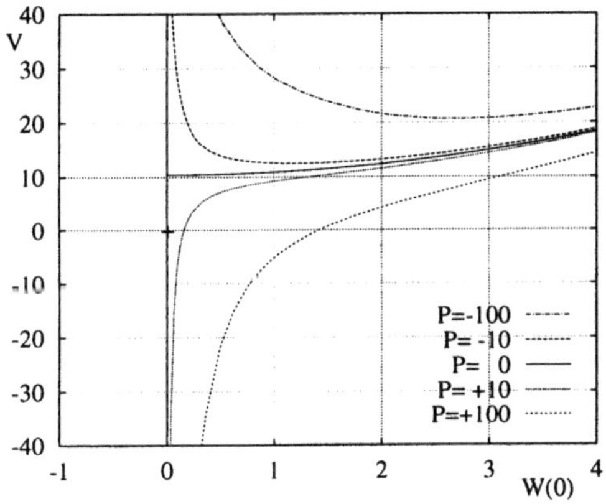


Fig. 1. Dimensionless deflexion $W(0)$ vs. dimensionless electric potential V.

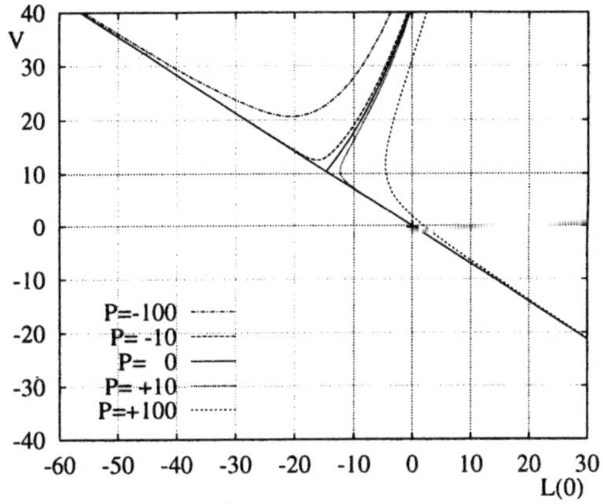


Fig. 2. Dimensionless in-plane stress $L(0)$ vs. dimensionless electric potential V.

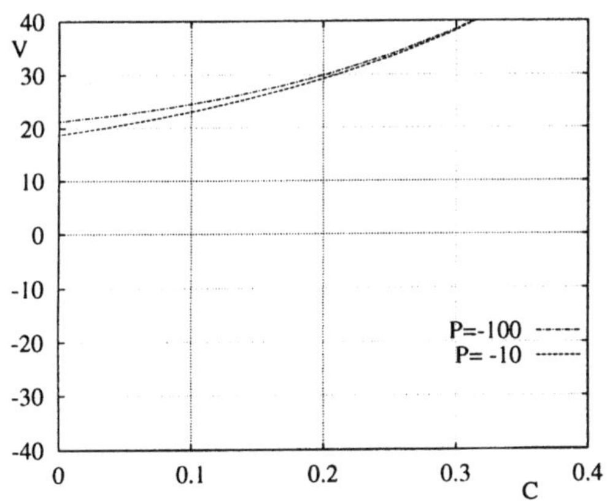


Fig. 3. Dimensionless width $C$ of the contact zone vs. dimensionless electric potential V.

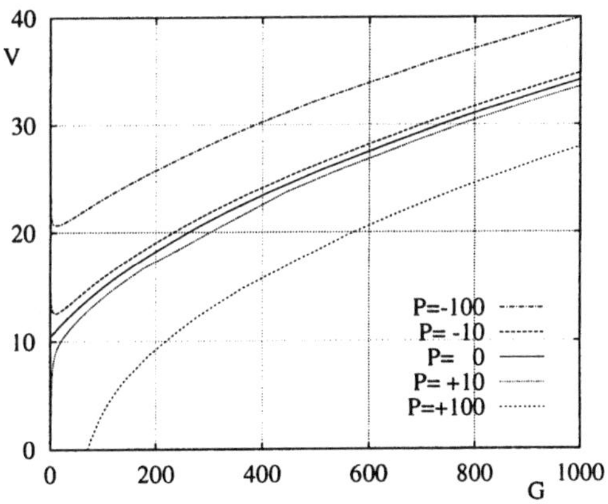


Fig. 4. Dimensionless energy-release rate $G$ vs. dimensionless electric potential V.

electric potential is sufficiently high (the threshold value increases with $P$), three equilibrium configurations are feasible: one is the undeformed configuration, the other two configurations lie on the dashed or dashed-dotted lines in figures 1-2. In particular, the equilibrium configurations lying on the increasing branch of those lines are stable and correspond to a debonded zone completely detached from the base plate. The equilibrium configurations on the decreasing branch are unstable: they entail unilateral contact over a peripheral annular zone, whose dimensionless width $C$ depends on $V$ and is plotted in figure 3.

The dimensionless energy-release rate $G$, as computed by equation (20), is plotted in figure 4. This quantity plays a central role in the analysis of the delamination-growth process.

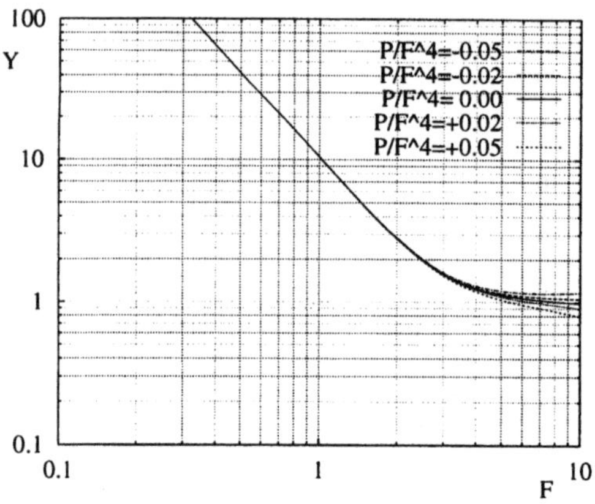


Fig. 5. Threshold curve of delamination.

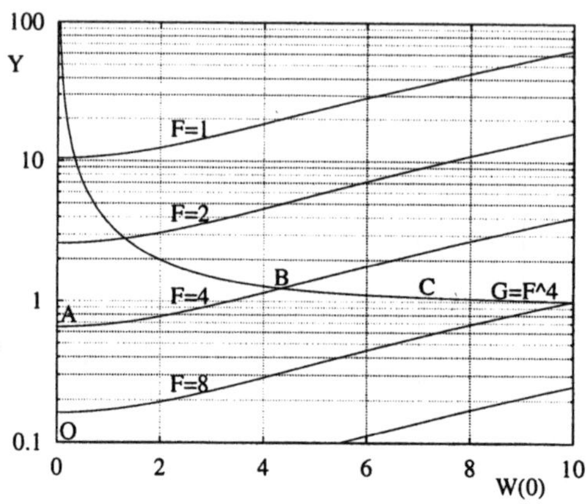


Fig. 6. Complete structural evolution. $P = 0$.

## 7. Delamination growth

In this work the growth behaviour of the delamination is studied by using the simple Griffith-Irwin criterion of critical fracture energy, instead of a more complex criterion that includes mode mixity (ratio of Mode II to Mode I) [17]. Thus, if $g^*$ is the critical fracture energy of the rigid-brittle bond between the piezoelectric layer and the base plate, an incipient delamination process is characterized by the equation

$$g = g^*. \tag{21}$$

For a given initial radius $R$ of the debonded zone, the dimensionless critical fracture energy is given by $G^* = R^4 g^*/[D(2h\eta)^2]$. Entering figure 4 with $G = G^*$, the dimensionless threshold value $V^*$ of the applied potential can be found on either one of the plotted curves, and hence the physical threshold value $v^*$ is easily computed.

Of course, the delamination process is "stable" if, under the condition (21), the *physical* threshold value $v^*$ increases with the radius $R$ (or the slenderness $\lambda$) of the debonded zone; otherwise a dynamical delamination process takes place when $v^*$ is reached.

In order to assess the stability characteristic of the delamination process, equation (21) is transformed into: $G = F^4$, where

$$F = \sqrt[4]{\frac{g^*(2h)^2}{D\eta^2}}\,\lambda.$$

Then, the relation between $v$ and $V$ is rewritten as $V/F^2 = Y$, where

$$Y = \frac{-\bar{e}_{31}}{2h\eta H(1+\nu)}\sqrt{\frac{D}{g^*}}\,v.$$

It is pointed out that $F$ and $Y$ are dimensionless quantities respectively proportional to the slenderness $\lambda$ and the physical applied potential $v$. Moreover, the respective proportionality coefficients do not change as the delamination proceeds (and $\lambda$ increases). Hence, the threshold curves of incipient delamination can be plotted in the $(Y, F)$ plane by using the equations: $F = \sqrt[4]{G}$ and $Y = V/\sqrt{G}$ (figure 5). It is emphasized that the curves in figure 5 are plotted at constant *physical* surface force $p$, which is proportional to $P/F^4$.

Since the curves in figure 5 are decreasing, it turns out that the delamination process is catastrophic, in the sense that when the applied difference of electric potential $v$ reaches its critical value $v^*$, a dynamic delamination process starts. On the other hand, a quasi-static delamination process could take place only if the applied electric potential $v$ would be decreased as the delamination proceeds. This fact is clearly shown in figure 6 (relevant to the case $p = P = 0$), where the complete structural evolution is plotted in the $(Y, W(0))$ plane. A typical evolution path is OABC: the OA branch represents the initial undeformed configuration, which is stable until the bifurcation point A, whose $Y$-coordinate depends on the initial slenderness of the debonded zone. Then, the structure follows the post-buckling branch AB, as the applied potential increases. The typical point of AB has coordinates $(W(0), V/F^2)$, where $V$ and $W(0)$ are given in table II. The delamination radius is still constant. At the point B the energy-release rate $g$ of the buckled structure equals the critical fracture energy $g^*$ of the bond (i.e., $G = F^4$): hence the delamination process starts. Then, the radius of the debonded zone increases quasi-statically at constant energy-release rate $g$ and the buckled structure follows the unstable branch BC (whose typical point has coordinates $(W(0), V/\sqrt{G})$, *provided that* the dynamic delamination process is prevented and the applied electric potential is gradually reduced.

## 8. Conclusions

The post-buckling behaviour of a circular debonded zone of a thin piezoelectric layer bonded on a rigid support was studied by means of a von Kármán-type plate theory. An applied transversal difference of electric potential was chosen as monotonic evolution parameter. The delamination process which may be induced by the buckling phenomenon was investigated via the classical Griffith-Irwin energy criterion.

The evolution parameter was found to be a decreasing function of the radius of the debonded zone, at constant energy-release rate. Hence, the quasi-static delamination process turned out to be unstable.

## Acknowledgments

The authors wish to thank Professor Antonio Grimaldi for a useful discussion on this paper. The financial support of the Italian National Research Council (CNR) is gratefully acknowledged.

## References

1. Bisegna, P. and Maceri, F. (1996) A consistent theory of thin piezoelectric plates. *J. Intell. Mater. Syst. Struct.* **7** 372-389.
2. Bisegna, P. (1997) A theory of piezoelectric laminates. *Rend. Mat. Acc. Lincei* (s. 9) **8** 137-165.
3. Bodner, S. R. (1954) The post buckling behavior of a clamped circular plate. *Quart. Appl. Math.* **12** 397-401.
4. Bottega, W. J. (1995) Separation failure in a class of bonded plates. *Composite Structures* **30** 253-269.
5. Bruno, D. and Grimaldi, A. (1990) Delamination failure of layered composite plates loaded in compression. *Int. J. Solids Structures* **26** 313-330.
6. Chai, H., Babcock, C. D. and Knauss, W. G. (1981) One dimensional modelling of failure in laminated plates by delamination buckling. *Int. J. Solids Structures* **17** 1069-1083.
7. Fichera, G. (1972) Boundary value problems in elasticity with unilateral constraints. In: Flügge, S. (ed.): *Handbuch der Physik*, Band VIa/2. Springer-Verlag, Berlin.
8. Friedrichs, K. O. and Stoker, J. J. (1941) The non-linear boundary value problem of the buckled plate. *Amer. J. Math.* **63** 839-888.
9. Friedrichs, K. O. and Stoker, J. J. (1942) Buckling of the circular plate beyond the critical thrust. *J. Appl. Mech.* **9** A7-A14.
10. Giannakopoulos, A. E., Nilsson, K.-F. and Tsamasphyros, G. (1995) The contact problem at delamination. *J. Appl. Mech.* **62** 989-996.
11. Ikeda, T. (1990) *Fundamentals of Piezoelectricity*. Oxford University Press, Oxford.
12. Kardomateas, G. A. (1993) The initial post-buckling and growth behaviour of internal delaminations in composite plates. *J. Appl. Mech.* **60** 903-910.
13. Ladevèze, P. (1992) A damage computational method for composite structures. *Computers and Structures* **44** 79-87.
14. Panagiotopoulos, P. D. (1985) *Inequality problems in mechanics and applications. Convex and nonconvex energy functions*. Birkhauser Verlag, Basel.
15. Point, N. and Sacco, E. (1996) A delamination model for laminated composites. *Int. J. Solids Structures* **33** 483-509.
16. Storåkers, B. and Andersson, B. (1988) Nonlinear plate theory applied to delamination in composites. *J. Mech. Phys. Solids* **36** 689-718.
17. Suo, Z. and Hutchinson, J. W. (1990) Interface crack between two elastic layers. *Int. J. Fracture* **43** 1-18.
18. Thompson, J. M. T. and Hunt, G. W. (1973) *A general theory of elastic stability*. John Wiley & Sons, London.
19. Timoshenko, S. P. and Gere, J. M. (1963) *Theory of elastic stability*. McGraw-Hill, Auckland.
20. Timoshenko, S. P. and Woinowsky-Krieger, S. (1970) *Theory of plates and shells*. McGraw-Hill, Auckland.
21. Yin, W.-L. (1985) Axisymmetric buckling and growth of a circular delamination in a compressed laminate. *Int. J. Solids Structures* **21** 503-514.

# A DELAMINATION MODEL. MATHEMATICAL PROPERTIES

NELLY POINT
*Département de Mathématiques - C.N.A.M.*
*292, Rue St. Martin - 75141 PARIS (FRANCE)*
*Email: point@vcnam.cnam.fr*

AND

ELIO SACCO
*Dipartimento di Ingegneria Industriale*
*Via Di Biasio 43 - 03043 CASSINO (ITALIA)*
*Email: sacco@img.unicas.it*

**Abstract.** In this paper some mathematical properties of a delamination model are studied. The laminate is schematized as two plates connected by a very special interface material. An interface constitutive model, based on the adhesion theory is introduced. The proposed model is governed by a functional which is neither smooth nor convex. The fundamental properties of this nonsmooth model are presented. Then a regularized interface model is constructed. The existence of a solution for the delamination problem obtained adopting the regularized interface model is proved. It is shown that this solution convergences to a solution of the nonsmooth initial delamination problem when the regularization parameters tend to 0. The lack of convexity of the functionals governing both the nonsmooth and the regularized problems makes this proof not straightforward.

## 1. Introduction

The study of the delamination is generally approached either by using the Elastic Fracture Mechanics [1, 2, 3, 4], or by introducing special constitutive laws for the interface material. This second approach often leads to a simpler model, both from a theoretical and computational point of view [5]. A model, based on the Frémond adhesive model [6, 7, 8, 9], has been recently presented in [10, 11]. Without external enforcement, it can lead



*P. Argoul et al. (eds.),*
*IUTAM Symposium on Variations of Domains and Free-Boundary Problems in Solid Mechanics*, 151–162.

to the Elastic Fracture Mechanics model or to interface material models, which are simply obtained as the regularization of the initial model [9].

The object of this paper is a mathematical investigation, in the framework of the small displacements and deformations theory, on some properties of the delamination model proposed in [10]. To remove the difficulty due to the non-differentiability, a regularized problem is introduced, which in fact corresponds to a relevant mechanical modelization of the delamination. The existence of a solution for this regularized problem is proved. Then the convergence of solution of the regularized problem to a solution of the initial one is proved.

## 2. The delamination model

Let $V$ be a laminated body realized by means of several thin laminae bonded together. It is supposed that delamination occurs only between two connected plates $V_1$ and $V_2$ . The surface of contact, denoted by $S$, is composed of $S_a$ and $S_d$ which are the part of the surface in adhesion and the debonded part, respectively.

The cartesian coordinate system $(x_1, x_2, x_3)$ is introduced such that the plane spanned by the axes $x_1, x_2$ contains the interface $S$. The displacement vector $u(x_1, x_2, x_3, t)$ is defined on $V \times R^+$, where $t$ denotes the time variable, and its restriction to $V_i$ is denoted by $u^{(i)}$ with $i = 1, 2$. The relative displacement vector of two points on the surface $S$, belonging to the plates $V_1$ and $V_2$ and characterized by the same in-plane coordinates, is defined as

$$s(x_1, x_2, t) = \overline{u}^{(2)}(x_1, x_2, t) - \overline{u}^{(1)}(x_1, x_2, t) \tag{1}$$

with $\overline{u}^{(i)}(x_1, x_2, t) = u^{(i)}(x_1, x_2, 0, t)$. The variable depends linearly on $\mathbf{u}$ : $s = Tu$, where $Tu$ represents the difference of the traces of $u^{(2)}$ and $u^{(1)}$ on $S$. The relative displacement $s$ can be decomposed in its normal $s_n$ and tangential $s_t$ components on the surface $S$, $s = s_n n + s_t$, where $s_n = s \bullet n$ with $n$ the unit vector oriented along the outward normal to $V_1$ on $S$.

A variable $\gamma(x_1, x_2, t)$ representing the damage state on the set $S \times R^+$ is introduced. This variable $\gamma$ is supposed to vary between 0 and 1, and is related to the intensity of adhesion parameter $\beta = 1 - \gamma$ proposed by Frémond [6, 7, 8]. It is assumed that the two plates are in adhesion in a point of the interface $S$ when $\gamma < 1$ in that point. When $\gamma = 1$, then the damage is complete. The proposed simple interface model is:

$$(1 - \gamma)s = 0 \quad 0 \leq \gamma \leq 1 \quad s_n \geq 0 \tag{2}$$

The first condition of (2) imposes that the relative displacement is zero until the damage is total. The last relation of (2) represents the condition of unilateral contact .

A very simple evolution law is considered here, based on the physical evidence that delamination cannot recede. Hence it is assumed that :

$$\dot{\gamma} \geq 0 \tag{3}$$

It can be proved that this simple and intuitive evolution law satisfies the Clausius-Duhem inequality on dissipation [10].

The problem of the delamination is formulated in variational form in the framework of the small displacements and strains theory. The sets $K$ and $A$ are introduced as:

$$K = \{(s, \gamma) : (1-\gamma)s = 0 \quad s_n \geq 0\} \qquad A = \{\gamma : \gamma_i \leq \gamma \leq 1\} \tag{4}$$

where $0 \leq \gamma_i \leq 1$ represents the initial damage state. Consequently $\gamma = \gamma_i = 1$ on $S_d$, the part of the surface which initially is not in adhesion.

In the following it is supposed that the external forces do not dependent on the evolution variable $t$ and inertial terms are neglected, hence a static problem is considered. The total potential energy $\pi$ of the system is the sum of the potential energy of the two loaded plates and of the potential energy of the interface:

$$\pi(u, \gamma) = \Lambda(u) + D(Tu, \gamma) \tag{5}$$

The potential energy of the two loaded plates is defined as

$$\Lambda(u) = \frac{1}{2} \sum_{i=1}^{2} \left\{ a^{(i)}(u^{(i)}, u^{(i)}) - f^{(i)}(u^{(i)}) \right\} \tag{6}$$

where $a^{(i)}$ is the symmetric bilinear form defined on $V_i$ , which quadratic form represents the double of the elastic energy of each plate of the body, $f^{(i)}$ the linear operator which characterizes the potential of the external forces.

The potential energy of the interface is:

$$\begin{aligned} D(Tu, \gamma) = D(s, \gamma) &= \int_{S_d} I_K(s, \gamma)|_{\gamma=1}\, ds + \int_{S_a} I_K(s, \gamma) ds \\ &+ \int_{S_a} I_A(\gamma) ds - \int_{S_a} (1-\gamma)\omega ds \end{aligned} \tag{7}$$

where $\omega$ is the Dupre's energy of adhesion [12], and the functions $I_K$ and $I_A$ are the indicator functions of the sets $K$ and $A$, respectively.

Finally, the delamination problem consists in finding the stationary point for the potential $\pi$, i.e. the solution state $(u^o, \gamma^o)$ is obtained by solving the problem:

$$\text{find } (u^o, \gamma^o) \text{ such that } 0 \in \partial\pi(u^o, \gamma^o) \tag{8}$$

## 3. Mathematical properties of the problem

In order to have a well-posed problem, it is assumed that the displacement is equal to zero on $\Gamma_{0i}$ $(i = 1,2)$, which is a subset of strictly positive measure of the boundary of the open set $V_i$. Let $H^1_{0i}$ denote the functions of $(H^1(V_i))^3$ whose traces on $\Gamma_{0i}$ are 0. The space $H^1_0$ is defined as:

$$H^1_0 = \left\{ u \text{ defined on } V_1 \cup V_2 \quad / \quad u^{(i)} \in H^1_{0i} \right\}$$

In this framework the quantity $s = Tu$, representing the difference of the traces of the variables $u^{(i)}$ on the interface $S$, is well defined and belongs to $(H^{1/2}(S))^3$. The following assumptions are made:

$a^{(i)}$ is a bounded bilinear form, coercive on $H^1_{0i}$ $(H1)$
$f^{(i)}$ is a bounded linear functional on $H^1_{0i}$ $(H2)$
$\omega$ is a positive constant $(H3)$

The problem (8) can be written in the following variational form:

find $(u^o, \gamma^o)$ in $H^1_0 \times L^2(S)$ such that for any admissible variation $(\delta u, \delta\gamma)$ in $H^1_0 \times L^2(S)$

$$\forall\ \delta u \in H^1_0 \qquad \sum_{i=1}^{2} \{a^{(i)}(u^{(i)o}, \delta u^{(i)}) - f^{(i)}(\delta u^{(i)})\} + \int_S t^o \bullet T\delta u\ ds = 0$$

$$\forall\ \delta\gamma \in L^2(S) \qquad \int_S (\omega - Y^o + q^o)\delta\gamma\ ds = 0$$

$$q^o \in \partial I_A(\gamma^o) \qquad (t^o, -Y^o) \in \partial I_K(s^o, \gamma^o) \tag{9}$$

The function $I_K$ is neither differentiable nor convex. Nevertheless, as any indicator function, it is subdifferentiable in any point where it is finite. Moreover, the particular structure of the set $K$ allows to explicit its subdifferential as the cartesian product of its subdifferentials with respect to each variable $s$ and $\gamma$, which is not true in general.

Hence $(t^o, -Y^o) \in \partial I_K(s^o, \gamma^o)$ is equivalent to $t^o \in \partial_s I_K(s^o, \gamma^o)$ and $-Y^o \in \partial_\gamma I_K(s^o, \gamma^o)$. Furthermore, it is easy to describe in detail, for each possible value taken by $(s^o, \gamma^o)$, the structure of the subdifferential of $I_K$.

**Proposition 1:** In each point of the surface $S_a$ the relationship $(t, -Y) \in \partial I_K(s, \gamma)$ leads to:

$$\begin{array}{llllllll}
\gamma \neq 1 & s = 0 & & \Longrightarrow & t_n \in R & t_t \in R^2 & Y = 0 \\
\gamma = 1 & s_n > 0 & & \Longrightarrow & t_n = 0 & t_t = 0 & Y \in R \\
\gamma = 1 & s_n = 0 & s_t \neq 0 & \Longrightarrow & t_n \in R^- & t_t = 0 & Y \in R \\
\gamma = 1 & s_n = 0 & s_t = 0 & \Longrightarrow & t_n \in R^- & t_t = 0 & Y = 0
\end{array} \tag{10}$$

where the vector $t$ is decomposed in its normal $t_n$ and tangential $t_t$ components on the surface $S$, such that $t = t_n n + t_t$.

From a mechanical point of view, $t$ represents the force per unit of area applied by the plate $V_1$ on $V_2$. The three last relations of (10) imply that when the delamination is complete, the stress on interface $S$ is only due to a frictionless unilateral contact.

The following proposition is a direct consequence of the second and the third equations of (9).

**Proposition 2:** In each point of the surface $S_a$ only the following three cases are possible for the solution of the problem (9):

$$\begin{array}{ll} \gamma^o = \gamma_i & q^o = Y^o - \omega \leq 0 \\ \gamma_i < \gamma^o < 1 & q^o = Y^o - \omega = 0 \\ \gamma^o = 1 & q^o = Y^o - \omega \geq 0 \end{array} \tag{11}$$

This proposition makes clear the relation of this model with the Fracture Mechanics. In fact, as it has been already shown [10], $Y^o$ is the energy release rate. The delamination occurs only when the energy release rate is greater or equal to the limit energy release rate $\omega$, which is in fact the Griffith limit energy. Furthermore, the Propositions 3 and 3 can be used to put in evidence some special features of the solution of equations (9).

**Proposition 3:** If $(u^o, \gamma^o)$ is a solution of problem (9) then either $\gamma^o = \gamma_i$ or $\gamma^o = 1$. Furthermore, in a point of $S_a$ where $\gamma^o = 1$, the relative displacement $s^o$ is different from 0.

This shows that the solution of problem (9) has an on-off character similar to the problem of Fracture Mechanics: in each point of $S$ the two plates are either in complete adhesion or completely delaminated.

## 4. The regularized model

The nondifferentiability of the functional $\pi$ may represent a strong difficulty in the treatment of the delamination problem, both from a mathematical and a computational point of view. A regularization of the total potential energy appears suitable. Let the scalars $\eta_n > 0$ and $\eta_t > 0$ be regularization parameters, the function $I_K$ is replaced with:

$$I_K^r(s,\gamma) = \frac{1}{2\eta_n}\left[(1-\gamma)(s_n^+)^2 + (s_n^-)^2\right] + \frac{1}{2\eta_t}(1-\gamma)\left\|s_t\right\|^2 \tag{12}$$

where $(\bullet)^+$ and $(\bullet)^-$ indicate the positive and negative parts of $(\bullet)$, respectively, such that $(\bullet) = (\bullet)^+ - (\bullet)^-$, and $\|\bullet\|$ denotes the norm of a vector.

Since $\gamma$ must be equal to 1 on $S_d$, an internal regularization of the function $I_A$ makes sense only on the initially adherent part denoted $S_a$. Hence, the indicator function $I_A$ is replaced on $S_a$ with:

$$I_A^r(\gamma) = r\frac{1-\gamma_i}{2}\int_0^{\nu(\gamma)} f(y)dy \tag{13}$$

where the scalar $r > 0$ is a regularization parameter and

$$f(\nu) = \frac{\nu}{1-|\nu|} \tag{14}$$

with

$$\nu(\gamma) = \frac{2}{1-\gamma_i}(\gamma - \frac{1+\gamma_i}{2}) \tag{15}$$

The derivative of the function $I_A^r$ is:

$$\forall\gamma \in ]\gamma_i, 1[ \qquad \frac{dI_A^r}{d\gamma}(\gamma) = r\ f(\nu(\gamma)) \tag{16}$$

Finally, the new regularized total potential energy $\pi^r$ is defined as:

$$\pi^r(u,\gamma) = \Lambda(u) + D^r(Tu,\gamma) \tag{17}$$

where the regularized adhesion energy $D^r$ can be written as:

$$\begin{aligned} D^r(Tu,\gamma) = D^r(s,\gamma) &= \int_S \left\{\frac{1}{2\eta_n}\left[(1-\gamma)(s_n^+)^2 + (s_n^-)^2\right] + \frac{1}{2\eta_t}(1-\gamma)\left\|s_t\right\|^2\right\} ds \\ &+ \int_{S_a} I_A^r(\gamma)ds - \int_{S_a}(1-\gamma)\omega ds \end{aligned} \tag{18}$$

The partial derivatives of $I_K^r(s,\gamma)$ with respect to $s$ and to $\gamma$, are:

$$\frac{\partial I_K^r}{\partial s}(s,\gamma) = \frac{1}{\eta_n}\left[(1-\gamma)s_n^+ + s_n^-\right]n + \frac{1}{\eta_t}(1-\gamma)s_t \tag{19}$$

$$\frac{\partial I_K^r}{\partial \gamma}(s) = -\frac{1}{2\eta_n}(s_n^+)^2 - \frac{1}{2\eta_t}\left\|s_t\right\|^2 \tag{20}$$

The corresponding regularized problem consists in:

find $(u^r, \gamma^r)$ in $H_0^1 \times L^2(S)$ with $\gamma^r = 1$ on $S_d$ /

$$\forall\, \delta u \in H_0^1 \qquad \sum_{i=1}^{2} \{a^{(i)}(u^{(i)r}, \delta u^{(i)}) - f^{(i)}(\delta u^{(i)})\} + \int_S t^r \bullet T\delta u \; ds = 0$$

$$\forall\, \delta\gamma \in L^2(S_a) \qquad \int_{S_a} (\omega - Y^r + q^r)\delta\gamma \; ds = 0$$

$$q^r = \frac{dI_A^r}{d\gamma}(\gamma^r) \qquad t^r = \frac{\partial I_K^r}{\partial s}(s^r, \gamma^r) \qquad -Y^r = \frac{\partial I_K^r}{\partial \gamma}(s^r) \tag{21}$$

It can be noted that equations (21) are similar to equations (9) with $q^o$, $t^o$ and $Y^o$ replaced by $q^r$, $t^r$ and $Y^r$, respectively.

The local form of the second equation of (21), the continuity and the strict monotonicity of the composed function $f \circ \nu$ from $]\gamma_i, 1[$ to $R$, allow to write on $S_a$:

$$\gamma^r = \nu^{-1} \circ f^{-1}\left(\frac{Y^r - \omega}{r}\right) \tag{22}$$

The damage parameter is an explicite function of the energy release rate $Y^r$ and hence, using equation (20) and the last one of (21), $\gamma^r$ can be written as function of the relative displacement $s^r$:

$$\gamma^r = \widehat{\gamma}(s^r) = \nu^{-1} \circ f^{-1}\left[\frac{1}{r}\left(\frac{1}{2\eta_n}(s_n^{r+})^2 + \frac{1}{2\eta_t}\left\|s_t^r\right\|^2 - \omega\right)\right] \tag{23}$$

## 5. Existence of solution of the regularized problem

A constructive method is used to prove the existence of the solution of the regularized problem. First the existence of the solution for the elastostatic problem corresponding to a fixed value of the damage function is given and the boundedness of this solution is proved. Then an iterative procedure is introduced and the convergence of a subsequence to a solution of the regularized problem is proved.

**Proposition 4:** For any $\gamma$ such that $0 \leq \gamma \leq 1$ almost everywhere on $S$, there exists a unique $u^\gamma$ solution of the following variational problem:

find $u^\gamma \in H_0^1$ /

$$\begin{aligned} \forall\, \delta u \in H_0^1 \qquad 0 \;=\; & \sum_{i=1}^{2} \{a^{(i)}(u^{(i)\gamma}, \delta u^{(i)}) - f^{(i)}(\delta u^{(i)})\} \\ + \; & \int_S \frac{1}{\eta_n} [(1-\gamma)s_n^{\gamma+} + s_n^{\gamma-}]\, n \bullet T\delta u \; ds \\ + \; & \int_S \frac{1}{\eta_t}(1-\gamma)s_t^\gamma \bullet T\delta u \; ds \end{aligned} \tag{24}$$

where $s^\gamma = Tu^\gamma$. Moreover this solution verifies the following estimates:

$$\left\| u^{(i)\gamma} \right\|_{(H^1(V_i))^3} \leq C \qquad (i = 1, 2) \tag{25}$$

$$\int_S \left\{ \frac{1}{\eta_n} \left[ (1-\gamma)(s_n^{\gamma+})^2 + (s_n^{\gamma-})^2 \right] + \frac{1}{\eta_t}(1-\gamma) \left\| s_t^\gamma \right\|^2 \right\} ds \leq C \tag{26}$$

where the constant $C$ is independent of the parameters $\eta_n$, $\eta_t$ and of the function $\gamma$.

A sequence $\{(u^{[n]}, \gamma^{[n]})\}$ is defined in the following way:

1. let $\gamma^{[0]} = \gamma_i$;
2. for a given $\gamma = \gamma^{[n-1]}$, let $u^{[n]}$ be the unique solution of equation (24);
3. let $\gamma^{[n]} = 1$ on $S_d$ and $\gamma^{[n]} = \widehat{\gamma}(s^{[n]})$ on $S_a$ with $\widehat{\gamma}$ given by equation (23);
4. with the new value $\gamma = \gamma^{[n]}$ go back to step 2.

The sequence $\{(u^{[n]}, \gamma^{[n]})\}$, obtained using the described procedure, verifies the following theorem:

**Theorem 5:** There exists a subsequence of $\{(u^{[n]}, \gamma^{[n]})\}$ still denoted by $\{(u^{[n]}, \gamma^{[n]})\}$ such that

- $\{u^{[n]}\}$ converges weakly to $u^\infty$ in $H_0^1$
- $\{s^{[n]}\}$ converges to $s^\infty = Tu^\infty$ strongly in $(L^2(S))^3$ and almost everywhere on $S$
- $\{\gamma^{[n]}\}$ converges almost everywhere to $\gamma^\infty$ with $\gamma_i \leq \gamma^\infty \leq 1$ and $\gamma^\infty$ is such that $\gamma^\infty = 1$ on $S_d$ and $\gamma^\infty = \widehat{\gamma}(s^\infty)$ on $S_a$.

The limit $(u^\infty, \gamma^\infty)$ is solution of the regularized problem (21), hence $\gamma^\infty = \gamma^r$ and $u^\infty = u^r$. Furthermore, $(u^r, \gamma^r)$ verifies the estimates (25) and (26) for $\gamma = \gamma^r$ and $u^\gamma = u^r$.

## 6. Convergence of the regularized solution

When the regularization parameters $r$, $\eta_n$ and $\eta_t$ tend to 0, the sequence of solutions $\{(u^r, \gamma^r)\}$ of the regularized problems (21) is expected to tend to a solution $(u^o, \gamma^o)$ of the initial nonsmooth problem (9). To prove this convergence the weak compactness of bounded sequences in Hilbert spaces is used again.

**Theorem 6:** Let $(u^r, \gamma^r)$ be a solution of problem (21). Under the hypotheses $(H1)-(H3)$, assuming that $\eta_n > 0$ and $\eta_t > 0$ tend to 0 for $r$ tending to 0, there exists a subsequence of $\{(u^r, \gamma^r)\}$ still denoted by $\{(u^r, \gamma^r)\}$ such that

- $\{u^r\}$ converges weakly to $u^o$ in $H_0^1$
- $\{s^r\}$ converges to $s^o = Tu^o$ strongly in $(L^2(S))^3$ and almost everywhere on $S$
- $\{\gamma^r\}$ converges to $\gamma^o$ in $L^\infty(S)$ weak* with $\gamma_i \leq \gamma^o \leq 1$ and $\gamma^o = 1$ on $S_d$
- $\{t^r\}$ converges to $t^o$ weakly in the dual space of $(H^{1/2}(S))^3$.

These limits verify:

$$\forall\ \delta u \in H_0^1 \qquad \sum_{i=1}^{2} \{a^{(i)}(u^{(i)o}, \delta u^{(i)}) - f^{(i)}(\delta u^{(i)})\} + \int_S t^o \bullet T\delta u\ ds = 0 \tag{27}$$

$$(1-\gamma^o)s^o = 0 \quad \gamma_i \leq \gamma^o \leq 1 \quad s_n^o \geq 0 \tag{28}$$

In order to complete the proof of the convergence of the sequence $\{(u^r, \gamma^r)\}$ to a solution of equations (9), it remains to prove the last relations of (9), or equivalently that the properties described in the Propositions 3, 3 and 3 are verified.

The Proposition 3 shows that the damage parameter in the initial non-smooth problem (9) has an on-off character. In fact, on $S_a$, the initially adherent area of $S$, $\gamma$ can only take two values: $\gamma_i$ (where no damage occurs) and 1 (where complete delamination occurs, and in this case the relative displacement must be different from 0). It is expected that the limit of the sequence $\{\gamma^r\}$ exibits the same feature. Recalling the equation (22), the function $\nu^{-1} \circ f^{-1}(y/r)$ defining the damage state on $S_a$, when $r \to 0$, tends to $\gamma_i$ for $y < 0$, to 1 for $y > 0$, and its limit has a discontinuity in $y = 0$. In the following proposition a partial result concerning the case when the limit $s^o$ of $\{s^r\}$ is different from zero, is obtained.

**Proposition 7:** On the subset $S^o = \{(x_1, x_2) \in S_a \ / \ s^o(x_1, x_2) \neq 0\}$ the sequence $\{Y^r\}$ tends almost everywhere to $+\infty$ and the sequence $\{\gamma^r\}$ tends almost everywhere to 1.

More generally, it is necessary to define the delaminated zone to known where the sequence $\{\gamma^r\}$ converges to 1 or to $\gamma_i$. As pointed out in Proposition 3, the sign of $Y^o - \omega$, limit of the sequence $\{Y^r - \omega\}$, defines the zone where delamination occurs. The next proposition shows in which sense this limit exists.

**Proposition 8:** The sequence $\{(1-\gamma^r)Y^r\}$ is a bounded sequence of positive functions in $L^1(S)$ hence it converges to a positive measure on $S$. On any subset of $S$ where $(1-\gamma^r)$ is bounded from below., the sequence $\{Y^r\}$

tends to a bounded positive measure.

Now, the rate of convergence of $\gamma^r$ to $\gamma_i$ or to 1 is studied. To this end, some subdomains of $S_a$ are initially defined as follows. For any fixed positive real numbers $\rho$ and $r$ let $S^r_{a\rho} \subset S_a$ and $S^r_{d\rho} \subset S_a$ be:

$$S^r_{a\rho} = \{(x_1, x_2) \in S_a \;/\; Y^r(x_1, x_2) - \omega \leq -\rho\} \tag{29}$$

$$S^r_{d\rho} = \{(x_1, x_2) \in S_a \;/\; Y^r(x_1, x_2) - \omega \geq \rho\} \tag{30}$$

It is a simple matter to verify that the sets introduced satisfy the properties:

- for $r \to 0$ the sequence of sets $\left\{\cap_{r \leq r^o} S^r_{a\rho}\right\}$, with $r^o > 0$, is increasing and converges to a set denoted $S_{a\rho}$ included in the adherent part of $S_a$ for the solution $(u^o, \gamma^o)$;
- for $r \to 0$ the sequence of sets $\left\{\cap_{r \leq r^o} S^r_{d\rho}\right\}$, with $r^o > 0$, is increasing and converges to a set denoted $S_{d\rho}$ included in the delaminated part of $S_a$ for the solution $(u^o, \gamma^o)$.

**Proposition 9:** On $S^r_{a\rho}$ and on $S^r_{d\rho}$, the following estimates hold:

$$\begin{aligned} 0 < 1 - \gamma^r < \frac{r}{2\rho} \qquad & \text{on } S^r_{a\rho} \\ 0 < \gamma^r - \gamma_i < \frac{r}{2\rho} \qquad & \text{on } S^r_{d\rho} \end{aligned} \tag{31}$$

The Proposition 6 implies the uniform convergence of $\{\gamma^r\}$ to 1 on any closed subset strictly included in some subset of $S_a$ and corresponding to the delaminated zone and to $\gamma_i$ on any closed subset strictly included in some subset of $S_a$. and correspnding to the adherent zone. This part is rather technical and hence ommitted here.

## 7. Convergences of $t^r$ and $Y^r$

The convergence of the gradient of a regularized functional to an element of the subdifferential of the initial nonsmooth one is a well-known result of convex analysis. The functional governing the delamination problem is not convex, and so this classical result does not apply. It is possible to prove that, at least locally, the relations $t^o \in \partial_s I_K(s^o, \gamma^o)$ and $-Y^o \in \partial_\gamma I_K(s^o, \gamma^o)$ are verified. To this end, it is necessary to give a local meaning to $t^o$, regarded as the limit of $t^r$.

The local regularity of $t^r$, depends on the regularity of the displacement field $u^r$. In fact, $t^r$ is the tension at the interface $S$, which is related to the stress tensor by the relation $t^r = \sigma^{(1)}n = -\sigma^{(2)}n$ on $S$. Since $\sigma$ is a linear function of the symmetric part of the gradient of $u^r$, the same property holds true for $t^r$ . The $L^2$ regularity of $t^r$ is ensured when $u^{r(i)} \in (H^{1+1/2})^3$, at least near the interface $S$. In fact, no global regularity results better than $(H^1(V_i))^3$ can be expected for $u^{r(i)}$. The lack of regularity results is due to the nonlinearity of the problem and to the boundary conditions of mixed type.

Nevertheless, it is possible to establish a local regularity result in any open set which does not contains points presenting difficulties (corners, change in the type of the boundary condition, delamination front of the initial nonsmooth problem).

**Theorem 11:** Let $\Theta$ be an open set included in $V$ and containing the closure of the open set $\vartheta$. Let $\vartheta$ be such that the subset $\vartheta \cap S$ is not empty and contained in the interior of $S$ and either in $S^r_{a\rho}$ or in $S^r_{d\rho}$ for some fixed positive constant $\rho$. Under the hypotheses $(H1)-(H3)$ and assuming that:

- the external forces are applied only on the boundary $\partial V \backslash \Gamma_0$,
- the elasticity coefficients $E_{ijhk} \in C^{0,1}(\overline{V}_i)$ with $i = 1, 2$,
- $\gamma^i \in C^{0,1}(\overline{V}_i)$ with $i = 1, 2$,

the solution $u^r$ of the regularized problem (21) verifies the following inequality:

$$\left\| u^{r(i)} \right\|_{(H^2(\vartheta \cap V_i))^3} \leq C \left\| u^{r(i)} \right\|_{(H^1(\Theta \cap V_i))^3} \qquad \text{with } i = 1, 2 \tag{32}$$

for $r$ small enough, e.g. $r < 2\alpha\rho \min(\eta_n, \eta_t)$ where $\alpha$ is the coercivity constant of the elasticity operator and $C$ is a constant independent of the regularization parameters.

It is important to note that for a problem with fixed values for $\eta_n$ and $\eta_t$ all the regularity results obtained hold for any $r$ small enough, even $r \to 0$. The case with $\eta_n$ and $\eta_t$ fixed has a clear mechanical meaning. In fact, the quantities $1/\eta_n$ and $1/\eta_t$ may represent the stiffnesses of the interface material.

It is possible to prove that the limit of the sequence $\{(t^r, -Y^r)\}$ evaluated at any point $(x_1, x_2) \in S_{a\rho}$ or $(x_1, x_2) \in S_{d\rho}$, for some strictly positive real number $\rho$, belongs to the subdifferential of $I_K(s^o(x_1, x_2), \gamma^o(x_1, x_2))$. In other words, the limit of $\{(t^r, -Y^r)\}$ belongs to the subdifferential of $I_K(s^o, \gamma^o)$ at least away from the delamination front.

## 8. Conclusions

A delamination model has been carried out. The interface model is governed by a nonsmooth functional involving two variables: the relative displacement of the joined plates and the damage parameter. A regularized model has been given. Some fundamental mathematical properties of the initial nonsmooth and the regularized delamination problems have been obtained. In particular, a constructive proof of the existence of a solution for the regularized problem has been given. The solutions of the regularized problem has been proved to converge to a solution of the initial nonsmooth problem.

## References

1. W.J. Bottega and A. Maewal, Delamination Buckling and Growth in Laminates. *J. Appl. Mech.* 50 (1983), 184-189.
2. H. Chai, and C.D. Babcock, Two-Dimensional Modelling of Compressive Failure in Delaminated Laminates. *J. Comp. Mater.* 19 (1985), pp. 67-98.
3. A.C. Garg, Delamination - A damage model in composite structures. *Engng. Fract. Mech.* 29 (1988), 557-584.
4. L.M. Kachanov, *Delamination Buckling of Composite Materials.* Kluwer Academic Publishers (1988).
5. O. Allix, P. Ladeveze, Interlaminar interface modelling for the prediction of the delamination. *Comp. Struct.* 22 (1992),pp 235-242.
6. M. Frémond, Contact Unilatéral avec Adhérence. *Unilateral Problems in Structural Analysis*", G. Del Piero and F. Maceri, Eds., Springer-Verlag (1985).
7. M. Frémond, Adhérence des Solides, *J. Méc. Théor. Appl.* 6, (1987), 383-407.
8. M.Frémond, Contact with Adhesion. *Topics in Nonsmooth Mechanics*, J.J. Moreau, P.D. Panagiotopoulos and G. Strang, Eds., Birkhäuser (1988).
9. N. Point, *Approche Mathématique de Problèmes à Frontières Libres. Application à des Exemples Physiques.* Thèse de Doctorat d'Etat es-Sciences Mathématiques de l'Université Paris XIII (1989).
10. N. Point, and E. Sacco, A delamination model for laminated composites. *Int. J. Solids Structures* 33, (1995), 483-509.
11. N. Point, and E. Sacco, Delamination of beams: A method for the evaluation of the strain energy release rates of DBC specimen. *Int. J. Fracture* 79, (1996), 225-247.
12. D. Maugis, and M. Barquins, Fracture mechanics and the adherence of viscoelastic bodies. *J. Phys. D., Appl. Phys.* 11 (1978), 1989-2023.

# DELAMINATION PREDICTION BY CONTINUUM DAMAGE MECHANICS

O. ALLIX*, L. GORNET, P. LADEVÈZE, D. LÉVÊQUE
*Laboratoire de Mécanique et Technologie*
*ENS de Cachan, CNRS, Université Paris VI*
*61 av. Président Wilson, 94230 Cachan, France*
* Also at Université d'Evry-Val-d'Essonne/ IUT GMP
2 cours Monségnieur Roméro, 91000 Evry Cédex -France

**Abstract.** The present study describes the basic principles of a general damage approach for laminates and demonstrates its prediction possibilities for simulating the complete fracture phenomenon in the case of the initiation and propagation of delamination. In this paper, the Damage Mechanics Approach is presented to solve delamination problems such as the propagation of an initiated crack or the initiation of delamination around holes. For the M55J/M18 carbon/epoxy material, numerical simulations are presented and compared with experimental results.

## 1. Introduction: the meso-modelling concept

Delamination often appears as the result of the interaction between different damage mechanisms, such as fiber breaking, transverse micro-cracking and the debonding of adjacent layers itself. Our aim herein is to build a bridge between damage mechanics and delamination by including all these damage mechanisms within the delamination analysis. To perform this, a damage meso-model, which allows us to predict both delamination initiation and propagation far more precisely, has been defined. Thanks to the proposed approach, both initiation and propagation are included in a single model.



*P. Argoul et al. (eds.),*
*IUTAM Symposium on Variations of Domains and Free-Boundary Problems in Solid Mechanics*, 163–171.

An initial step, which has been conducted in other studies, was to model the laminate as a stacking sequence of non-linear layers and non-linear interlaminar interfaces (see Figure 1). At the layer level, the inner damage mechanisms are taken into account by means of internal damage variables. These damage variables are prescribed to be uniform throughout the thickness of each ply which defines what we call a damage meso-model (Ladevèze, 1986). This play a crucial role for a mesh-independent damage prediction (Ladevèze, 1992,). The single-layer model and its identification, including damage and inelasticity, were previously developed by Ladevèze and Le Dantec (1992).

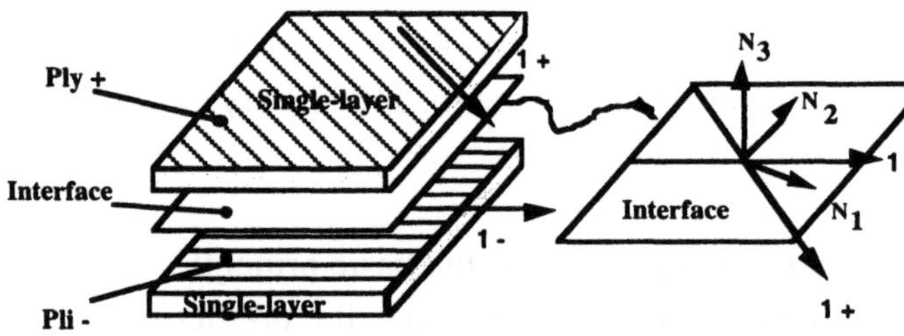


*Figure 1.* Laminate meso-modelling.

The interlaminar interface is a two-dimensional entity which ensures traction and displacement transfer from one ply to another. Its mechanical behaviour depends on the angles between the fibers of two adjacent layers. Its primary interest is to allow the modelling of more or less progressive degradation of the interlaminar connection (Allix et al. 1985, Allix 1992, Allix and Ladevèze 1996). One consequence of this meso-model is that only two types of cracks are being considered: delamination cracks and orthogonal cracks which extend across the entire thgickness of the layer.
In this paper, attention is being focused on the the interface model and on its use in solving delamination problems: Edge Delamination Tension problems, Fracture Mechanics specimens, holed plates. For several materials, numerical simulations are presented and compared with experimental results.

## 2. Interlaminar Interface Modelling

The effect of the deterioration of the interlaminar connection on its mechanical behaviour is taken into account by means of internal damage variables. The energy per unit area proposed by Allix and Ladevèze (1992) is:

$$E_D = \frac{1}{2}\left[\frac{<-\sigma_{33}>_+^2}{k_3^0} + \frac{<\sigma_{33}>_+^2}{k_3^0(1-d_3)} + \frac{\sigma_{32}^2}{k_2^0(1-d_2)} + \frac{\sigma_{31}^2}{k_1^0(1-d_1)}\right] \quad (1)$$

where $k_i^0$ is an interlaminar stiffness value and $d_i$ the internal damage indicator associated with its Fracture Mechanics mode, while i corresponds to an orthotropic direction $N_i$ of the interface (Figure 1). These directions are defined as the bisectors of the angle formed by the fibre directions of the adjacent layers. The damage energy release rates are introduced as:

$$Y_{d3} = \frac{1}{2}\frac{<\sigma_{33}>_+^2}{k_3^0(1-d_3)^2};\ Y_{d1} = \frac{1}{2}\frac{\sigma_{31}^2}{k_1^0(1-d_1)^2};\ Y_{d2} = \frac{1}{2}\frac{\sigma_{32}^2}{k_2^0(1-d_2)^2} \quad (2)$$

In what follows, an "isotropic" damage evolution law is described. The damage evolution law is assumed to be governed by means of an equivalent damage energy release rate of the following form:

$$\underline{Y}(t) = \sup\Big|_{\tau\leq t}\left[\left((Y_{d3})^\alpha + (\gamma_1 Y_{d1})^\alpha + (\gamma_2 Y_{d2})^\alpha\right)^{1/\alpha}\Big|_\tau\right] \quad (3)$$

The evolution of the damage indicators is thus assumed to be strongly coupled. $\gamma_1$, $\gamma_2$ and $\alpha$ are material parameters. A simple damage evolution law is then defined as follows:

$$d_3 = d_1 = d_2 = \left[\frac{n}{n+1}\frac{<\underline{Y}-Y_o>_+}{Y_c-Y_o}\right]^n \text{ if } d_3 < 1,\ d_3 = d_1 = d_2 = 1 \text{ otherwise} \quad (4)$$

wherein a critical value $Y_c$ and a threshold value $Y_o$ have been introduced. High values of the "n" case correspond to a brittle interface. In Allix and Corigliano (1996), it is shown that the Fracture Mechanics associated with this damage evolution law is:

$$\left(\frac{G_I}{G_{cI}^p}\right)^\alpha + \left(\frac{G_{II}}{G_{cII}^p}\right)^\alpha + \left(\frac{G_{III}}{G_{cIII}^p}\right)^\alpha = 1 \quad (5)$$

with:

$$G_{cI}^p = Y_c;\ G_{cII}^p = \frac{Y_c}{\gamma_1};\ G_{cIII}^p = \frac{Y_c}{\gamma_2} \quad (6)$$

## 3. Fracture Mechanics Tests

### 3.1 Experimental part

The tests of crack propagation in interlaminar fracture specimens conducted in this work and developed at the AÉROSPATIALE facility in Suresnes (France) are the pure-mode I DCB (Double-Cantilever Beam) Test (De Charentay et al., 1984), the pure-mode II ENF (End-Notched Flexure) Test (Davies et al., 1990) and a mixed-mode test: the MMF (Mixed-Mode Flexure) Test (Russell and Street, 1985). These tests were conducted on M55J/M18 material laminates. In this work, we focus our attention on only the prediction of unidirectional laminate fracture mechanics tests. The mean thickness of a single ply is on the order of 0.1 mm. The tests have been conducted on an INSTRON testing machine at ambient temperature. The displacement rate was fixed at 2 mm/min in the DCB test and at 1 mm/min in the ENF and MMF tests. In these two latter bending tests, the total useful length was 180 mm.
The critical energy release rates are usually reduced by means of Linear Elastic Fracture Mechanics. The links between Linear Elastic Fracture Mechanics and Damage Mechanics serve to identify the parameters of the interfacial modelling relative to those critical rates. In the DCB test, the mode I critical energy release rate measured for the unidirectional material was $\mathcal{G}^{p}_{cI} = 113\ \mathrm{J/m^2}$. This value was obtained by derivation of the experimental compliance of the specimen. The mode II critical energy release rate measured in the ENF test provides us with the value of $\gamma_1 = 0.37$. The value of the $\alpha$ parameter was found by means of the mixed-mode test: $\alpha = 1.6$ for the 0°/0° interface.

Remark: In another work (Allix et al. 1997), we emphasize the necessity of taking account of the dissipative phenomena inside the layers, especially when $\theta \neq 0°$, because the measured energy release rate cannot be attributed to the debonded interface only.

### 3.2 Comparison between predicted values and experimental results for a 0°/0° interface case

With the previous measured values of the energy release rate and making use of relations (6) we are able to identify an interface model and then to simulate the response of the different tests. These finite element predictions have been conducted with the help of ENDO-STRAT-EF, a three dimensional non-linear finite element code that includes the meso-model (Gornet et al., 1997). Numerical results are mesh-independent; the basic factor behind this is the presence of internal lengths in the meso-modeling. In this work, the same finite element mesh was used for all laminate Fracture Mechanics specimen test predictions. This mesh consisted of 678 elements for each of both arm (layers) and 226 special elements for the interface (Beer, 1985). Here, we will only show the prediction for the DCB and MMF tests.

Figure 2 depicted the correlation obtained between experimental results and finite element predicted values in the case of a DCB specimen. The MMF test specimen is shown in Figure 3. In this mixed mode test, the mode I is dominant. Experimental results and finite element predicted values also exhibit good correlation (Figure 4).

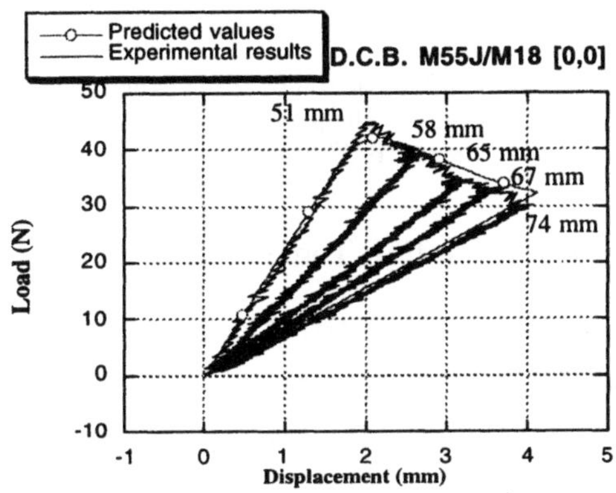


*Figure 2*. Prediction of a D.C.B. test. Comparison between experimental results and predicted values. The initial crack closure is a= 50 mm. The growth of the delaminattion area at the end of the test is 23mm.

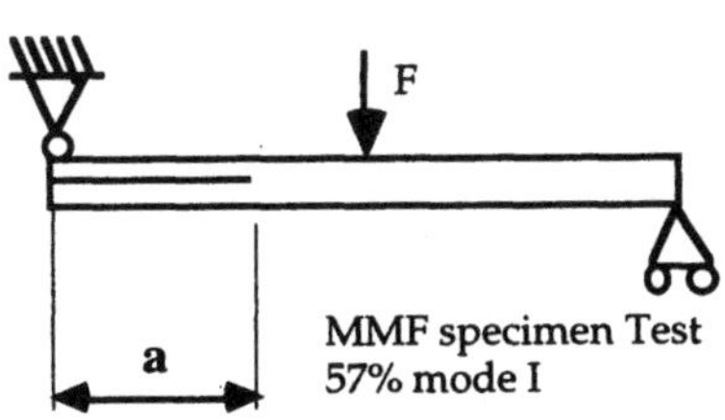


*Figure 3.* M.M.F. specimen test. The initial crack length has been denoted by "a".

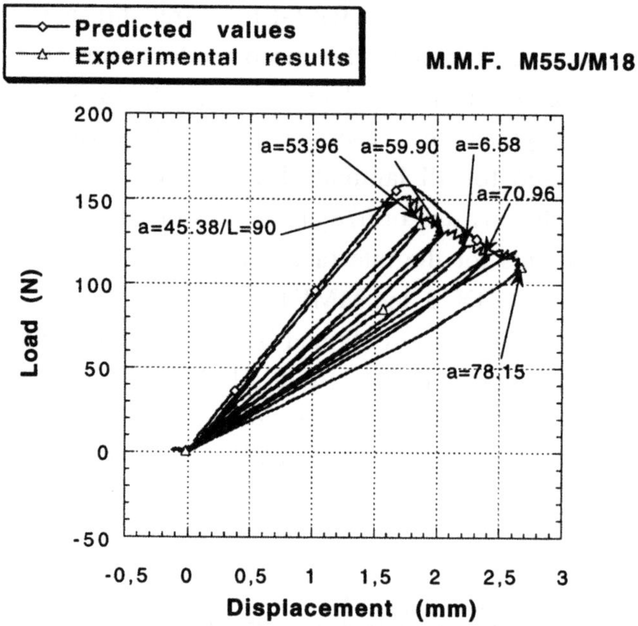


*Figure 4.* Prediction of an M.M.F. test . Comparison between experimental results and predicted values. The initial crack closure is a=45mm. The growth of the delaminattion area at the end of the test is 32.77mm.

## 4. Example of an Edge Delamination Tension specimen

The specimen under consideration is a $(0_3,\pm45_2,90)s$. The edge is straight and the problem to solve can be set as a generalised plain strain problem in a strip perpendicular to the edge. In the previous test the delamination starts on the O°/O° interface and after this initiation point the load can still increase, the maximum value of the applied tension being around four times the initiation load (Allix et al. 1997). Without including the inner layer mechanisms into account in our Finite Element Analysis the initiation and maximum load coincide. Introducing the inner damage mechanisms (namely the transverse cracking), the delamination propagation becomes then stable under increasing tension up to the fiber rupture of the 0° plies (Gornet et al. 1997). This example shows the necessity of including all the damage mechanisms into the delamination analysis even for quite simple specimens. Figure 5 depictes the state of damage before the final failure of the specimen.

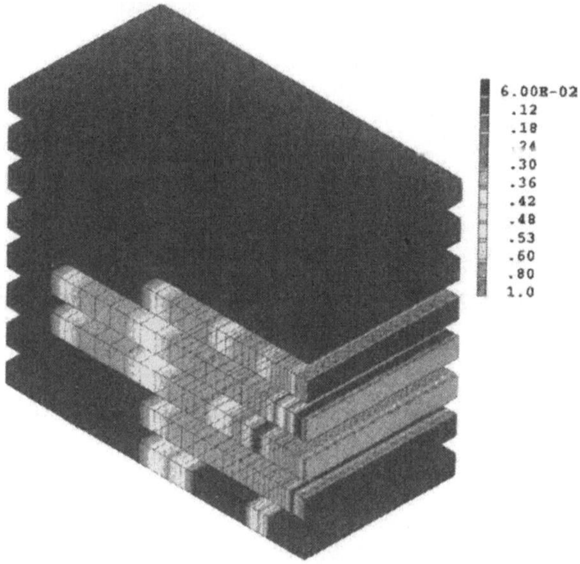


Figure 5. Simulation of the damage state of a $(0_3,\pm45_2,90)$s E.D.T. specimen before the final failure.

## 5. Damage identification using plates with holes

More reliable delamination tests would be those on laminated plates with a circular hole. In fact, for such specimens, the delamination crack initiation is reproducible and the growth is often stable within a certain range. The idea herein is to use the first parameter set identified previously as initial data for a global identification using plates with holes. We are just at the beginning of this procedure. Also of interest in this test is the valuable information it providesin terms of the shape and size of the delaminated area revealed by means of X-ray photography (Figure 6). The difficulty herein is that, due to the complexity of the state of stresses, the interpretation of this test requires complex computations (Figure 7). For such tests, a specialized software DSDM (Delamination Simulation by Damage Mechanics) has been previously developed (Allix 1992).

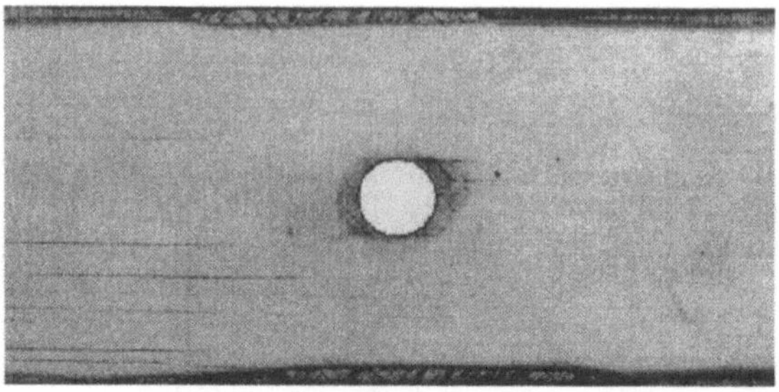

*Figure 6.* $[0_3/\pm45_2/90]_s$ X-ray photograph (86% of the rupture load).

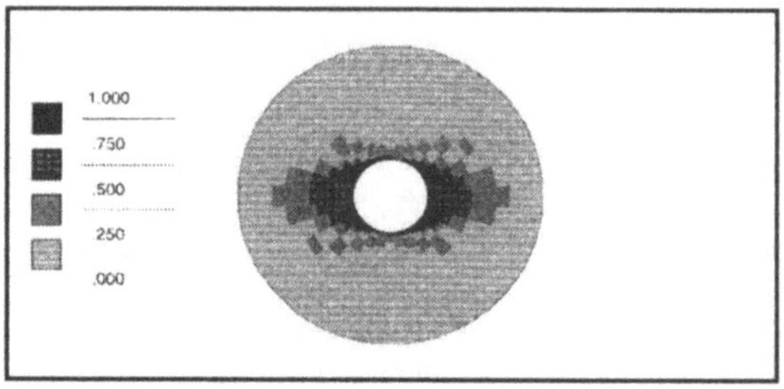


*Figure 7.* $d_3$ indicator computed at the 0°/+45° interface in the $[0_3/\pm45_2/90]_s$ laminate.

## 6. Conclusions

This work represents the first step towards a global identification of an interface damage model dedicated to delamination prediction. An initial set of parameters was determined using standard initiation and propagation tests conducted on M55J/M18 material specimens. It has been shown that experimental results and finite element predicted values exhibit strong correlation for classical fracture mechanics tests. Lastly, an initial comparison between computation and a holed-specimen test in tension has revealed the interest of using such an approach as a standard identification test.

## 7. References

Allix, O. (1992) Damage analysis of delamination around a hole. *New Advances in Computational Structural Mechanics*, P. Ladevèze, O. C. Zienkiewicz eds., Elsevier Science Publishers B. V., pp 411-421.

Allix, O., Gilletta D. and Ladevèze, P. (1985) Mechanical behaviour of elementary constituents of laminates, Proceedings I.C.C.M.M. V, San Diego, August 1985.

Allix, O. and Corigliano, A. (1996) Modeling and simulation of crack-propagation in mixed-modes interlaminar fracture specimens. In *International Journal of Fracture* 77, pp. 11-140.

Allix, O. and Ladevèze, P. (1992) Interlaminar interface modelling for the prediction of delamination, *Comp. Struct.*, 22:235-242.

Allix, O. and Ladevèze, P. (1996) Damage mechanics of interfacial media: basic aspects, identification and application to delamination. In *Damage and Interfacial Debonding in Composites*. Studies in applied Mechanics, 44, Eds Allen D. and Voyiadjis G., Elsevier, pp. 167-88.

Allix O., Ladevèze P., Lévêque D., Perret L. (1997) Identification and validation of an interface damage model for delamination prediction, *Computational Plasticity*, eds Owen D.R.J., Oñate E. & Hinton E., Barcelone, pp. 1139-1147.

Beer, G. (1985). An isoparamétric joint/interface element for finite element analysis, Int. J. for Numer. Methods Eng., 21, pp. 585-600.

De Charentenay, F.-X., Harry, J.M., Prel, Y.J. and Benzeggagh, M.L. (1984) Characterizing the Effect of Delamination Defect by Mode I Delamination Test. *Effect of Defects in Composite Materials, ASTM STP 836*, pp 84-103.

Davis P. 1990. Measurement of GIc and GIIc in carbon/epoxy composites, *Comp. Sci. & Technol.*, 39, pp. 193-205.

Gornet, L. Hochard, C., Ladevèze, P. and Perret, L. (1997), Proceedings *Damage and Failure of Interfaces*, DFI-1, Vienna, Austria, Sept 22-24.

Ladevèze, P. (1986) Sur la mécanique de l'endommagement des composites, Proceedings of the Fifth French National Colloquium on Composite Materials, JNC 5, C. Bathias & D. Menkès eds, Pluralis Publication, Paris, pp. 667-683.

Ladevèze, P. (1992) Towards a fracture theory, Proceedings of the third International Conference on Computational Plasticity Part II, D.R.J. OWEN, E. ONATE, E. HINTON ed, Pineridge press, Cambridge U.K., pp. 1369-1400.

Ladevèze, P. and Le Dantec, E. (1992) Damage modelling of the elementary ply for laminated composites, *Comp. Sci. & Technol.*, 43:257-267.

Russell, A.J. and Street., K.N. (1985) Moisture and temperature effects on the mixed-mode delamination fracture of unidirectional graphite/epoxy, *Delamination and Debonding of Materials, ASTM STP 876*, W.S. Johnson, Ed., pp 349-370.

# DYNAMICAL SYSTEMS FOR NON-ISOTHERMAL PHASE SEPARATION

ALAIN DAMLAMIAN
*Centre de Mathématiques, Ecole Polytechnique, URA 169 CNRS*
*91128 Palaiseau Cedex, France*

AND

NOBUYUKI KENMOCHI
*Department of Mathematics, Faculty of Education, Chiba University*
*1-33 Yayoi-chō, Inage-ku, Chiba, 263 Japan*

**Abstract.** A non-isothermal model for phase separation in a binary mixture is considered. In this paper we try the subdifferential approach to the model, based on the recent development of nonlinear operator $-\Delta\alpha$ defined on the dual space the Sobolev space $H^1(\Omega)$.

## 1. Introduction

Various models for nonlinear phenomena accompanied with free boundaries and phase transitions, which will be called simply phase transition models in this paper, have been proposed and actively analyzed from both the physical and mathematical point of views since 1970. In the recent development on this subject the nonlinear operator theoretic approach might be noteworthy and one expects efficient application of the abstract theory on nonlinear semigroups and subdifferentials of convex functions established so far in order to investigate not only the existence and uniqueness of solutions of models but also the asymptotic stability for the associated dynamical processes. In treating mathematically our phase transition model, some difficulties arise from its structural features as mentioned below:

- The model is a system of fully nonlinear parabolic PDEs which includes singularity and degeneracy at the same time.



*P. Argoul et al. (eds.),*
*IUTAM Symposium on Variations of Domains and Free-Boundary Problems in Solid Mechanics*, 173–180.

– Each PDE in the system has its own function space in which the PDE is well-posed. However, in general, it would not be easy to find function spaces in which the good structure is still kept, when two PDEs are linked together.

Recently, an idea was proposed in [7,8,9] in order to overcome these difficulties; actually it is to use nonlinear (multivalued) differential operator $-\Delta\alpha$ defined in the dual space $H^1(\Omega)^*$ of the Sobolev space $H^1(\Omega)$ in modeling the energy balance equation, where $\alpha$ is a maximal monotone graph in $R \times R$. The main objective of this paper is to expound the subdifferential approach to phase transition models with some recent results on the operator $-\Delta\alpha$.

*Notation.* For a general (real) Banach space $Y$, we denote by $Y^*$ the dual space. Given a proper l.s.c. convex function $\psi$ on $Y$, we denote by $D(\psi)$ the effective domain and by $\partial_{Y,Y^*}\psi$ the subdifferential from $Y$ into $Y^*$, and the sets $D(\partial_{Y,Y^*}\psi) := \{v \in Y : \partial_{Y,Y^*}\psi(v) \neq \emptyset\}$ and $R(\partial_{Y,Y^*}\psi) := \bigcup_{v\in Y} \partial_{Y,Y^*}\psi(v)$ are respectively the domain and range of $\partial_{Y,Y^*}\psi$. In particular, if $Y$ is a Hilbert space and $Y^*$ is identified with $Y$, then $\partial_{Y,Y^*}\psi$ is denoted simply by $\partial_Y\psi$.

In this paper, let $\Omega$ be a bounded domain in $R^N$ with smooth boundary $\Gamma := \partial\Omega$, and let $V := H^1(\Omega)$ be the standard Sobolev space with norm $||v||$ defined by

$$||v|| := \left\{\int_\Omega |\nabla v|^2 + n_o \int_\Gamma |v|^2\right\}^{\frac{1}{2}},$$

where $n_o$ is a fixed positive constant. We simply denote by $\langle\cdot,\cdot\rangle$ the duality pairing between $V^*$ and $V$; as usual we have $V \subset L^2(\Omega) \subset V^*$ with compact and densely defined injections. Also, we denote by $F$ the duality mapping from $V$ onto $V^*$ which is defined by

$$\langle Fv, z\rangle := \int_\Omega \nabla v \cdot \nabla z + n_o \int_\Gamma vz, \qquad \forall v, z \in V.$$

Clearly $F$ is an isometric operator from $V$ onto $V^*$. Given $f \in L^2(\Omega)$ and $h \in L^2(\Gamma)$, an element $f^* \in V^*$ is uniquely determined by

$$\langle f^*, z\rangle = \int_\Omega fz + \int_\Gamma hz, \qquad \forall z \in V, \tag{1.1}$$

and "$Fv = f^*$" is formally equivalent to "$-\Delta v = f$ *in* $\Omega$, $\frac{\partial v}{\partial n} + n_o v = h$ *on* $\Gamma$ ", which holds in the variational sense. It is easy to check that $V^*$ is a Hilbert space endowed with inner product $(\cdot,\cdot)_*$ given by

$$(v, z)_* := \langle v, F^{-1}z\rangle, \qquad \forall v, z \in V^*. \tag{1.2}$$

Moreover we denote by $\Delta_N$ the Laplacian with homogeneous Neumann boundary condition, i.e., given $q \in L^2(\Omega)$,

$$\Delta_N v = q \Longleftrightarrow \Delta v = q \ \ in\ L^2(\Omega), \quad \frac{\partial v}{\partial n} = 0 \ \ in\ H^{\frac{1}{2}}(\Gamma),$$

with $D(\Delta_N) = \{v \in H^2(\Omega); \frac{\partial v}{\partial n} = 0 \ in\ H^{\frac{1}{2}}(\Gamma)\}$.

## 2. Operator $-\Delta\alpha$

In this section, let $\alpha$ be a maximal monotone graph in $R \times R$ such that the domain $D(\alpha) := \{r \in R; \alpha(r) \neq \emptyset\}$ has the non-empty interior. Fix a primitive $\hat{\alpha}$ of $\alpha$; the subdifferential of $\hat{\alpha}$ is $\alpha$ in $\mathbf{R}$, namely $\partial_R \hat{\alpha} = \alpha$. Now, for this $\hat{\alpha}$ we define a functional $j_o$ by

$$j_o(z) := \begin{cases} \int_\Omega \hat{\alpha}(z) & if\ z \in L^1(\Omega)\ and\ \hat{\alpha} \in L^1(\Omega), \\ +\infty & otherwise. \end{cases}$$

It is easy to see that the restriction of $j_o$ onto $L^2(\Omega)$, denoted by the same notation $j_o$, is proper, l.s.c. and convex in $L^2(\Omega)$, and its subdifferential $\partial_{L^2(\Omega)} j_o$ is given by $\partial_{L^2(\Omega)} j_o(v) = \{z \in L^2(\Omega); z \in \alpha(v)\ a.e.\ on\ \Omega\}$. But, when $j_o$ is considered on $V^*$, it is not l.s.c. on $V^*$, in general, without any restriction. Under the condition that

$$\liminf_{|r| \to +\infty} \frac{\hat{\alpha}(r)}{|r|} = +\infty, \tag{2.1}$$

it was shown in [2] that its restriction onto $V^*$, denoted by $j_o$ again, is proper, l.s.c. and convex on $V^*$ and $D(\partial_{V^*,V} j_o) \subset L^1(\Omega)$, and moreover

$$\partial_{V^*,V} j_o(v) = \{z \in V; z \in \alpha(v)\ a.e.\ on\ \Omega\}, \quad \forall v \in D(\partial_{V^*,V} j_o). \tag{2.2}$$

In the general case we have:

**Theorem 1.** (cf. [7]) *There is a proper l.s.c. convex function $j$ on $V^*$ such that $j = j_o$ on $L^2(\Omega)$. Furthermore, if $v \in D(\partial_{V^*,V} j) \cap L^2(\Omega)$, then $v \in D(\partial_{L^2(\Omega)} j_o)$ and*

$$\begin{aligned} \partial_{V^*,V} j(v) &= \partial_{L^2(\Omega)} j_o(v) \cap V \\ &= \{z \in V; z \in \alpha(v)\ a.e.\ on\ \Omega\}. \end{aligned}$$

The main point of the proof of Theorem 1 is to consider the closure of the epigraph of the restriction of $j_o$ onto $L^2(\Omega)$ in $V^* \times R$ and to show that this closure gives the epigraph of a required proper l.s.c. convex function on $V^*$. For simplicity, we denote by $\partial_* j$ the subdifferential $\partial_{V^*} j$ of $j$ in the Hilbert space $V^*$ with inner product given by (1.2). It is easy to see that $\partial_* j(v) = F\partial_{V^*,V} j(v)$ for any $v \in D(\partial_* j)$. Hence, from Theorem 1 it follows that for any $v \in D(\partial_* j) \cap L^2(\Omega)$

$$\partial_* j(v) = \{Fz; z \in V, z \in \alpha(v) \ a.e. \ on \ \Omega\}.$$

Accordingly, the initial-boundary problem

$$\begin{cases} u_t - \Delta\alpha(u) \ni f(x,t) & in\ Q := \Omega \times (0,T), \\ \dfrac{\partial \alpha(u)}{\partial n} + n_o \alpha(u) \ni h(x,t) & on\ \Sigma := \Gamma \times (0,T) \\ u(\cdot, 0) = u_o & on\ \Omega \end{cases} \tag{2.3}$$

can be reformulated in the form

$$\begin{cases} u'(t) + \partial_* j(u(t)) \ni f^*(t) & in\ V^*,\ 0 < t < T, \\ u(0) = u_o, \end{cases} \tag{2.4}$$

where $f^*(t)$ is the element of $V^*$ determined from $f(t) \in L^2(\Omega)$ and $h(t) \in L^2(\Gamma)$ in the same way as (1.1).

**Remark 2.1.** Various choice of maximal monotone graph $\alpha$ represents a large number of physical processes. For instance:

**(a)** When $\alpha$ is Lipschitz continuous on $\mathbf{R}$ and satisfies that $\hat{\alpha}(r) \geq c_o|r|^2$, $r \in R$ for some positive constant $c_o$, (2.3) is the enthalpy formulation of the multi-phase Stefan problem (cf. [6]).

**(b)** When $\alpha(u) = -\frac{1}{u}$ for $u > 0$, (2.3) was recently discussed as a Stefan-like problem with singular heat flux in [4,5].

**Remark 2.2.** In the general case of $\alpha$ (without condition (2.1)), $D(\partial_* j)$ is big enough; actually it contains singular measures on $\Omega$ (cf. [3]). Therefore, the solution of (2.4) is not necessarily that of (2.3), although (2.4) always has one and only one solution, for any general initial datum $u_o$ in $V^*$.

## 3. Phase separation model with constraint

For instance, let us consider a binary alloy with components A and B which occupies $\Omega$. Phenomenologically we see that the domain $\Omega$ is divided into three regions (of pure A, pure B and mixture) which vary in time by phase separation. We are interested in the dynamics of these time-dependent regions. This sort of dynamics can be described by absolute temperature field $\theta := \theta(x,t)$ and order parameter $w := w(x,t)$ which is formulated for instance by

$$w(x,t) := 2w_A(x,t) - 1, \quad (x,t) \in Q,$$

where $w_A(x,t)$ is the local concentration of component A at position $x \in \Omega$ and time $t$. Note here that $-1 \le w \le 1$ because of $0 \le w_A \le 1$, and the state of the material is respectively of pure A, of pure B and of mixture at $(x,t)$, if $w(x,t) = 1$, $w(x,t) = -1$ and $-1 < w(x,t) < 1$. In this paper, consider a phase separation process in which the free energy functional $F_\Omega(w,\theta)$ is given by

$$F_\Omega(w,\theta) := \int_\Omega \left\{ \frac{\kappa\theta}{2}|\nabla w|^2 + \tau(\theta) + \theta(\hat{\beta}(w) + \hat{g}(w)) + \lambda(w) \right\}, \tag{3.1}$$

where $\kappa$ is a (small) positive constant, the coefficient of interfacial energy; $\tau(\theta)$ is a smooth concave function for $\theta > 0$, for example $\tau(\theta) := -c_1\theta\log\theta$ with constant $c_1 > 0$; $\hat{\beta}$ is a proper l.s.c. convex function on $\mathbf{R}$ with $D(\hat{\beta}) = [-1,1]$, for example the indicator function of the interval $[-1,1]$; $\hat{g}(\cdot)$ is a smooth function on $\mathbf{R}$, for example a function with double wells; $\lambda(\cdot)$ is a smooth convex function, for example a linear or quadratic function. In the same approach as in [1, 10] we derive from (3.1) the following energy and mass balance equations

$$[\rho(\theta) + \lambda(w)]_t - \Delta(-\frac{1}{\theta}) = f(x,t) \quad in\ Q, \tag{3.2}$$

$$w_t - \Delta\left\{-\kappa\Delta w + \tilde{\beta} + g(w) + \frac{\lambda'(w)}{\theta}\right\} = 0, \quad \tilde{\beta} \in \beta(w), \quad in\ Q, \tag{3.3}$$

where $\rho(\theta) := \tau(\theta) - \theta\tau'(\theta)$, $\beta := \partial_R\hat{\beta}$, $g := \hat{g}'$. These equations are considered with the following initial and boundary conditions:

$$\frac{\partial}{\partial n}\left(-\frac{1}{\theta}\right) + n_o\left(-\frac{1}{\theta}\right) = h(x,t) \quad on\ \Sigma, \tag{3.4}$$

$$\frac{\partial w}{\partial n} = 0, \quad \frac{\partial}{\partial n}\left\{-\kappa\Delta w + \tilde{\beta} + g(w) + \frac{\lambda'(w)}{\theta}\right\} = 0 \quad on\ \Sigma, \tag{3.5}$$

$$\theta(\cdot,0) = \theta_o, \quad w(\cdot,0) = w_o \quad in\ \Omega, \tag{3.6}$$

where $f \in L^2(0,T;L^2(\Omega))$, $h \in L^2(0,T;L^2(\Gamma))$, and $\theta_o, w_o \in L^2(\Omega)$ are given as data. System (3.2)-(3.6) is referred as (PSC). We are going to reformulate our phase separation model (PSC) as a single evolution equation in the product space

$$X := \begin{matrix} V^* \\ \times \\ L^2(\Omega)_o \end{matrix}$$

where $L^2(\Omega)_o := \{z \in L^2(\Omega); \int_\Omega z = 0\}$; we denote by $\pi_o$ the projection from $L^2(\Omega)$ onto $L^2(\Omega)_o$; $X$ is a Hilbert space with inner product $(\cdot,\cdot)_X$ defined by

$$(U_1, U_2)_X := (e_1, e_2)_* + \int_\Omega w_1 w_2, \quad \forall U_i := \begin{pmatrix} e_i \\ w_i \end{pmatrix} \in X, \quad i = 1, 2.$$

To this end we use such a proper l.s.c. convex function $j$ on $V^*$ as in the previous section, corresponding to a maximal monotone graph $\alpha$ in $R \times R$ which is given as follows:

$$\alpha(u) := \left\{ -\frac{1}{\theta}; \theta \in \rho^{-1}(u) \cap (0, +\infty) \right\}$$

with the (possibly multivalued) inverse $\rho^{-1}$ of $u = \rho(\theta)$, assuming that

$$\rho > 0 \ on \ (0, +\infty), \quad \lim_{\theta \downarrow 0} \rho(\theta) = 0. \tag{3.7}$$

Since $\tau$ is concave, we see that $\rho$ is non-decreasing on $(0, +\infty)$, which implies together with (3.7) that $\alpha$ is a maximal monotone graph in $R \times R$. Since the operator $-\Delta_N$ is invertible in $L^2(\Omega)_o$, with a new variable $e := \rho(\theta) + \lambda(w)$, our system (PSC) can be written in the form

$$\frac{d}{dt} A \begin{pmatrix} e(t) \\ w(t) \end{pmatrix} + \begin{pmatrix} F\tilde{\alpha}(t) \\ -\kappa\Delta_N w(t) + \pi_o[\tilde{\beta}(t) - \tilde{\alpha}(t)\lambda'(w(t))] \end{pmatrix} + \begin{pmatrix} 0 \\ \pi_o g(w(t)) \end{pmatrix} = \begin{pmatrix} f^*(t) \\ 0 \end{pmatrix} \tag{3.8}$$

$$in \ X, \quad t \in (0, T),$$

$$\tilde{\alpha} \in \alpha(e - \lambda(w)), \quad \tilde{\beta} \in \beta(w) \ \ in \ Q,$$

$$\begin{pmatrix} e(0) \\ w(0) \end{pmatrix} = \begin{pmatrix} e_o \\ w_o \end{pmatrix}, \quad e_o := \rho(\theta_o) + \lambda(w_o),$$

where

$$A := \begin{pmatrix} I & 0 \\ 0 & (-\Delta_N)^{-1} \end{pmatrix},$$

$f^*(t)$ is an element in $V^*$ determined from $f(t) \in L^2(\Omega)$ and $h(t) \in L^2(\Gamma)$ in the same way as (1.1); we note that $\tilde{\alpha} := -\frac{1}{\theta} \in \alpha(e - \lambda(w))$.

## 4. Evolution equation associated with (PSC)

We use the same notation as in the previous section. Fixing a primitive $\hat{\alpha}$ of $\alpha$, we consider the proper, l.s.c. convex function $j$ on $V^*$ given by Theorem

2 corresponding to $\hat{\alpha}$, and define a function $\varphi$ on $X$ by putting

$$\varphi\begin{pmatrix} e \\ w \end{pmatrix} := j(e - \lambda(w)) + \int_\Omega \left\{ \frac{\kappa}{2}|\nabla w|^2 + \hat{\beta}(w) \right\}, \quad \forall \begin{pmatrix} e \\ w \end{pmatrix} \in X.$$

Then we have:

**Theorem 2.** (cf. [8,9]) *$\varphi$ is proper, l.s.c. and convex on $X$. Moreover, if $\binom{e}{w} \in D(\partial_X \varphi)$ and $\binom{\tilde{e}}{\tilde{w}} \in \partial_X \varphi\binom{e}{w}$, then the following statements hold:*

*(1) $\tilde{e} \in \partial_* j(e - \lambda(w))$, hence there is $\tilde{\alpha} \in \partial_{V^*,V} j(e - \lambda(w))$ such that $\tilde{e} = F\tilde{\alpha}$. Furthermore, if $e \in L^2(\Omega)$, then $\tilde{\alpha} \in \alpha(e - \lambda(w))$ a.e. on $\Omega$.*

*(2) $w \in D(\Delta_N)$ and there is $\tilde{\beta} \in L^2(\Omega)$ such that*

$$\tilde{\beta} \in \beta(w) \ \text{ a.e. on } \Omega, \quad \tilde{w} = -\kappa \Delta_N w + \pi_o[\tilde{\beta} - \tilde{\alpha}\lambda'(w)],$$

*where $\tilde{\alpha}$ is the same function as in (1).*

By virtue of Theorem 2 we see that the second term of (3.8) is included in $\partial_X \varphi\binom{e(t)}{w(t)}$ and therefore our system (PSC) is reformulated, via transformation

$$U := \begin{pmatrix} e \\ w \end{pmatrix}, \qquad e = \rho(\theta) + \lambda(w),$$

as an evolution equation of the form

$$\text{(E)} \qquad (AU)'(t) + \partial_X \varphi(U(t)) + G(U(t)) \ni \ell(t) \quad \text{in } X, \quad t \in [0, T],$$

subject to initial condition

$$(A^{\frac{1}{2}}U)(0) = A^{\frac{1}{2}}U_o, \qquad U_o := \begin{pmatrix} e_o \\ w_o \end{pmatrix}, \tag{4.1}$$

where

$$G(U) := \begin{pmatrix} 0 \\ \pi_o g(w) \end{pmatrix} \quad for\ U := \begin{pmatrix} e \\ w \end{pmatrix} \in X, \quad \ell(t) := \begin{pmatrix} f^*(t) \\ 0 \end{pmatrix}$$

and $A := \begin{pmatrix} I & 0 \\ 0 & (-\Delta_N)^{-\frac{1}{2}} \end{pmatrix}$ is the fractional power of $A$ in $X$; note that $A^{\frac{1}{2}}$ is positive, continuous, linear and selfadjoint in $X$, since so is $A$. By a solution $U$ of (E) we mean that $A^{\frac{1}{2}}U \in W^{1,2}(0, T; X)$ and $\ell(t) - G(U(t)) - (AU)'(t) \in \partial_X \varphi(U(t))$ for a.e. $t \in [0, T]$.

**Theorem 3.** (cf. [9]) *Assume that*

$$\frac{h}{n_o} \in \alpha(\tilde{h}) \text{ a.e. in } Q \text{ for some } \tilde{h} \in L^2(0, T; L^2(\Gamma)), \ \begin{pmatrix} e_o \\ w_o \end{pmatrix} \in D(\varphi), \ e_o \in L^2(\Omega).$$

*Then there exists one and only one solution* $U := \binom{e}{w}$ *of (E) satisfying initial condition (4.1) such that e and w are respectively weakly continuous* $[0,T]$ *into* $L^2(\Omega)$ *and* $V$. *Moreover, the couple of functions* $\{\theta, w\}$, *with* $\theta = -\frac{1}{\tilde{\alpha}}$, *is a solution of (PSC) in the variational sense, where* $\tilde{\alpha}(t)$ *is the function in the expression of* $\partial_X \varphi\binom{e(t)}{w(t)}$ *given by Theorem 2.* See [9] for the detail proof of Theorem 3.

## References


1. H.W. Alt and I. Pawlow, Existence of solutions for non-isothermal phase separation, Adv. Math. Sci. Appl. **1**(1992),319-409.
2. H. Brézis, Monotonicity methods in Hilbert spaces and some applications to nonlinear partial differential equations, *Contributions to Nonlinear Functional Analysis*, ed. E. Zarantonello, Academic Press, New York, 1971.
3. H. Brézis, Intégral convexes dans les espaces de Sobolev, Israel J. Math. **13**(1972, 9-23.
4. P. Colli and G. Savaré, Time-discretization of Stefan problems with singular flux, pp. 16-28, in *Free Boundary Problems, Theory and Applications*, Pitman Research Notes Math. Ser. **Vol. 363**, Longman, 1996.
5. P. Colli and J. Sprekels, Stefan problems and the Penrose-Fife phase field models, Adv. Math. Sci. Appl., to appear.
6. A. Damlamian, Some results on the multi-phase Stefan problems, Communs. Partial Diff. Eqns. **2**(1977), 1017-1044.
7. A. Damlamian and N. Kenmochi, Evolution equations generated by subdifferentials in the dual space of $H^1(\Omega)$, submitted to Discrete and Continuous Dynamical Systems.
8. A. Damlamian and N. Kenmochi, Evolution equations associated with non-isothermal phase transitions, in *Functional Analysis and Global Analysis*, Lecture Notes Math., Springer-Verlag, Singapore, to appear.
9. A. Damlamian and N. Kenmochi, Evolution equations associated with non-isothermal phase separation: Subdifferential approach, preprint, 1997.
10. O. Penrose and P.C. Fife, Thermodynamically consistent models of phase-field type for the kinetics of phase transitions, Physica D **43** (1990), 44-62.

# GLOBAL SOLUTION TO THE PENROSE–FIFE PHASE FIELD MODEL WITH SPECIAL HEAT FLUX LAWS

PIERLUIGI COLLI
*Dipartimento di Matematica "F. Casorati"*
*Università di Pavia, Via Ferrata 1, I–27100 Pavia, Italy*
PHILIPPE LAURENÇOT
*Institut Elie Cartan - Nancy, Université de Nancy I*
*BP 239, F–54506 Vandœuvre les Nancy Cedex, France*
AND
JÜRGEN SPREKELS
*Weierstraß-Institut für Angewandte Analysis und Stochastik*
*WIAS, Mohrenstraße 39, D–10117 Berlin, Germany*

**Abstract.** The phase field model introduced by Penrose and Fife is considered for diffusive phase transitions with non conserved order parameter. Different motivations lead to investigate the case when the heat flux is the gradient of some function of the absolute temperature $\vartheta$ behaving like $1/\vartheta$ as $\vartheta$ approaches 0 and like $-\vartheta$ as $\vartheta \nearrow +\infty$. Uniqueness is proved for the solution of related initial and boundary value problems.

## 1. Introduction

This note is concerned with the study of the following initial–boundary value problem in the cylindrical domain $Q := \Omega \times (0,T)$, where $\Omega \subset I\!R^N$ $(N \leq 3)$ is a bounded domain with a smooth boundary $\Gamma$ and $T > 0$. Find the pair $(\vartheta, \chi) : Q \to I\!R^2$ satisfying

$$\partial_t(\vartheta + \lambda(\chi)) - \Delta\alpha(\vartheta) = g \quad \text{in } Q, \tag{1}$$

$$\mu\chi_t - \nu\Delta\chi + \beta(\chi) + \sigma'(\chi) \ni -\frac{\lambda'(\chi)}{\vartheta} \quad \text{in } Q, \tag{2}$$

$$-\partial_n\alpha(\vartheta) = \gamma\alpha(\vartheta) - h, \quad \partial_n\chi = 0 \quad \text{in } \Sigma := \Gamma \times (0,T), \tag{3}$$



*P. Argoul et al. (eds.),*
*IUTAM Symposium on Variations of Domains and Free-Boundary Problems in Solid Mechanics*, 181–188.

$$\vartheta(\cdot,0)=\vartheta_0,\quad \chi(\cdot,0)=\chi_0 \quad \text{in } \Omega, \tag{4}$$

with $\partial_t = \partial/\partial t$, $\Delta$ being the Laplacian, and $\partial_n$ denoting the outward normal derivative to $\Gamma$. Regarding the data, we have that

$$\mu,\ \nu,\ \gamma \text{ are positive constant coefficients,} \tag{5}$$

$$\lambda,\ \sigma \text{ are two } C^2 \text{ functions such that } \lambda',\ \lambda'',\ \sigma'' \in L^\infty(I\!R), \tag{6}$$

$$\alpha : (0,+\infty) \to I\!R \text{ is } C^1, \text{ strictly increasing, and concave,} \tag{7}$$

$$\beta : I\!R \to 2^{I\!R} \text{ is a maximal monotone graph with } 0 \in \beta(0), \tag{8}$$

$$g \in L^2(Q), \quad h \in L^2(\Sigma), \quad \vartheta_0 \in L^2(\Omega), \quad \chi_0 \in H^1(\Omega). \tag{9}$$

The system of partial differential equations (1–2) provides a quite general version of the phase–field model proposed by Penrose and Fife [9] for the kinetics of phase transitions. This model describes the evolution of a material exhibiting two different phases (e.g., solid–liquid in melting phenomena or solid–solid in martensitic tranformations) in terms of the absolute temperature $\vartheta : Q \to (0,+\infty)$ and of the order parameter $\chi : Q \to I\!R$ (representing, for instance, the local concentration of one of the phases). Within (1–4), it is actually considered the standard case when the order parameter is non–conserved, and significant examples of $\lambda$, $\beta$, and $\sigma$ are discussed, e.g., in [1, 5, 10] besides [9].

The term $-\Delta\alpha(\vartheta)$ in the energy balance (1) comes from applying the heat flux law $q = -\nabla\alpha(\vartheta)$. Now, let us observe that whatever $\alpha$ obeying (7) makes the system (1–2) consistent with the second principle of thermodynamics. Several papers have been devoted to the investigation of (1–4) or variations of it (see [2, 4–6, 10] and references therein) under the common position that $\alpha(\vartheta) = -k_0/\vartheta$ for some constant $k_0 > 0$ (henceforth $k_i$, $i \in I\!N$, denote positive constant coefficients). It is plain that this assumption (quite unusual in the framework of heat conduction) offers some advantages, due to the presence of the factor $1/\vartheta$ in the right hand side of (2) and to the penalization introduced for the forbidden value $\vartheta = 0$. Taking instead the classical Fourier law (which corresponds to $\alpha(\vartheta) = k_1\vartheta$) causes serious difficulties to the analysis, as one can easily imagine, and the only existence result we know [7] is addressed to the particular situation of $\beta(\chi) = \chi^3$.

However, despite the possibility of carrying out a nice mathematical treatment, constitutive relations like $\alpha(\vartheta) = -k_0/\vartheta$ might look acceptable only for low and intermediate temperatures, while the behaviour for high temperatures is not satisfactory. Indeed, in this case no sort of coerciveness is furnished by $-\Delta\alpha(\vartheta)$ as $\vartheta$ becomes larger and larger. In order to avoid this drawback, some attempt has been done in [3], where the quoted law is complemented by a memory term in $\vartheta$, and especially in [1], in which existence of weak solutions is proved for a wide class of nonlinearities $\alpha$. Noticing that the framework of [3] allows only certain phase transitions processes due to the convexity property required for $\lambda$, let us comment on the approach of [1]. There, the expected conditions $\alpha(\vartheta) \searrow -\infty$ as $\vartheta \searrow 0$ and $\alpha(\vartheta) \nearrow +\infty$ as $\vartheta \nearrow +\infty$ are prescribed along with some restrictions on $\alpha'$, namely,

$$\alpha'(r) \sim k_2 r^{-2p} \text{ when } r \searrow 0, \qquad \alpha'(r) \sim k_3 r^{-2q} \text{ when } r \nearrow +\infty, \tag{10}$$

for two values $p \geq 1/2$ and $0 \leq q \leq 1/2$.

Now, it is clear that, among all the possible choices of $\alpha$, the most interesting ones are related to the case $p = 1$, $q = 0$ and find

$$\alpha(\vartheta) = -\frac{k_2}{\vartheta} + k_3\vartheta \quad \text{ or } \quad \alpha'(\vartheta) = \max\left\{k_2, k_3\vartheta^2\right\}\vartheta^{-2}$$

as rather natural examples. In fact, it is the purpose of the present paper to examine this reduced setting more in detail and complete the results of [1] by showing further regularity and uniqueness of the solution. Then, in addition to (7) we suppose that there exist two positive constants $d_0$, $d_\infty$ and a function $\ell \in C^1(0, +\infty)$ such that

$$-r^{-1} = d_0\alpha(r) + \ell(r) \quad \forall\, r > 0, \quad \ell' \in L^\infty(0, +\infty), \tag{11}$$

$$\lim_{r \nearrow +\infty} d_\infty \alpha'(r) = 1. \tag{12}$$

Note that, letting $p = 1$ and $q = 0$, (11) implies the first of (10) for $k_2 = 1/d_0$ and (12) coincides with the second of (10) if $k_3 = 1/d_\infty$. As a consequence of (7) and (11–12), $\alpha'$ is decreasing and bounded from below by $1/d_\infty$. Hence, indicating by $\rho$ the inverse function of $\alpha$, it turns out that $\rho$ is a $C^1$ strictly increasing, convex, Lipschitz continuous function defined in all of $I\!R$.

By readily exploiting the conclusions of [1], in this note we obtain the existence and uniqueness theorem listed below, whose proof is developed in the next Sections. More precisely, Section 2 deals with uniqueness, while Section 3 is devoted to collect the necessary a priori estimates, recalling those of [1] and deriving the others.

For the sake of convenience, we set $V := H^1(\Omega)$, identify $H := L^2(\Omega)$ with its dual space $H'$, and note that $V \subset H \subset V'$ with dense and compact injections. Also, let $(\cdot, \cdot)$ stand both for the duality pairing between $V'$ and $V$ and for the inner product in $H$. Finally, let us specify the bilinear form

$$a(v_1, v_2) := \int_\Omega \nabla v_1 \cdot \nabla v_2 + \gamma \int_\Gamma v_1 v_2, \qquad v_1, v_2 \in H^1(\Omega),$$

(compare (1) and the first boundary condition in (3) with the subsequent (20)) and come to the desired statement.

**Theorem 1** *Assume that (5–9), (11–12), and*

$$g \in W^{1,1}(0,T;L^2(\Omega)), \qquad h \in W^{1,1}(0,T;L^2(\Gamma)), \tag{13}$$

$$\vartheta_0 > 0 \quad a.e.\ in\ \Omega, \qquad u_0 := \alpha(\vartheta_0) \in H^1(\Omega), \tag{14}$$

$$\chi_0 \in H^2(\Omega), \qquad \partial_n \chi_0 = 0 \quad a.e.\ in\ \Gamma, \tag{15}$$

$$\chi_0 \in D(\beta) \quad a.e.\ in\ \Omega, \qquad \beta^0(\chi_0) \in L^2(\Omega) \tag{16}$$

*hold, where $D(\beta) \subseteq I\!R$ is the domain of $\beta$ and the single–valued graph $\beta^0$ represents the minimal section of $\beta$. Then there exists a unique pair $(\vartheta, \chi)$ satisfying*

$$\vartheta \in W^{1,\infty}(0,T;V') \cap C^0([0,T];H) \cap L^2(0,T;V), \tag{17}$$

$$\vartheta > 0 \quad a.e.\ in\ Q, \qquad u := \alpha(\vartheta) \in L^\infty(0,T;V), \tag{18}$$

$$\chi \in W^{1,\infty}(0,T;H) \cap H^1(0,T;V) \cap L^\infty(0,T;H^2(\Omega)), \tag{19}$$

$$((\vartheta + \lambda(\chi))_t\,(t), v) + a(u(t), v) = \int_\Omega g(t)v + \int_\Gamma h(t)v \qquad \forall\, v \in V, \quad a.e.\ in\ (0,T), \tag{20}$$

$$\chi \in D(\beta) \quad a.e.\ in\ Q, \qquad \partial_n \chi = 0 \quad a.e.\ in\ \Sigma, \tag{21}$$

$$\mu \chi_t - \nu \Delta \chi + \xi + \sigma'(\chi) = -\,\lambda'(\chi)\vartheta^{-1} \quad a.e.\ in\ Q, \tag{22}$$

$$for\ some\ \ \xi \in L^\infty(0,T;H)\ \ such\ that\ \ \xi \in \beta(\chi) \quad a.e.\ in\ Q, \tag{23}$$

$$\vartheta(0) = \vartheta_0, \quad \chi(0) = \chi_0 \quad in\ H. \tag{24}$$

## 2. Uniqueness

We start by inferring one property of the positive function $\rho = \alpha^{-1}$, besides those already mentioned in the Introduction. In view of (7) and (11–12), we remark that

$$d_1 \left(1 + r^2\right)^{-1} \le \rho'(r) \le d_\infty \qquad \forall\, r \in I\!R, \tag{25}$$

for some constant $d_1 > 0$ fulfilling $d_1 \alpha' \le 1 + |\alpha|^2$ in $(0, +\infty)$. Indeed, the existence of such a constant $d_1$ is ensured by the continuity and boundedness of the function $\alpha'(r)/\left(1 + |\alpha(r)|^2\right)$, $r \in (0, +\infty)$ (check the limits as $r \searrow 0$ and $r \nearrow +\infty$).

Let now $(\vartheta_i, u_i, \chi_i, \xi_i)$, $i = 1, 2$, be two quadruples fulfilling (17–24), that is, $(\vartheta_1, \chi_1)$ and $(\vartheta_2, \chi_2)$ are two solutions of problem (1–4) in the sense of Theorem 1. First, we take the difference of the equalities (20), integrate from 0 to $s \in (0,T)$, and choose $v = (u_1 - u_2)(s)$. Thanks to (18) and (6), we have that

$$\int_\Omega (\rho(u_1) - \rho(u_2))(s) U(s) + a\Big(\int_0^s U(t)dt, U(s)\Big) \le \int_\Omega \|\lambda'\|_{L^\infty(I\!R)}\, |\mathcal{X}(s)|\, |U(s)| \tag{26}$$

where $U := u_1 - u_2$ and $\mathcal{X} := \chi_1 - \chi_2$. Then, we subtract the two equations (22) and test by $\mathcal{X}$. Observe that $(\xi_1 - \xi_2)(\chi_1 - \chi_2) \geq 0$ almost everywhere in $Q$ because of (23) and (8). Since (cf. (11)) $-\vartheta_i^{-1} = d_0 u_i + \ell(\rho(u_i))$ for $i = 1, 2$, from (21), (6), (11), and (25) we deduce that

$$\mu(\mathcal{X}_t(s), \mathcal{X}(s)) + \nu \int_\Omega |\nabla \mathcal{X}(s)|^2 \leq L \int_\Omega (|\mathcal{X}(s)| + (d_0 + L d_\infty)|U(s)|)\, |\mathcal{X}(s)|$$
$$+ \int_\Omega L(d_0 |u_2(s)| + |\ell(\rho(0))| + L d_\infty |u_2(s)|) |\mathcal{X}(s)|^2 \tag{27}$$

for a.a. $s \in (0, T)$, with $L$ denoting the maximum among the Lipschitz constants of $\sigma'$, $\lambda$, $\lambda'$, and $\ell$. Next, we add (27) to (26) and integrate over $(0, t)$, with $t \in (0, T]$. Noting that the form $a(\cdot, \cdot)$ induces a norm on $V$ and applying the mean value theorem along with (25), it is straightforward to determine two positive constants $C_1$ and $C_2$ such that

$$\int_0^t \int_\Omega \frac{d_1 |U|^2}{1 + |u_1|^2 + |u_2|^2} + C_1 \left\| \int_0^t U(s) ds \right\|_V^2 + \frac{\mu}{2} \|\mathcal{X}(t)\|_H^2$$
$$+ \nu \int_0^t \int_\Omega |\nabla \mathcal{X}|^2 \leq C_2 \int_0^t \int_\Omega ((1 + |u_2|)|\mathcal{X}|^2 + |U|\, |\mathcal{X}|)\,. \tag{28}$$

for any $t \in [0, T]$. On account of (18), the Hölder inequality and the continuous embedding $H^1(\Omega) \subset L^4(\Omega)$ yield

$$\int_0^t \int_\Omega |u_2|\, |\mathcal{X}|^2 \leq C_3 \|u_2\|_{L^\infty(0,T;V)} \int_0^t \|\mathcal{X}(s)\|_V \|\mathcal{X}(s)\|_H \, ds,$$

so that the elementary Young inequality enables us to achieve

$$C_2 \int_0^t \int_\Omega (1 + |u_2|)|\mathcal{X}|^2 \leq \frac{\nu}{4} \int_0^t \int_\Omega |\nabla \mathcal{X}|^2$$
$$+ C_4 \left(1 + \|u_2\|_{L^\infty(0,T;V)}^2\right) \int_0^t \|\mathcal{X}(s)\|_H^2 \, ds, \tag{29}$$

where the constants $C_3$ and $C_4$ depend only on $\Omega$, $C_2$, and $\nu$. Moreover, as the space dimension $N$ is not larger than 3, we can exploit the compactness of the injection $H^1(\Omega) \subset L^4(\Omega)$ and apply also a well–known inequality (see, e.g., [8, Thm. 16.4, p. 102]) to estimate

$$C_2 \int_0^t \int_\Omega |U|\, |\mathcal{X}| \leq C_2 \int_0^t \int_\Omega \frac{|U|}{\sqrt{1 + |u_1|^2 + |u_2|^2}} \sqrt{1 + |u_1|^2 + |u_2|^2}\, |\mathcal{X}|$$
$$\leq \frac{1}{2} \int_0^t \int_\Omega \frac{d_1 |U|^2}{1 + |u_1|^2 + |u_2|^2} + C_5 \int_0^t \|\mathcal{X}(s)\|_H^2 \, ds$$
$$+ C_5 \sum_{i=1}^2 \|u_i\|_{L^\infty(0,T;L^4(\Omega))}^2 \int_0^t \|\mathcal{X}(s)\|_{L^4(\Omega)}^2 \, ds$$

$$\leq \frac{1}{2}\int_0^t\int_\Omega \frac{d_1|U|^2}{1+|u_1|^2+|u_2|^2} + \frac{\nu}{4}\int_0^t\int_\Omega |\nabla \mathcal{X}|^2$$
$$+ C_6\left(1+\sum_{i=1}^{2}\|u_i\|^2_{L^\infty(0,T;V)}\right)\int_0^t \|\mathcal{X}(s)\|_H^2\, ds \tag{30}$$

for two other constants $C_5$, $C_6$. At this point, combining (28–30) and then applying the Gronwall lemma, we find that $|U| = 0$ a.e. in $Q$ and $\mathcal{X}(t) = 0$ for all $t \in [0,T]$. On account of (20) and (22), this entails that the two quadruples $(\vartheta_1, u_1, \chi_1, \xi_1)$ and $(\vartheta_2, u_2, \chi_2, \xi_2)$ must necessarily coincide.

## 3. Existence

The following preparatory result is a consequence of the analysis carried out in [1] (cf. especially Theorem 2.3 and Remark 5.6).

**Lemma 2** *Under the assumptions (5–9), (11–12), (14), and (16), there exist* $\vartheta \in H^1(0,T;V')\cap L^\infty(0,T;H)\cap L^2(0,T;V)$, $u := \alpha(\vartheta) \in L^2(0,T;V)$, $\chi \in H^1(0,T;H) \cap L^\infty(0,T;V) \cap L^2(0,T;H^2(\Omega))$ *satisfying (20–22) and (24), for* $\xi \in L^2(Q)$ *such that* $\xi \in \beta(\chi)$ *a.e. in* $\Omega$.

Therefore, to complete the proof of Theorem 1 it remains to verify the additional regularity stated in (17–19). For this aim, it suffices to derive further uniform bounds on the solution of a suitable approximating problem. The main procedure consists in taking $v = u_t$ in (20) and multiplying the time derivative of (22) by $\chi_t$. Unfortunately, these steps cannot be performed on the regularization scheme proposed in [1]. However, in the setting of this note (cf. (11)), we can rewrite (22–23) as

$$\mu\chi_t - \nu\Delta\chi + \beta(\chi) \ni -\sigma'(\chi) + \lambda'(\chi)(d_0 u + \ell(\rho(u))) \quad \text{a.e. in } Q, \tag{31}$$

so that here a possible approximation could simply be given by inserting the term $\varepsilon u_t$ in the left hand side of (1) (to get $u_t$ and $-\Delta u$ both in $L^2(Q)$) and by substituting $\beta$ with its Yosida regularization $\beta_\varepsilon$ in (31) (to have the component $\chi$ smooth). Proving that the approximating problem admits one and only one (cf. Section 2) solution looks quite standard, and then one can pass to the limit as $\varepsilon \searrow 0$ like in [5], for instance.

That having been said, let us argue directly on (20) and (31) assuming $\beta$ Lipschitz continuous (whence $\xi = \beta(\chi)$) and paying attention to avoid the use of $\|\beta'\|_{L^\infty(I\!R)}$ (obviously $\beta'$ denotes the derivative of $\beta$) in the formal estimates. Moreover, it is understood that $\chi_t(0) = z_0$ with

$$z_0 := \Delta\chi_0 - \beta^0(\chi_0) - \sigma'(\chi_0) + \lambda'(\chi_0)(d_0 u_0 + \ell(\vartheta_0)) \in H$$

because of (31), (24), (14–16), (6), and (11) (we remark that $\beta_\varepsilon(y)$ converges to $\beta^0(y)$ as $\varepsilon \searrow 0$ whenever $y \in D(\beta)$). In short, the precise computation we do is $d_0 \left(\int_0^t (20)|_{v=u_t}\right) + \int_0^t \int_\Omega (31)_t \times \chi_t$. Thanks to (13), (6), (11), and (25), arguing as in (27–28) and observing that $\beta' \geq 0$, one sees there is a cancellation and manages to infer that

$$
\begin{aligned}
d_0 \int_0^t \int_\Omega \rho'(u)|u_t|^2 &+ d_0 C_1 \|u(t)\|_V^2 + \frac{\mu}{2} \|\chi_t(t)\|_H^2 \\
+ \nu \int_0^t \int_\Omega |\nabla \chi_t|^2 &\leq d_0 a(u_0, u_0) + \frac{\mu}{2} \|z_0\|_H^2 \\
&+ C_7 \max_{0 \leq s \leq T} \left(\|g(s)\|_H + \|h(s)\|_{L^2(\Gamma)}\right)\left(\|u(t)\|_V + \|u_0\|_V\right) \\
&+ \frac{C_7}{2} \int_0^t \left(\|g_t(s)\|_H + \|h_t(s)\|_{L^2(\Gamma)}\right)\left(1 + \|u(s)\|_V^2\right) ds \\
&+ C_8 \int_0^t \int_\Omega (1 + |u|)\, |\chi_t|^2 + L^2 \int_0^t \int_\Omega \rho'(u)|u_t|\, |\chi_t| \qquad (32)
\end{aligned}
$$

for $t \in (0, T)$, where the constant $C_7$ depends only on $\Omega$, and $C_8$ relies on $L$, $d_0$, $|\ell(\rho(0))|$, and $d_\infty$. Now, examine the right hand side of (32). The first term containing $C_7$ can be treated via the Young inequality, which is again helpful, along with (25), to obtain

$$
L^2 \int_0^t \int_\Omega \rho'(u)|u_t|\, |\chi_t| \leq \frac{d_0}{2} \int_0^t \int_\Omega \rho'(u)|u_t|^2 + \frac{d_\infty L^4}{2d_0} \int_0^t \int_\Omega |\chi_t|^2. \qquad (33)
$$

Recalling that $H^1(\Omega)$ is compactly embedded into $L^4(\Omega)$, due to the Hölder inequality and [8, formula (16.16), p. 102] there is a constant $C_9$ such that

$$
\begin{aligned}
C_8 \int_0^t \int_\Omega |u|\, |\chi_t|^2 &\leq \int_0^t \|u(s)\|_{L^4(\Omega)} \|\chi_t(s)\|_H \|\chi_t(s)\|_{L^4(\Omega)}\, ds \\
&\leq \frac{\nu}{2} \int_0^t \int_\Omega |\nabla \chi_t|^2 + C_9 \int_0^t \left(1 + \|u(s)\|_V^2\right) \|\chi_t(s)\|_H^2\, ds. \qquad (34)
\end{aligned}
$$

Then, we can collect (32–34) and remark that $\|\chi_t\|_{L^2(Q)}$ and $\|u\|_{L^2(0,T;V)}$ are already under control, by virtue of Lemma 2. Thus, in view of (13–14), if we set

$$
S(t) := d_0 \int_0^t \int_\Omega \rho'(u)|u_t|^2 + d_0 C_1 \|u(t)\|_V^2 + \frac{\mu}{2} \|\chi_t(t)\|_H^2 + \nu \int_0^t \int_\Omega |\nabla \chi_t|^2,
$$

it turns out that $(1/2)S(t) \leq C_{10} + \int_0^t R(s)S(s)ds$ for some constant $C_{10}$ and some known function $R \in L^1(0, T)$. Hence, the Gronwall lemma allows us to conclude that (18) holds and that $\chi_t \in L^\infty(0, T; H) \cap L^2(0, T; V)$ (this would mean that the related approximating sequences are bounded in the respective norms!). Moreover, (19) can be recovered by applying a monotonicity argument. Indeed, on account of (18), (6), (11), and (25), the inclusion (31) entails that $-\Delta\chi + \beta(\chi)$ equals an element $f$ of $L^\infty(0,T;H)$. Therefore, multiplying by $-\Delta\chi$ and integrating only in space, by (21) we get

$$\frac{1}{2}\,\|\Delta\chi(t)\|_H + \int_\Omega \beta'(\chi)|\nabla\chi|^2 \le \frac{1}{2}\,\|f\|_{L^\infty(0,T;H)}$$

for a.a. $t \in (0,T)$, so that the classical regularity theory for elliptic problems (refer to [8]) yields $\chi \in L^\infty(0,T;H^2(\Omega))$. Finally, (17) follows from a comparison in (20) for we have that (and this was not stated in Lemma 2) $\vartheta \in C^0([0,T];H)$ by interpolation (cf, e.g., [8, Thm. 3.1, p. 19]).

**Remark 3** *An open question concerns the possibility of deriving $L^\infty$ estimates for u. Such estimates would immediately imply that also $\vartheta$ and $1/\vartheta$ are uniformly bounded almost everywhere in Q and would open the way to other regularity results. A promising technique, already employed successfully in [10, 6, 4], consists in implementing suitable Moser iteration procedures.*

## References

1. Colli, P. and Laurençot, Ph. (1998) Weak solutions to the Penrose–Fife phase field model for a class of admissible heat flux laws, *Physica D*, **Vol. no. 111**, pp. 311–334.
2. Colli, P. and Sprekels, J. (1995) On a Penrose–Fife model with zero interfacial energy leading to a phase–field system of relaxed Stefan type, *Ann. Mat. Pura Appl. (4)*, **Vol. no. 169**, pp. 269–289.
3. Colli, P. and Sprekels, J. (1998) Weak solution to some Penrose–Fife phase–field systems with temperature–dependent memory, *J. Differential Equations*, **Vol. no. 142**, pp. 54–77.
4. Horn, W., Laurençot, Ph. and Sprekels, J. (1996) Global solutions to a Penrose–Fife phase–field model under flux boundary conditions for the inverse temperature, *Math. Methods Appl. Sci.*, **Vol. no. 19**, pp. 1053–1072.
5. Kenmochi, N. and Niezgódka, M. (1993/94) Systems of nonlinear parabolic equations for phase change problems, *Adv. Math. Sci. Appl.*, **Vol. no. 3**, pp. 89–117.
6. Laurençot, Ph. (1994) Solutions to a Penrose–Fife model of phase–field type, *J. Math. Anal. Appl.*, **Vol. no. 185**, pp. 262–274.
7. Laurençot, Ph. (1998) Weak solutions to a Penrose–Fife model with Fourier law for the temperature, *J. Math. Anal. Appl.*, **Vol. no. 219**, pp. 331–343.
8. Lions, J.L. and Magenes, E. (1972) *Non-homogeneous boundary value problems and applications, Vol. I*. Springer–Verlag, Berlin.
9. Penrose, O. and Fife, P.C. (1990) Thermodynamically consistent models of phase field type for the kinetics of phase transitions, *Physica D*, **Vol. no. 43**, pp. 44–62.
10. Sprekels, J. and Zheng, S. (1993) Global smooth solutions to a thermodynamically consistent model of phase–field type in higher space dimensions, *J. Math. Anal. Appl.*, **Vol. no. 176**, pp. 200–223.

# DAMAGE AND GRADIENT OF DAMAGE IN TRANSIENT DYNAMICS

B. NEDJAR
*Laboratoire de Mécanique et Technologie,*
*ENS Cachan/CNRS/Université Paris VI,*
*61, avenue du Président Wilson, 94235, Cachan, France.*

**Abstract.** This paper is concerned with the modelling of progressive degradation in transient dynamics. Within the framework of the continuum damage mechanics, a model involving the gradient of damage is investigated. This model together with the equations of motion in dynamics are issued from a new formulation of the principle of virtual power recently proposed.

## 1. Introduction

Continuum damage theories are based on the thermodynamics of irreversible processes where damage is described by internal quantities. Isotropic damage formulations are extensively employed in the literature because of their simplicity, efficiency and adequacy for many practical applications.

For transient dynamic problems, the consequences of shocks and impacts on structures made with ductile or brittle materials, such as concrete, are of engineering importance and then need to be analysed. This is the aim of this paper where we use a model and a formalism of the equations of motion issued from a new formulation of the principle of virtual power recently proposed in [3, 4].

Damage in a solid results from microscopic movements. In this theory, it is decided to take into account the power of these microscopic movements in the power of the internal forces. This later is chosen to depend, besides on the strain rates, also on the damage rate which is clearly related to the microscopic movements. Furthermore, it is assumed that it depends also on the gradient of the damage rate to account for the microscopic interactions between the material points.



*P. Argoul et al. (eds.),*
*IUTAM Symposium on Variations of Domains and Free-Boundary Problems in Solid Mechanics,* 189–196.

The remainder of this paper is organized as follows: first, a brief description of the damage theory we use is given. Then a rate independent/rate dependent damage model is outlined. Next, the numerical implementation within the finite element formulation is discussed and a numerical example is given.

## 2. Equations of motion and constitutive laws

In this section, the formulation of the damage theory we use is recalled. For a complete treatment, we refer to [3, 4, 7].

Let the scalar $\beta(\mathbf{x}, t)$ be a damage quantity with value 1 when the material is undamaged and value 0 when it is completely damaged. For a domain $\Omega$, we choose the power of the internal forces $P_i$ as:

$$P_i(\Omega, \mathbf{u}, \frac{d\beta}{dt}) = -\int_\Omega \sigma : \mathbf{D}(\mathbf{u}) d\Omega - \int_\Omega (B\frac{d\beta}{dt} + \mathbf{H}.\mathbf{grad}\frac{d\beta}{dt}) d\Omega \tag{1}$$

where $\sigma$ is the stress tensor and $\mathbf{D}(\mathbf{u})$ the strain rates tensor ($\mathbf{u}$ is the macroscopic velocity). The two non-classical quantities are: $B$, the internal work of damage and $\mathbf{H}$, the flux vector of internal work of damage.

We choose the power of the acceleration forces $P_a$ as [3, 4]:

$$P_a(\Omega, \mathbf{u}, \frac{d\beta}{dt}) = \int_\Omega \rho \frac{d\mathbf{u}}{dt}.\mathbf{u} d\Omega + \int_\Omega \underline{\rho} \frac{d^2\beta}{dt^2}\frac{d\beta}{dt} d\Omega \tag{2}$$

where $\rho$ is the density. The quantity $\underline{\rho}\frac{d^2\beta}{dt^2}$ stands for the acceleration forces of the microscopic links; $\underline{\rho}$ is proportional to their mass. For the applications, we take $\underline{\rho} = \alpha\rho$, where $\alpha$ is a material parameter which can be zero valued if we decide to neglect the inertia effects of the microscopic movements.

The principle of virtual power gives two sets of movement equations:

$$div\sigma + \mathbf{f} = \rho\frac{d\mathbf{u}}{dt} \; in \; \Omega \qquad \sigma.\mathbf{n} = \mathbf{F} \; on \; \partial\Omega \tag{3}$$

$$div\mathbf{H} - B = \alpha\rho\frac{d^2\beta}{dt^2} \; in \; \Omega \qquad \mathbf{H}.\mathbf{n} = 0 \; on \; \partial\Omega \tag{4}$$

where *div* is the divengence operator and $\mathbf{n}$ is the outward normal unit vector to $\Omega$. $\mathbf{f}$ and $\mathbf{F}$ are the volumetric and surfacic external forces, respectively. Equations (4) are new and non-classical. They describe the damage, *i.e.* the microscopic movements in $\Omega$.

Within the framework of continuum thermodynamics [5], the state of the material is characterized by its free energy $\Psi = \Psi(\epsilon, \beta, \mathbf{grad}\beta)$, where $\epsilon$ is the strain tensor.

As in [5], we postulate the existence of a pseudo-potential of dissipation $\Phi$, and we assume that there are only dissipative phenomena with respect to the damage. The constitutive laws are then given by, [3, 4]:

$$\sigma = \frac{\partial \Psi}{\partial \epsilon} \qquad \mathbf{H} = \frac{\partial \Psi}{\partial (\mathbf{grad}\beta)} \qquad B = \frac{\partial \Psi}{\partial \beta} + \frac{\partial \Phi}{\partial \dot{\beta}} \tag{5}$$

where $(\dot{})$ denotes the time derivative.

## 3. Outlines of a rate independent/rate dependent model

It is mainly observed that damage within a solid is produced by the extensions into the material when a certain threshold is achieved. Such a situation is observed in concrete. Another characteristic of this material is its softening behaviour, *i.e.* a gradual reduction of the load carrying capacity with increasing deformation.

We propose in this section a rate independent / rate dependent damage model (a variant of those proposed in [4, 7]). The choices for $\Psi$ and $\Phi$ are given by:

$$\Psi = \frac{1}{2}\beta\epsilon : \mathbf{C} : \epsilon \; + \; \frac{k}{2}(\mathbf{grad}\beta)^2 \; + \; I(\beta) \tag{6}$$

$$\begin{aligned}\Phi = \frac{1}{2}c\dot{\beta}^2 - W\dot{\beta} - \frac{1}{2}\dot{\beta}\, \{2\mu\epsilon^- : \epsilon^- + \lambda((tr[\epsilon])^-)^2 \\ +(1-\beta^n)[2\mu\epsilon^+ : \epsilon^+ + \lambda((tr[\epsilon])^+)^2]\} + I_-(\dot{\beta})\end{aligned} \tag{7}$$

where the $tr$ denotes the trace operator and : the double contracted product.

The first term of $\Psi$ constitutes the simplest model where damage is coupled to elasticity. $\mathbf{C}$ is the symmetric rank four elastic tensor at the undamaged state ($\lambda$ and $\mu$ are the Lamé parameters). In the second term, $k$ is a material parameter which measures the influence of damage at a material point on the damage of its neighbourhood.

In the expression of $\Phi$, the state variables $\epsilon$ and $\beta$ are considered as parameters. The quantity $W$ is the initial damage threshold expressed in terms of volumetric energy and the exponent $n$ is a material parameter that controls the softening behaviour of the material, with $n \geq 0$. $c$ is the viscosity parameter of damage. Its influence is shown in [7].

In the literature, several models introduce rate dependent constitutive relations. Besides on the physical motivations, the regularizing role of viscosity avoids the ill-posedness of the boundary value problem due to strain softening, as shown in [1, 2, 8].

However, rate independent damage evolution can also be considered. In fact, one can check that if $c = 0$, the function $\Phi$ is positively homogeneous of degree one.

The functions $\langle . \rangle^+$ and $\langle . \rangle^-$ are respectively the positive part and the negative part of the scalar $\langle . \rangle$. The positive part $\epsilon^+$ and the negative part $\epsilon^-$ of the strain tensor are obtained after diagonalisation. The functions $I$ and $I_-$ are the indicator functions of the intervals $[0, 1]$ and $]-\infty, 0]$, respectively. With these functions, the free energy and the pseudo-potential of dissipation have their physical values for any actual or physical value of $\beta$ and $\dot{\beta}$ [3, 4].

In the domain $\Omega$, the equations of motion are obtained by using (6), (7) and (5) into (3) and (4). We get then:

$$\begin{aligned} \rho\ddot{\mathbf{u}}(t) - div(\beta\mathbf{C} : \epsilon) = \mathbf{f}(t) \quad & \text{in } \Omega \\ \sigma.\mathbf{n} = \mathbf{F}(t) \quad & \text{on } \partial\Omega \end{aligned} \tag{8}$$

$$\begin{aligned} \alpha\rho\ddot{\beta} + c\dot{\beta} - k\Delta\beta + \partial I(\beta) + \partial I_-(\dot{\beta}) \ni W - \beta^n S(\epsilon) \quad & \text{in } \Omega \\ k\frac{\partial\beta}{\partial\mathbf{n}} = 0 \quad & \text{on } \partial\Omega \end{aligned} \tag{9}$$

where $\Delta$ is the Laplacian and where we have used the notation,

$$S(\epsilon) = \frac{1}{2}[2\mu\epsilon^+ : \epsilon^+ + \lambda((\langle tr[\epsilon]\rangle^+)^2]$$

This coupled problem is solved by taking into account the intial conditions for the displacements and damage fields. Eqn.(9) describes the evolution of damage. Its right hand side is the source of damage governed by extensions. The sub-differentials $\partial I$ and $\partial I_-$ are reactions which force $\beta$ to remain between 0 and 1 and $\dot{\beta}$ to be negative.

## 4. Finite element formulation and resolution algorithm

The variational formulations of the local forms of the governing equations play a central role in the numerical resolution of the coupled problem at hand.

Let us note that if the indicator functions $I$ and $I_-$ are approximated by smooth functions, the sub-differentials $\partial I$ and $\partial I_-$ are approximated by classical derivatives. However, these approximations will not be done and their representations will be forced.

The two governig equations (8) and (9) are equivalent to:

$$\forall\eta, \int_\Omega \eta\rho\ddot{\mathbf{u}}d\Omega + \int_\Omega \epsilon(\eta) : \beta\mathbf{C} : \epsilon(\mathbf{u})d\Omega = \int_\Omega \mathbf{f}.\eta d\Omega + \int_{\partial\Omega} \mathbf{F}.\eta d\Gamma \tag{10}$$

$$\begin{aligned} \forall\psi, \qquad & \int_\Omega \psi\alpha\rho\ddot{\beta}d\Omega \quad + \quad \int_\Omega \psi c\dot{\beta}d\Omega \\ & \int_\Omega k\mathbf{grad}\psi.\mathbf{grad}\beta d\Omega \quad + \quad \int_\Omega \psi S(\epsilon)\beta^n d\Omega = \int_\Omega W\psi d\Omega \end{aligned} \tag{11}$$

where $\eta$ and $\psi$ are admissible variations of displacements and damage.

The development of an algorithm consistent with the coupled problem is given as follows. As the damage is considered as a variation of the mechanical characteristics, at each global iteration, the stiffness matrix of each element is recomputed. Thus, for a given intermediate state of damage $\beta^i$, the matrix form of eqn.(10) is given by:

$$\mathbf{M}.\ddot{\mathbf{U}}(t) + \mathbf{K}(\beta^i).\mathbf{U}(t) = \mathbf{F}(t) \tag{12}$$

where $\mathbf{M}$ is the mass matrix, $\mathbf{K}(\beta^i)$ is the global secant stiffness, $\mathbf{U}$ is the vector of nodal displacements and $\mathbf{F}$ is the vector of the applied external forces.

In the other hand, given an intermediate state of deformation, the matrix representation of eqn.(11) is given by:

$$\mathbf{m}.\ddot{\beta}(t) + \mathbf{c}.\dot{\beta}(t) + \mathbf{k}.\beta(t) + \mathbf{P}(\beta(t), \mathbf{U}(t)) = \mathbf{f} \tag{13}$$

where $\mathbf{m}$ is the mass matrix related to the internal links, $\mathbf{c}$ is the viscosity matrix of damage, $\mathbf{k}$ is the matrix of influence of damage, $\mathbf{P}(\beta, \mathbf{U})$ is the nonlinear term function of the state of deformation and $\beta$ is the vector of nodal values of damage.

### 4.1. TIME-STEPPING SCHEME - TEMPORAL DISCRETISATION

The finite difference scheme used for the time discretisation of the mechanical part, eqn.(12), is the implicit Newmark scheme. For the damage evolution, eqn.(13), the time discretisation depends on whether or not the inertia effects of the microscopic movements are neglected. If so ($\alpha = 0$), the time stepping uses a $\theta$-method scheme (implicit or semi- implicit), else, if they are not neglected ($\alpha \neq 0$), it uses the implicit (or explicit) Newmark scheme. However, if there is no viscosity and no inertia effects ($\alpha = 0$ and $c = 0$), the damage evolution is instantaneous and then no time stepping is needed.

### 4.2. COUPLED PROBLEM AND ALGORITHM

Starting from an equlibrated states of deformation and damage at time $t_m$: $\mathbf{U}_m$ and $\beta_m$, the proposed algorithm to solve for the new time step $t_{m+1}$ is summarized as follows:

1. The solution $\beta_m$ is taken as a first approximation: $\beta^0_{m+1} = \beta_m$,
2. Compute $\mathbf{K}(\beta^i_{m+1})$, $\mathbf{M}$ and $\mathbf{F}_{m+1}$,
3. Solve for the displacements, eqn.(12):$\mathbf{U}^{i+1}_{m+1}$ ,
4. Solve for the damage field, eqn.(13): $\beta^{i+1}_{m+1}$,
5. Check for the convergence of the coupled problem by means of

classical residual forces: $\mathbf{R}_{m+1}^{i+1}$,
if $||\mathbf{R}_{m+1}^{i+1}|| >$ *tolerance* go to step 2,
else, the convergence is obtained for $\mathbf{U}$ and $\beta$, go to the next time step.

## 5. A numerical example

The algorithmic treatment outlined above has been implemented in an extended version of the CESAR-LCPC finite element code [6].

A bidimensional analysis, with the plane strain assumption, of a strain softening bar problem in tension is considered. The straight bar is $200mm$ long with a cross section A of 5x5$mm$ ($Fig$.1). This analysis has been accomplished by assuming a *rate independent* constitutive relation ($c = 0$), and negligeable inertia effects for the damage evolution ($\alpha = 0$).

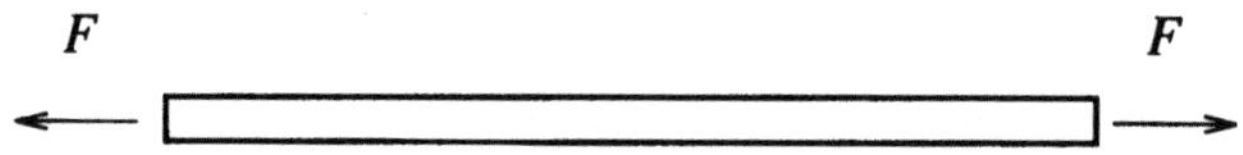


*Figure 1.* Strain softening bar in tension.

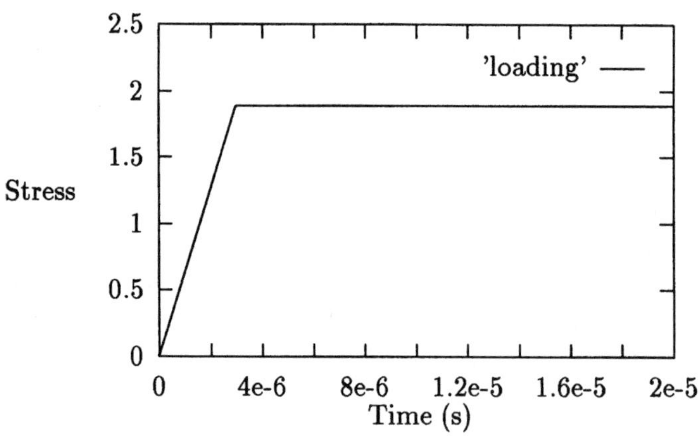


*Figure 2.* Loading history.

The material characteristics used for the model are: $E = 35000MPa$, $\nu = .18$, $W = 6.10^{-5}MPa$, $n = 2/3$, $k = 0.01MPa.mm^2$ and the mass density $\rho = 2400Kg/m^3$.

With these characteristics, the stress-strain behaviour in tension performed in statics is shown in $Fig$.3. The tensile strenght is $f_t = 2.1MPa$.

The bar is set into motion by a step load such that $F = 0.9(Af_t)$ at both ends ($Fig$.2). The computation has been performed with 200 linear

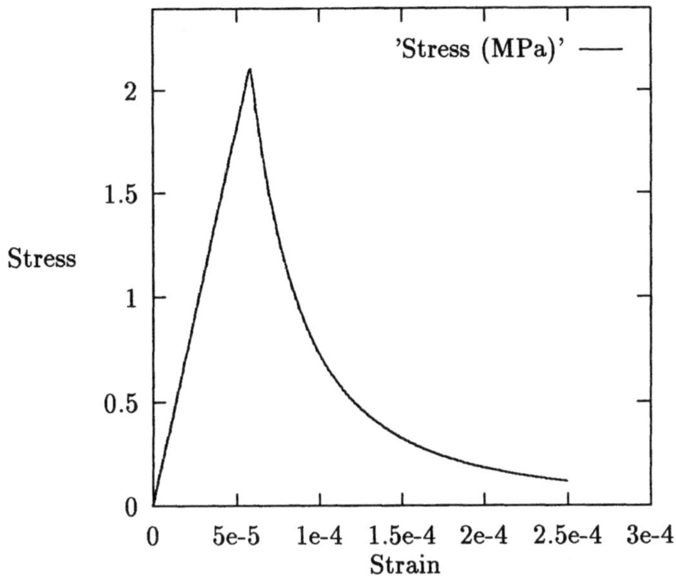


*Figure 3.* Stress-Strain behaviour in tension.

quadrilateral finite elements along the bar on two rows, and a time step $\Delta t = 2.10^{-7}s$ was used for the implicit time discretisation.

The stress waves produced by the step-loads at each end are elastic until they meet in the middle of the bar. At this time, the combined stress level exceeds the tensile strenght, damage develops and a strain softening process is initiated with the emergence of a localization zone. $Fig.4$ shows the damage profiles at different times along the direction of the bar: $\beta(x,t)$. The bar is broken after 3.6 $10^{-5}s$ because $\beta$ reaches 0 at the middle.

## 6. Conclusion

Within the framework of continuum damage mechanics, a model involving the gradient of damage has been investigated in transient dynamics. The equations of motion are issued from a new formulation of the principle of virtual power recently proposed, and the damage model follows the classical continuum thermodynamics.

The finite element formulation of the coupled nonlinear problem has been briefly outlined and a numerical example was given. The damage model proposed can describe both rate dependent and rate independent damage evolutions. The algorithmic treatment of this later does not need the addition of an artificial viscosity for computational purposes.

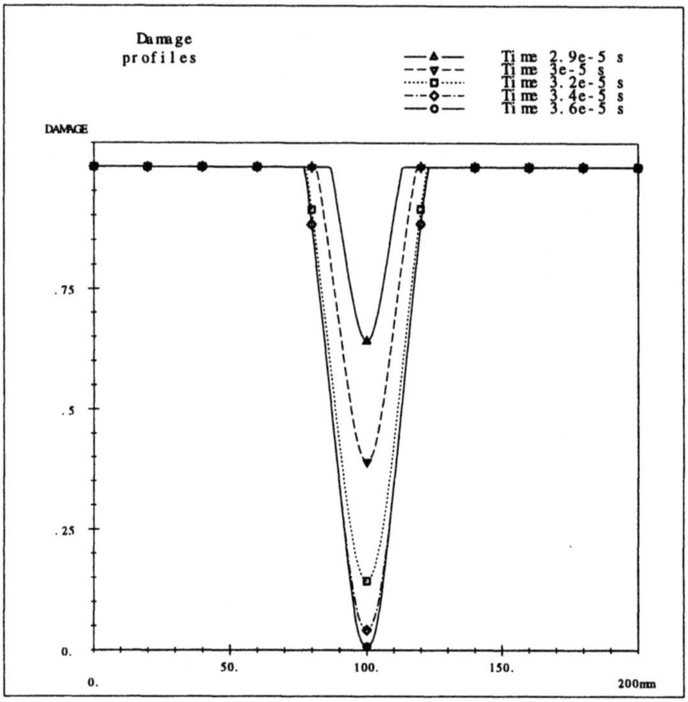


*Figure 4.* Damage profiles at different times.

## 7. Acknowledgements

We thank Professor M. Frémond for his many helpful discussions.

## References

1. Comi, C and Perego, U. (1997). On visco-damage models for concrete at high strain rates. *in Proc. of Computational Plasticity*, D.J.R. Owen, E. Oñate and E. Hinton (eds.), Barcelona, Pineridge Press, 1551-1555.
2. Dubé, J.F., Pijaudier-Cabot, G. and La Borderie, C. (1994). Rate dependent damage model for concrete in dynamics.*ASCE J. Eng. Mech.*, **122**, 939-947.
3. Frémond, M. and Nedjar, B. (1993). Endommagement et principe des puissances virtuelles.*C. R. Acad. Sci. Paris*, **317** (7), serie II, 857-864.
4. Frémond, M. and Nedjar, B. (1996). Damage, gradient of damage and principle of virtual power. *Int. J. Solids Struc.*, **33** (8), 1083-1103.
5. Germain, P., Nguyen, Q.S. and Suquet, P. (1983). Continuum thermodynamics. *Jour. Appl. Mech., ASME*, **50**, 1010-1021.
6. Humbert, P. (1989). CESAR-LCPC, un code général de calcul par éléments finis. *Bull. Liais. Labo. des Ponts et Chaussées*, **160**, 112-116.
7. Nedjar, B. (1995). *Mécanique de l'endommagement. Théorie du premier gradient et application au béton.* Doctorat de l'Ecole Nationale des Ponts et Chaussées, Paris.
8. Needleman, A. (1987). Material rate dependance and mesh sensitivity in localization problems. *Comp. Meth. Appl. Mech. Engng.*, **67**, 68-85.

# CRACKS IN FRACTURE MECHANICS : A TIME INDEXED FAMILY OF ENERGY MINIMIZERS

*Griffith's theory revisited*

G. A. FRANCFORT AND J.-J. MARIGO
*LPMTM (UPR-CNRS 9001)*
*Institut Galilée, Université Paris-Nord*
*93430 Villetaneuse, France*

## 1. Introductory remarks

Brittle fracture mechanics is classically thought of as operating under various restrictive premises, two of which seem both drastic and unrealistic: no crack will appear unless a crack is already present; cracks propagate along predefined trajectories. Our main goal is to do away with the above, while departing as little as possible from the sanctity of Griffith's criterion. In a nutshell, the following will be achieved:

1. crack initiation, and its subsequent evolution till complete failure of the loaded sample,
2. complete freedom in the mechanical and geometric characteristics of the sample,
3. boundary cracks,
4. unilateral contact if needed.

There is, of course a price to pay, at least for the time being. The proposed formulation does not know well how to handle any kind of loadings, whether body loadings or surface tractions; thus the only admissible boundary conditions are on the displacement field. In the text all such boundary conditions are referred to as "loadings" . The formulation is also, as of yet, unable to model a continuous time evolution of the cracking process, except in the case of linearly increasing boundary displacements; it can however predict the cracking process for any kind of time-discretized boundary displacement.

Without further ado, we briefly present, in the next section, our model; we then implement it in a one-dimensional setting. The interested reader is kindly referred to (Francfort and Marigo, 1997) for a more detailed and thorough exposition of the model.



*P. Argoul et al. (eds.),*
*IUTAM Symposium on Variations of Domains and Free-Boundary Problems in Solid Mechanics*, 197–202.

## 2. The model

In all that follows, $\Omega$ denotes a bounded domain of $\mathbf{R}^N$ with $N = 1, 2,$ *or* 3.

Our main ingredient, the potential cracks, can be, quite simply, any compact set; we impose no restrictions of size or shape. Note however that the actual cracks are always of codimension 1, *i.e.* of finite $N-1$-Hausdorff dimension. Following Griffith's fundamental idea, see (Griffith, 1920), the associated surface energy is of the Griffith's type, that is, upon denoting by $k(x)$ the (nonnegative) energy density per unit area required to create a crack at that point,

$$E_s(\Gamma) = \int_\Gamma k(x) dH^{N-1}(x). \tag{1}$$

Note that a dependence of $k$ upon $n$, the "normal" to the crack $\Gamma$ could be introduced at the expense of further specifying the geometric structure of $\Gamma$.

Our second ingredient is the "loading"; as mentioned before, only displacements can be imposed on part of the boundary of our sample. We denote by $U$ the time-dependent displacement field applied to part of the boundary $\partial_d\Omega$. Assume that $\Gamma$ denotes the crack at a given time; we have to distinguish between the uncracked part of the boundary, *i.e.*, $\partial_d\Omega\backslash\Gamma$, on which we impose $U$, and the cracked part of the boundary on which nothing is imposed if unilateral contact is neglected, while unilateral conditions of some type must be imposed otherwise.

Our third ingredient is the actual material. We will assume that the domain $\Omega$ is occupied by a linearly elastic material. It should be pointed out that the extension of our model to finite elasticity can be easily performed (the underlying mathematical structure of the problem is actually simpler in the latter setting); see (Fonseca and Francfort, 1995) on a mathematical treatment of fracture in such a setting. The elasticity matrix $A(x)$ satisfies $\alpha\varepsilon\cdot\varepsilon \le A(x)\varepsilon\cdot\varepsilon \le \beta\varepsilon\cdot\varepsilon$, for any symmetric $\varepsilon$.

To each "loading" and potential crack we associate the corresponding elastic energy, $E_d(\Gamma, U)$, defined as

$$E_d(\Gamma, U) = inf\left\{\int_{\Omega\backslash\Gamma}\frac{1}{2}A(x)\varepsilon(u)\cdot\varepsilon(u)dx; \quad u = U \ on \ \partial_d\Omega\backslash\Gamma\right\}. \tag{2}$$

If the crack $\Gamma$ is smooth enough, as well as $\partial_d\Omega\backslash\Gamma$, then the infimum is classically attained for a displacement field $u_\Gamma$, solution to the elastic equilibrium on $\Omega\backslash\Gamma$; if not, then $E_d(\Gamma, U)$ should be understood as a generalized compliance because the associated elastic equilibrium may be ill-posed. In any case it can be conjectured that the actual crack will exhibit more regularity, which will in turn permit a writing of the elastic equilibrium on the uncracked part of the domain. It should be said however that no such regularity has been proved at this point; we do not want to dwell any further on this issue because a precise discussion necessarily involves a detailed explanation of the functional analytic foundation of this formulation, which would distract us from our main goal in the present work, that is, a concise exposition of the proposed model. The interested reader is referred to (Fonseca and Francfort, 1995) and references therein for further inquiries; see also (Ambrosio, 1990) or (De Giorgi *et al.*, 1989) for an exposition of the mathematical results available in the setting of image segmentation as well as linear antiplane elasticity.

Our fourth ingredient is the total energy, which we define to be

$$E(\Gamma, U) = E_s(\Gamma) + E_d(\Gamma, U). \tag{3}$$

It now remains to postulate an evolution law for the cracking process. This is done firstly for a sequence of set times $0 = t_0 < t_1 < \cdots < t_i < \cdots$; upon denoting by $U_i$ and $\Gamma_i$ the corresponding "loadings" and resulting cracks, we propose the following principle:

*Assume that $\Gamma_0$, ...,$\Gamma_{i-1}$,are known, then the crack $\Gamma_i$ at time $t_i$ will be such that*

$$\Gamma_{i-1} \subset \Gamma_i, \tag{4}$$

$$E(\Gamma_i, U_i) \leq E(\Gamma, U_i) \quad \forall \Gamma \supset \Gamma_{i-1}. \tag{5}$$

The first inclusion translates the irreversible character of the fracturing process while the second expresses a principle of least energy: the preferred state, among all admissible cracks, *i.e.* , among all cracks that include the actual crack at the previous time step, is that which produces the smallest total energy.

As it stands, the proposed evolution law allows for any kind of "loading"; we emphasize however our current inability at producing a time-continuous version of this evolution law. This is why the continuous analogue is only proposed for increasing "loadings", that is, for loadings $U$ of the form

$$U(t) = \begin{cases} tU_0 & , \quad t \geq 0, \\ 0 & , \quad t < 0. \end{cases} \tag{6}$$

The time $t \in (-\infty, +\infty)$ should be understood as a loading parameter, rather than an actual time, as is always the case when investigating quasistatic evolutions; in this respect, the only relevant feature of time is its monotonically increasing character and, as such, any increasing function of $t$ would do. The associated evolution law for the cracking process becomes

- **No initial crack** (any initial condition would do) :

$$\Gamma^-(0) = \emptyset; \tag{7}$$

- **Irreversibility of the process** :

$$\Gamma(t) \textit{ increases with } t; \tag{8}$$

- **First least energy principle** :

$$E(\Gamma(t), tU_0) \leq E(\Gamma, tU_0), \qquad \forall \Gamma \supset \Gamma^-(t); \tag{9}$$

- **Second least energy principle** :

$$E(\Gamma(t), tU_0) \leq E(\Gamma(s), tU_0), \qquad \forall s, \quad 0 \leq s \leq t, \tag{10}$$

where $\Gamma^-(t) = \overline{\bigcup_{s<t} \Gamma(s)}$.

If the three first relations are an exact transcription of the time-discretized model in a continuous framework, the fourth relation is new. It can be formally deduced from the discrete version for increasing "loadings" (see (Francfort and Marigo, 1997)); it is not contained in the third relation but does provide an additional restriction which proves essential in preventing too large a set of possible solutions (cf. section 2 below).

Note also that the evolution law guarantees that $\Gamma(0) = \emptyset$, at least when $k(x) > 0, H^{N-1} - a.e.$ on $\Omega$. Indeed, since, at time 0, the "loading" is 0, the associated compliance is also 0, so that $E(\Gamma(0), 0) = E_s(\Gamma(0))$ and (9) yields $E_s(\Gamma(0)) \leq E_s(\emptyset)$. Since the reverse inequality is a direct consequence of irreversibility, that implies that $\Gamma(0) \subset \emptyset$ up to a set of null $H^{N-1}$-Hausdorff measure.

We refer the reader to (Francfort and Marigo, 1997) for a detailed presentation of the properties of the proposed evolution model. The following features are particularly noteworthy and should be compared to Griffith's theory results, see (Liebowitz, 1968) :

1. *A crack will appear at some finite time, provided that some elastic energy is stored in the sound loaded body,* i.e. *provided that* $E_d(\emptyset, U_0) \neq 0$;
2. *Cracking will stop only when there will be no more elastic energy inside the body;*
3. *If the crack follows a smooth enough space-trajectory and if it further advances (with time) along that trajectory in a regular manner,* i.e. , *if the* $H^{N-1}$*-Hausdorff measure of the crack is an absolutely continuous function of* $t$*, then the crack evolution obeys the classical Griffith's law;*
4. *But even if the space-trajectory is smooth, the crack evolution may differ from that predicted by Griffith's law if it propagates with (a) jump(s) of* **finite** *length;*

5. *If the "initial" displacement field corresponding to elastic equilibrium in the absence of any cracks (or more generally in the presence of the initial crack) exhibits singular points, then a crack — of finite or infinitesimal length — will form at such points, either right away or at finite time, according to the strength of the singularity of the displacement field at such a point;*
6. *In the absence of singularities on the "initial" displacement field, then, either the sample will remain crack free — whether $E_d(\emptyset, U_0) = 0$ —, or a crack of* **finite** *length will appear at a finite non-zero time.*

## 3. The one-dimensional case

Consider a one-dimensional bar $\Omega = (0, L)$ of cross-sectional area $S$ made of a heterogeneous material with Young's modulus $E(x)$. Assume that the end displacements are respectively $u(0) = 0$ and $u(L) = \varepsilon t$ where $\varepsilon = \pm 1$.

If the bar is crack-free, its "surface" energy is 0, while its bulk energy is

$$E(\emptyset, \varepsilon t) = \frac{1}{2}\bar{E}S\frac{t^2}{L}, \qquad \textit{with} \quad \frac{1}{\bar{E}} = \frac{1}{L}\int_0^L \frac{1}{E(x)}dx. \tag{11}$$

For any crack $\Gamma$ with finite surface energy, *i.e.* any countable set $\mathcal{T}$ of points in $[0, L]$ (we implicitly assume that $k$ is continuous on $[0, L]$ and that $k(x) > \alpha, x \in \bar{\Omega}$, for some $\alpha > 0$),

$$E(\Gamma, \varepsilon t) = E_s(\Gamma) = S \sum_{x \in \mathcal{T}} k(x). \tag{12}$$

Introduce the critical time

$$t_r = \sqrt{\frac{2k_{min}L}{\bar{E}}}, \qquad \textit{with} \quad k_{min} = \min_{x \in [0,L]} k(x). \tag{13}$$

Then, the bar remains crack-free for all $t$'s with $t < t_r$. Indeed note that $\Gamma(0) = \emptyset$; if there exists $t < t_r$ such that $\Gamma(t) \neq \emptyset$, then, since (12) and (10) give $E(\Gamma(t), \varepsilon t) = E_s(\Gamma(t)) \leq E(\emptyset, \varepsilon t)$, we should have

$$k_{min}S \leq \sum_{x \in \Gamma(t)} k(x)S = E_s(\Gamma(t)) \leq \frac{1}{2}\bar{E}S\frac{t^2}{L} < \frac{1}{2}\bar{E}S\frac{t_r^2}{L} = k_{min}S, \tag{14}$$

which is impossible.

Further, the bar is cracked as soon as $t > t_r$. Indeed, if such is not the case, by (9) we should have $E(\Gamma, \varepsilon t) \geq E(\emptyset, \varepsilon t), \forall \Gamma\,'s$, *i.e.*,

$$k(x)S \geq \frac{1}{2}\bar{E}S\frac{t^2}{L} > \frac{1}{2}\bar{E}S\frac{t_r^2}{L} = k_{min}S, \quad \forall x \in \bar{\Omega}, \tag{15}$$

which is impossible.

Thirdly, since (10) gives $E(\Gamma(t_r), \varepsilon t_r) \leq E(\emptyset, \varepsilon t_r) = k_{min} S$, while (9) and (12) imply

$$E(\Gamma(t_r), \varepsilon t) \geq E(\Gamma(t), \varepsilon t) = E_s(\Gamma(t)) \geq k_{min} S, \quad \forall t > t_r, \tag{16}$$

by passing to the limit when $t \to t_r$, we obtain

$$E(\Gamma(t_r), \varepsilon t_r) = \lim_{t \to t_r} E_s(\Gamma(t)) = k_{min} S. \tag{17}$$

But by using once more (10), we have, when $t_r < s < t$ :

$$k_{min} S \leq E_s(\Gamma(t)) = E(\Gamma(t), \varepsilon t) \leq E(\Gamma(s), \varepsilon t) = E_s(\Gamma(s)). \tag{18}$$

By passing to the limit when $s \to t_r$ and using (17), we get $E_s(\Gamma(t)) = k_{min} S, \forall t > t_r$, which is only possible if $\Gamma(t)$ is reduced to a single point $x$ such that $k(x) = k_{min}$. Then, for any $t > t_r$, $\Gamma(t) = \{x\}$. We have thus proved that *the bar remains uncracked until time $t_r$, at which it suddenly breaks at a single point $\{x\}$ where $k$ reaches its minimum.*

Remark that, in the setting of unilateral contact, it is easily verified that the bar will not break under compression but that it will break at time $t_r$ under traction. The contrast with Griffith's theory of fracture is striking : our formulation predicts "crack" initiation where Griffith's cannot.

As a final note, the reader's attention is drawn to a peculiar byproduct of the formulation. The end displacement, at the time of breakage, grows like the square root of the length of the bar so that the average force per unit length at that time decreases like the square root of the length of the bar. This effect, which is equally present in the classical theory of fracture mechanics, is due to the interplay between surface and bulk energies which in turn produces a characteristic length. The physical plausibility of such a scaling might be deemed debatable by the physically inclined reader.

## References

AMBROSIO, L. Existence theory for a new class of variational problems. *Arch. Rat. Mech. Anal. 111* (1990), 291–322.

DE GIORGI, E., CARRIERO, M., AND LEACI, A. Existence theorem for a minimum problem with free discontinuity set. *Arch. Rat. Mech. Anal. 108* (1989), 195–218.

FONSECA, I., AND FRANCFORT, G. A. Relaxation in $BV$ versus quasiconvexification in $W^{1,p}$; a model for the interaction between fracture and damage. *Calculus of Variations 3* (1995), 407–446.

FRANCFORT, G. A., AND MARIGO, J.-J. Stable damage evolution in a brittle continuous medium. *Eur. J. Mech., A/Solids 12*, 2 (1993), 149–189.

FRANCFORT, G. A., AND MARIGO, J.-J. Griffith's theory of brittle fracture revisited. In preparation, 1997.

GRIFFITH, A. The phenomena of rupture and flow in solids. *Phil. Trans. Roy. Soc. London CCXXI-A* (1920), 163–198.

LIEBOWITZ, H., Ed. FRACTURE *: An Advanced Treatise*, vol. II : Mathematical Fundamentals. Academic Press, New York, London, 1968.

# One-Dimensional Ductile-Brittle Transition, Yielding, and Structured Deformations

GIANPIETRO DEL PIERO
*Facoltà di Ingegneria, Via Saragat 1, 44100 Ferrara*

## 1. Introduction

The alternative between two types of rupture, one due to a gradual loss (ductile fracture or strain-softening), and one due to a sudden drop (brittle fracture or catastrophic failure), of the bearing capacity of the body, has received large attention in recent literature. Several one-dimensional models have been studied, mainly consisting of bars divided into two pieces with different physical properties. It is the competition between these different properties which gives rise to stress-strain curves exhibiting one or the other type of failure.

In most of these models [2, 7, 8, 9], strain-softening is a constitutive property of the material, and fracture is obtained as an instability of the solution of the equilibrium problem. In a model proposed by Carpinteri [3] and founded on Barenblatt's *cohesive crack* theory [1], failure inside a linear elastic body is due to pre-existing cohesive cracks. The presence of a crack of measure zero instead of a band of finite volume renders the formation of fractures dependent upon a *scale factor,* a fact that has received wide experimental confirmation.

The present communication refers to some research in progress [5], based upon the use of *structured deformations,* introduced by D.R. Owen and the present author to study the mechanics of bodies undergoing changes in internal structure [6], and on some recent work of Choksi and Fonseca [4] on the definition and on the mathematical properties of the energy associated with such deformations. I show that, prescribing only the forms of the bulk energy of the material constituting the bar and of the energy of the cohesive cracks, and characterizing the equilibrium configurations as local minima of the total energy, it is possible to reproduce some phenomena, such as the formation of a crack or a change in the internal structure of the material, and to predict the conditions of their initiation. The form of the bulk energy need not allow for strain softening, and no initial cracks, inhomogeneities, imperfections, or weakening bands are needed. The formation of a crack may cause either ductile or brittle fracture, and the type of fracture is influenced by the total length of the bar, which therefore plays the role of a scale factor.

According to the theory developed in [6], changes in internal structure are attributed to microcracks. The alternative between macroscopic cracks and microcracks depends upon the analytic expression of the energy of the cohesive cracks; in what follows, I show that a concave energy determines either brittle or ductile fracture, depending on the scale factor, whereas an energy with a convex behavior near the origin determines a change in internal structure.



*P. Argoul et al. (eds.),*
*IUTAM Symposium on Variations of Domains and Free-Boundary Problems in Solid Mechanics,* 203–210.

## 2. The bar with concave fracture energy

Consider a bar of length $l$, with strain energy of the form

$$E(f) = \int_0^l w(f'(x))\,dx + \sum_{z\in\Gamma(f)} \theta([f(z)]) , \tag{2.1}$$

where $f$ is the axial deformation of the bar, $f'$ is the deformation gradient, $[f(z)]$ is the jump of $f$ at $z$, and $\Gamma(f)$ denotes the set of all jump points of $f$. A jump of $f$ represents a fracture, and $[f(z)]$ will be called the *fracture opening*. To exclude interpenetration of matter, it is convenient to make the assumptions

$$f'(x) > 0 \quad \text{a.e. in } (0, l) , \qquad [f(z)] \geq 0 \quad \forall z\in\Gamma(f) , \tag{2.2}$$

which ensure the injectivity of $f$. The functions $w$ and $\theta$ are the *bulk energy density* and the *fracture energy*, respectively; they are assumed to be twice differentiable, with $w$ strictly convex and $\theta$ strictly concave

$$w'' > 0 , \qquad \theta'' < 0 , \tag{2.3}$$

and they are supposed to satisfy

$$w(1) = w'(1) = 0 , \qquad \theta(0+) = 0 , \qquad \theta'(0+) > 0 . \tag{2.4}$$

The bar is subject to a prescribed elongation $\beta$, so that

$$\beta l = \int_0^l f'(x)\,dx + \sum_{z\in\Gamma(f)} [f(z)] \tag{2.5}$$

is the length of the bar after deformation. The problem of characterizing the equilibrium configurations of the bar reduces to finding the relative minima of the functional $E$ under the constraint (2.5). It can be solved with classical techniques of Calculus of Variations [5, Sect.5]; in fact, the analysis of the first variation shows that, at equilibrium, the deformation gradient is constant, all fractures have the same opening, and the derivatives of $w$ and $\theta$ take the same value:

$$w'(f') = \theta'([f]) . \tag{2.6}$$

It also shows that the formation of a crack at a point not belonging to $\Gamma(f)$ involves an increase of energy only if

$$\theta'(0+) \geq \theta'([f]) . \tag{2.7}$$

This condition rules out as unstable all those deformations for which $f'$ exceeds the value

$$f'_1 := (w')^{-1}\, \theta'(0+) \tag{2.8}$$

at any point of the bar. Indeed, because $w'$ is increasing, it is invertible and its inverse

$(w')^{-1}$ is decreasing; therefore, by (2.6) and (2.7),

$$f' = (w')^{-1}\,\theta'([f]) \;\leq\; (w')^{-1}\,\theta'(0+) = f'_1 \,. \tag{2.9}$$

In the following, by an *equilibrium configuration* I mean a deformation $f$ with $f'$ constant and not greater than $f'_1$, and with a number N≥0 of jumps, each one of amplitude $[f] = h(f')$, with $h := (\theta')^{-1}w'$.

Among all equilibrium configurations, of interest are those which are *metastable*, i.e., which are local minimizers of the energy. Two necessary conditions for metastability come from the analysis of the second variation [5]; the first one is

$$w''(f') + l^{-1}\,\theta''(h(f')) \geq 0\,. \tag{2.10}$$

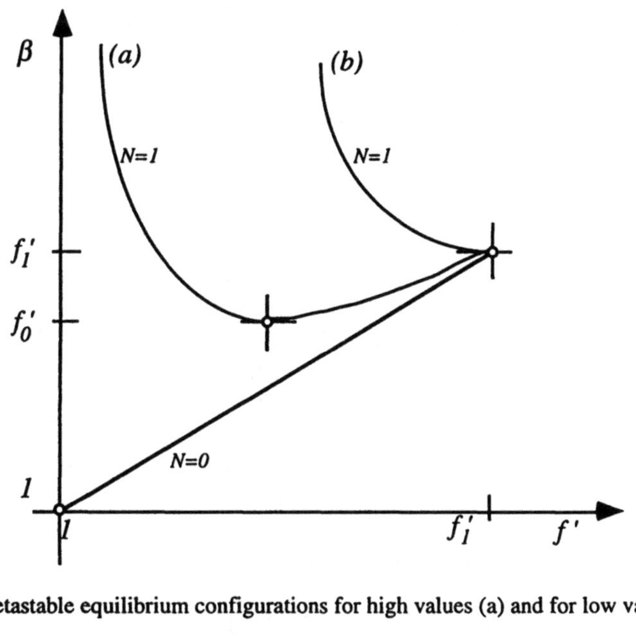


Fig.1. Metastable equilibrium configurations for high values (a) and for low values (b) of $l$.

Because $w''$ is positive and $\theta''$ is negative, for each fixed $f'$ this condition is satisfied for small values, and violated for large values, of $l$. Thus, depending on $l$, the same deformation $f'$ may correspond either to a metastable or to an unstable equilibrium configuration. In other words, $l$ plays here the role of a scale factor.

The second necessary condition for metastability is that (see formula (3.6) below) *a metastable equilibrium configuration may have at most one jump point.* This condition establishes a simple relation between $f'$ and $\beta$ at equilibrium. Indeed, Equation (2.5) yields

$$\beta = f' \tag{2.11}$$

for equilibrium configurations with N=0, and

$$\beta = f' + l^{-1}\,h(f') \tag{2.12}$$

for equilibrium configurations with N=1. The graphs of these two equations are plotted in Fig.1. It can be proved that, by Inequality (2.10), the second equation determines a metastable equilibrium configuration only at those $f'$ at which the slope of the graph is negative [5].

Note that the curve N=1 is influenced by the scale factor $l$; in the figure, the curves relative to two different values of $l$ are shown. In both, for $\beta > f'_1$ there is only one equilibrium configuration, with N=1. Moreover, if we denote by $f'_0$ the minimum of the curve N=1, we see that for large values of $l$ there are one equilibrium configuration with N=0 for $\beta < f'_0$, and three equilibrium configurations, one with N=0 and two with N=1, one of which unstable, for $\beta$ between $f'_0$ and $f'_1$. Consequently, if we let $\beta$ grow from the initial value $\beta$=1, there is a point in $(f'_0, f'_1)$ at which the equilibrium solution jumps from the curve N=0 to the curve N=1, with a drastic reduction of $f'$. This behavior reproduces the phenomenon of *strain localization,* usually associated with the process of fracturing.

For low values of $l$, the part of the curve with negative slope may disappear, and this can be viewed as a transition from brittle to ductile fracture, the two types of fracture being characterized by a jump and by a gradual decrease of $f'$, respectively.

## 3. Effects of convexity near the origin

Let us see how the preceding conclusions change if, instead of assuming $\theta$ concave, we assume that $\theta''$ is positive in an interval $(0, \lambda_1)$ near the origin and negative outside. For convenience, we also assume that there is a $\lambda_2 > \lambda_1$ such that $\theta'(\lambda_2) = \theta'(0+)$ (see Fig.2). Note that the case of a concave $\theta$ coincides with the limit case $\lambda_2 = \lambda_1 = 0$.

Under these assumptions, the analysis of the first and second variations of $E$ yields the same necessary conditions for equilibrium and metastability as above, and therefore the definition of an *equilibrium configuration* remains unchanged. However, Inequality (2.7) is now violated by all $[f]$ in $(0, \lambda_2)$, and therefore all fracture openings smaller than $\lambda_2$ are unstable.

For an equilibrium configuration, the energy is

$$E(f) \;=\; l\,w(f') + \mathrm{N}\;\theta(h(f'))\,, \tag{3.1}$$

and the condition (2.5) takes the form

$$\beta = f' + l^{-1}\,\mathrm{N}\,h(f')\;. \tag{3.2}$$

Take a perturbation $\eta$ with $\eta' = 0$, for which

$$E(f+\eta) \;=\; E(f) + \sum_{z\in\Gamma(\eta)\cap\Gamma(f)} \big(\theta(h(f')+[\eta(z)]) - \theta(h(f'))\big) \;+\; \sum_{z\in\Gamma(\eta)\setminus\Gamma(f)} \theta([\eta(z)])\,, \tag{3.3}$$

and, again by (2.5),

$$\sum_{z\in\Gamma(\eta)} [\eta(z)] \;=\; 0\;. \tag{3.4}$$

Consider the expansion

$$E(f+\eta) = E(f) + (\theta'(h(f')) - \theta'(0+)) \sum_{z\in\Gamma(\eta)\cap\Gamma(f)} [\eta(z)] +$$

$$+ \frac{1}{2}\theta''(h(f')) \sum_{z\in\Gamma(\eta)\cap\Gamma(f)} [\eta(z)]^2 + \frac{1}{2}\theta''(0+) \sum_{z\in\Gamma(\eta)\setminus\Gamma(f)} [\eta(z)]^2 + \dots \tag{3.5}$$

of $E$. Note that the intersection $\Gamma(\eta)\cap\Gamma(f)$ cannot be empty; indeed, if all $z$ belong to $\Gamma(\eta)\setminus\Gamma(f)$ then all $[\eta(z)]$ must be positive by $(2.2)_2$, in contradiction with (3.4). If $\Gamma(\eta)\cap\Gamma(f)$ consists of a single point, say $z_o$, then again by (3.4) the jump $[\eta(z_o)]$ must be negative. Because for $f'<f'_1$ we have $\theta'(h(f')) < \theta'(0+)$, it follows that $E(f+\eta) > E(f)$ for sufficiently small $\eta$.

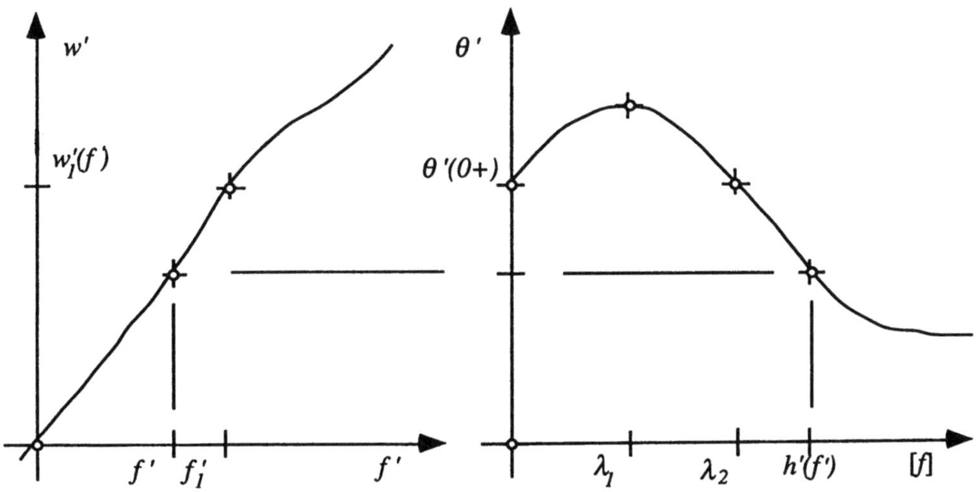


Fig.2. Graphs of the functions $w'$, $\theta'$ and construction of the function $h$ for surface energy with convexity near the origin.

If we now assume that $f$ has at least two jump points and take $\Gamma(\eta)\subset\Gamma(f)$, we have

$$E(f+\eta) = E(f) + (\theta'(h(f')) - \theta'(0+)) \sum_{z\in\Gamma(\eta)} [\eta(z)] + \frac{1}{2}\theta''(h(f')) \sum_{z\in\Gamma(\eta)} [\eta(z)]^2 + \dots, \tag{3.6}$$

and we see that the first variation is zero by (3.4) and the second variation is negative because $\theta''(h(f'))$ is negative. Thus, an equilibrium configuration with N>1 is unstable.

Of the remaining equilibrium configurations, those with N=0 are limited to the range $0 < \beta < f'_1$, and those with N=1 are limited to $\beta > f'_0$, with $f'_0$ the minimum of the function $\beta = \tilde{\beta}(f')$ determined by Equation (3.2) with N=1. But in the case of convexity near the origin we have

$$\tilde{\beta}(f'_1) = f'_1 + l^{-1}\lambda_2 > f'_1, \tag{3.7}$$

and therefore $f'_0$ can be greater than $f'_1$. If this occurs, there are no stable equilibrium configurations for $f'_1<\beta<f'_0$.

Let us see if it is possible to define some *generalized solution* covering this range of $\beta$. For every given N and $\beta$, let $f_{N,\beta}$ be the equilibrium configuration with N jumps and with total length $\beta$. Then, $f'_{N,\beta}$ is the solution of the equation

$$f'_{N,\beta} + \mathrm{N}\, l^{-1}\, h(f'_{N,\beta}) = \beta\,, \tag{3.8}$$

and the corresponding energy is

$$E(f_{N,\beta}) = l\, w(f'_{N,\beta}) + \mathrm{N}\, \theta(h(f'_{N,\beta}))\,. \tag{3.9}$$

Because $f'_{N,\beta}$ is less or equal to $f'_1$ for all N by condition (2.7), the term $(\mathrm{N}\, l^{-1}\, h(f'_{N,\beta}))$ in Equation (3.8) remains bounded when $\mathrm{N}\to\infty$ and, consequently, $h(f'_{N,\beta})\to 0$. But $h = (\theta')^{-1}w'$, and $(\theta')^{-1}w'(f'_{N,\beta})\to 0$ implies $f'_{N,\beta}\to f'_1$. Thus, by the continuity of $\theta$ and by Equation (3.8) we get

$$\lim_{\mathrm{N}\to\infty} (\mathrm{N}\, \theta(h(f'_{N,\beta}))) = \lim_{\mathrm{N}\to\infty} (\mathrm{N}\, \theta'(0+)\, h(f'_{N,\beta})) =$$

$$= \lim_{\mathrm{N}\to\infty} (l\, \theta'(0+)(\beta - f'_{N,\beta})) = l\, \theta'(0+)(\beta - f'_1)\,. \tag{3.10}$$

Using the continuity of $w$ and recalling that $\theta'(0+) = w'(f'_1)$ by Equation (2.8), we get

$$E_\infty(\beta) := \lim_{\mathrm{N}\to\infty} E(f_{N,\beta}) = l\,(w(f'_1) + \theta'(0+)(\beta - f'_1))\,. \tag{3.11}$$

Thus, the sequence $\mathrm{N}\mapsto f_{N,\beta}$ converges in energy when $\mathrm{N}\to\infty$. Moreover, $E_\infty(\beta)$ is the infimum of the whole sequence, provided that $\beta$ is sufficiently small. Indeed, by the convexity of $w$ and by (3.8),

$$E_\infty(\beta) - E(f_{N,\beta}) = l\,(w(f'_1) - w(f'_{N,\beta}) + (\beta - f'_1)\, w'(f'_1)) - \mathrm{N}\, \theta(h(f'_{N,\beta})) \le$$

$$\le l\,(\beta - f'_{N,\beta})\, w'(f'_1) - \mathrm{N}\, \theta(h(f'_{N,\beta})) = \mathrm{N}\,((h(f'_{N,\beta}))\, w'(f'_1) - \theta((h(f'_{N,\beta}))))\,, \tag{3.12}$$

and, after replacing $w'(f'_1)$ by $\theta'(0+)$,

$$E_\infty(\beta) - E(f_{N,\beta}) = \mathrm{N} \int_0^{h(f'_{N,\beta})} (\theta'(0+) - \theta'(x))\, dx\,. \tag{3.13}$$

Because $\theta'$ is increasing near the origin, it follows that $E_\infty(\beta) \le E(f_{N,\beta})$ if $h(f'_{N,\beta})$ does not exceed a critical value $K$ corresponding to the equal-area rule shown in Fig.3. Since the only metastable solution may occur for N=1, it follows that $E_\infty(\beta)$ is the infimum in the class of all equilibrium configurations if

$$h(f'_{1,\beta}) < K\,. \tag{3.14}$$

In the graph of Fig.1, $h(f'_1,\beta)$ is the distance, at the given $\beta$, between the curve N=0 and the curve N=1 for $l = 1$. Thus, for all $\beta$ for which this distance in less than $K$, $E_\infty(\beta)$ is indeed the infimum of the energies of the equilibrium configurations.

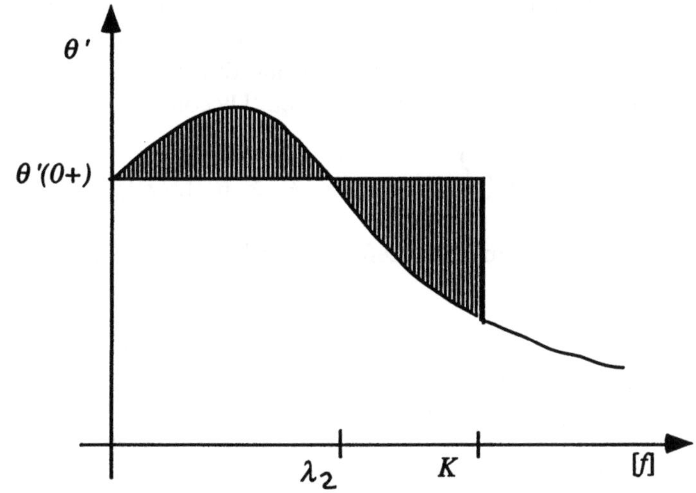


Fig.3. Equal-area rule for the determination of *K*.

## 4. Discussion

The energy $E_\infty(\beta)$ can be associated with a *structured deformation* in the sense of [6]. In the present one-dimensional context, such a deformation is a pair $(g, G)$, with $g$ a deformation of the type considered in the preceding sections, and $G$ a scalar field over $(0, l)$, obeying the condition $G \leq g'$ and called the *deformation gradient without disarrangements*. The difference $(g'-G)$ is called the *deformation due to micro-disarrangements*, and is a measure of change in the internal structure of the material.

For a fracture energy with convexity near the origin, it can be shown that the sequence $N \mapsto f_{N,\beta}$ converges uniformly to the function $x \mapsto \beta x$, while the sequence of the derivatives $f'_{N,\beta}$ converges to $f'_1$ as explained in the preceding section.

The energy $E_\infty(\beta)$ can be associated with the structured deformation $(g, G)$, with $g(x) = \beta x$ and $G(x) = f'_1$. It describes a configuration whose macroscopic deformation gradient is $\beta$, while the bulk undergoes the deformation $f'_1$, the remaining deformation $(\beta - f'_1)$ being due to changes in internal structure. It is then reasonable to identify the stress with the derivative $w'(f'_1)$ of the bulk energy density.

Consider a bar subjected to a growing elongation $\beta$ starting from $\beta = 1$. At first, we have $g'=G=\beta$, and we say that the bar undergoes a *simple deformation*, which can be described by the function $g$ alone. When $\beta$ attains the value $f'_1$, a *structured deformation* takes place: there is further elongation, but the stress stays constant at the value $w'(f'_1)$. According to formula (3.10), the energy $E_\infty(\beta)$ is composed of two terms, a constant term due to bulk deformation, plus the energy spent for the change of internal structure.

This model can be used to describe the phenomenon of *yielding*. It is remarkable that this comparatively rich model has been obtained with very few assumptions on the structure of the energy functional.

It is also remarkable that the energy $E_\infty(\beta)$, as given by Equation (3.10), coincides with the expression of the energy of a structured deformation, evaluated in [4] under more restrictive assumptions on the energy fracture. In one dimension, and for the given structured deformation, the expression given in [4, Sect.2] becomes

$$I(\beta, f'_1) \;=\; l\; inf\big( \int_0^1 w(h'(x))\,dx \;+\; \sum_{z\in\Gamma(h)} \theta_0([h(z)])\big)\,, \tag{4.1}$$

with the infimum to be evaluated among all deformations $h$ which satisfy the conditions

$$\int_0^1 h'(x)\,dx + \sum_{z\in\Gamma(h)} [h(z)] = \beta\,, \qquad \int_0^1 h'(x)\,dx \;=\; f'_1\,, \tag{4.2}$$

with $\theta_0$ the function $x\mapsto x\theta'(0+)$. It is easy to show that the infimum is indeed $E_\infty(\beta)$; in fact, the replacement of $\theta$ by $\theta_0$ and the rescaling of the domain from $(0, l)$ to $(0, 1)$ do not alter the analysis of the minima of $E$ made in the preceding sections and the corresponding conclusions.

ACKNOWLEDGEMENT. This research has been supported by a CNR grant within the Progetto Strategico *Problemi della progettazione complessa.*

## References

1. BARENBLATT, G.I., *The formation of equilibrium cracks during brittle fracture: general ideas and hypotheses. Axially symmetric cracks,* Appl. Math. Mech. (PMM), 23, 622-636, 1959.
2. BILLARDON, R., & J.G.S. PATIÑO, *One-dimensional softening and stability due to continuum damage,* Proc. 9th Brazilian Congr. on Mech. Engineering, 1987.
3. CARPINTERI, A., *Cusp catastrophe interpretation of fracture instability,* J. Mech. Phys. Solids 37, 567-582, 1989.
4. CHOKSI, R., & I. FONSECA, *Bulk and interfacial energy densities for structured deformations of continua,* Arch. Rat. Mech. Analysis, forthcoming.
5. CHOKSI, R., G. DEL PIERO, I. FONSECA, & D. OWEN, *Structured deformations as energy minimizers in some models of fracture and hysteresis,* forthcoming.
6. DEL PIERO, G., & D.R. OWEN, *Structured deformations of continua,* Arch. Rat. Mech. Analysis 124, 99-155, 1993.
7. NEEDLEMAN, A., *On the competition between failure and instability in progressively softening solids,* J. Appl. Mech. 58, 294-296, 1991.
8. SCHREYER, H.L., & Z. CHEN, *One-dimensional softening with localization,* J. Appl. Mech. 53, 791-797, 1986.
9. TRUSKINOVSKY, L., *Fracture as a phase transition,* pre-print.

# APPROXIMATION RESULTS FOR A CLASS OF QUASISTATIC CONTACT PROBLEMS INCLUDING ADHESION AND FRICTION

M. COCU*, L. CANGEMI AND M. RAOUS
*Laboratoire de Mécanique et d'Acoustique - C.N.R.S.*
*31, Chemin Joseph Aiguier, 13402 Marseille Cedex 20*
* *L.M.A. - C.N.R.S. and Université de Provence*

## 1. Introduction

This paper is concerned with a continuum thermodynamic and mathematical formulation of an interface law including unilateral contact, adhesion and friction. The model is based on the notion of material boundary associated to the interface and its derivation follows from the principle of virtual power and the principles of thermodynamics (Cangémi *et al.*, 1996 a). Adhesion and friction are strongly coupled and adhesion is characterized by a new variable, the intensity of adhesion $\beta$ introduced by Frémond. We consider a quasistatic unilateral contact problem for which we present a variational formulation. A time discretization is adopted and we prove that if the friction coefficient is sufficiently small then the incremental formulation that can be derived from this discretization has a unique solution. Finally an application to an indentation problem is mentioned.

## 2. Mechanical and Thermodynamic Formulation

### 2.1. THE CLASSICAL EQUATIONS AND CONDITIONS

Let $\Omega^1$ and $\Omega^2$ be two disjoint domains of $I\!R^d$, $d = 2,3$ , occupied by two elastic bodies with sufficiently regular boundaries $\Gamma^\alpha = \bar{\Gamma}^\alpha_U \cup \bar{\Gamma}^\alpha_\phi \cup \bar{\Gamma}_c$, $\alpha = 1,2$. On $\Gamma^1_\phi$, $\Gamma^2_\phi$ external force densities $\phi^1$, $\phi^2$ are respectively applied and on $\Gamma^1_U$ and $\Gamma^2_U$ (with $mes(\Gamma^\alpha_U) > 0$, $\alpha = 1,2$) the displacements are given. On $\Omega^1 \cup \Omega^2$ the volume force densities $f^1$ and $f^2$ are applied.
Assuming small deformations, the contact between the two bodies is defined by a point-to-point correspondence. The relative displacement between cou-



*P. Argoul et al. (eds.),*
*IUTAM Symposium on Variations of Domains and Free-Boundary Problems in Solid Mechanics*, 211–218.

pled particles on $\Gamma_c$ is denoted by $[u] = u^1 - u^2$.
The contact boundary $\Gamma_c$ will be considered in what follows as a material boundary with a specific thermodynamic behaviour (Frémond, 1987; Klarbring, 1990). By applying the principle of virtual power to the whole system $\Omega^1 \cup \Omega^2 \cup \Gamma_c$, the following equilibrium equations can be deduced ($\alpha$ =1, 2) :

$$div\, \sigma^\alpha(u) + f^\alpha = 0 \;\; in \quad \Omega^\alpha \tag{1}$$

$$\sigma^\alpha(u)n^\alpha = \phi^\alpha \;\; on \quad \Gamma^\alpha_\phi \tag{2}$$

$$u^\alpha = 0 \;\; on \quad \Gamma^\alpha_U \tag{3}$$

$$\sigma^1(u)n^1 = -\sigma^2(u)n^2 = -R \;\; on \quad \Gamma_c, \tag{4}$$

where $\sigma$ is the stress tensor and $R$ the contact force on $\Gamma_c$.

We make the classical decomposition into normal and tangential components for the contact forces and the displacement vector on $\Gamma_c$:

$$\begin{cases} [u] = [u_N]\, n + [u_T] & where \quad [u_N] = [u].n \\ R = R_N\, n + R_T & where \quad R_N = R.n \end{cases} \tag{5}$$

with $n = n^1 = -n^2$, where $n^\alpha$ is the outward unit normal vector to $\Gamma^\alpha$ , $\alpha = 1, 2$.

The constitutive laws will be obtained using thermodynamic considerations for the bodies and the material interface.

## 2.2. STATE LAWS

In addition to the specific internal energy e and the specific entropy s, we define on $\Gamma_c$ a surface density of internal energy E and a surface density of entropy S. The internal energy and the entropy corresponding to a subdomain $\mathcal{D} \subset \Omega^1 \cup \Omega^2 \cup \Gamma_c$ can then be written as :

$$\mathcal{E} = \int_{\mathcal{D}} \rho e\, dx + \int_{\partial\mathcal{D}\cap\Gamma_c} E\, ds \;\; \text{and} \;\; \mathcal{S} = \int_{\mathcal{D}} \rho s\, dx + \int_{\partial\mathcal{D}\cap\Gamma_c} S\, ds. \tag{6}$$

The classical state variable is the infinitesimal strain tensor $\epsilon = (\epsilon_{ij})$. Following Frémond, we introduce two other variables, the jump [u] on $\Gamma_c$ and the intensity of adhesion $\beta$, considered to be an internal variable. We introduce the Helmholtz free energies $\psi = e - sT$ in $\Omega^1 \cup \Omega^2$ and $\Psi = E - ST$ on $\Gamma_c$ , where $T$ is the absolute temperature. The decomposition of the forces into their reversible and irreversible parts :

$$\begin{array}{ll} \sigma = \sigma^r + \sigma^{ir} & in\, \Omega^1 \cup \Omega^2 \\ \left.\begin{array}{l} -\sigma_N = R_N = R^r_N + R^{ir}_N \\ -\sigma_T = R_T = R^r_T + R^{ir}_T \end{array}\right\} & on\, \Gamma_c, \end{array}$$

will be used and $\{\sigma^r, R^r, -G_\beta\}$ will represent the thermodynamic forces associated to the state variables $\{\epsilon, [u], \beta\}$. The specific forms of the free energy potentials are :

– *for each body* $\Omega^1$ *and* $\Omega^2$

$$\psi(\epsilon) = \frac{1}{2\rho} A^{\alpha}_{ijkl} \epsilon_{ij} \epsilon_{kl} \,,$$

where $A^{\alpha}_{ijkl}$ $(1 \leq i, j, k, l, \leq 3, \ \alpha = 1, 2)$ are the elasticity coefficients.

– *on the interface* $\Gamma_c$

$$\Psi([u_N], [u_T], \beta) = \frac{C_N}{2} [u_N]^2 \beta^2 + \frac{C_T}{2} [u_T]^2 \beta^2 - w\,\beta + \\ + I_{\tilde{K}}([u]) + I_P(\beta) \,,$$

where $\tilde{K} = \{v;\ v_N \leq 0\}$ and $P = \{\gamma; 0 \leq \gamma \leq 1\}$.

The adhesion is governed here by a compliance law, where $C_N$ and $C_T$ are material characteristic coefficients. For each state of adhesion $\beta$ belongs to $]0, 1]$ and the interface is able to develop an elastic strength, proportional to the jump $[u]$. The product $w\,\beta)$ represents the surface energy of adhesion.

The state laws are then written as follows :

In $\Omega^1 \cup \Omega^2$

$$\sigma^r = \rho \frac{\partial \psi(\epsilon)}{\partial \epsilon} \,. \tag{7}$$

On $\Gamma_c$

$$(R^r_N, R^r_T) \in \partial_{[u]} \Psi([u_N], [u_T], \beta) \text{ and } - G_\beta \in \partial_\beta \Psi([u_N], [u_T], \beta). \tag{8}$$

### 2.3. COMPLEMENTARY LAWS

In accordance with the two principles of thermodynamics we can write the dissipation of the system $\Omega^1 \cup \Omega^2 \cup \Gamma_c$ in an isothermic evolution as follows (Cangémi *et al.*, 1996 a) :

$$\left\{ \begin{array}{l} d = (\sigma - \sigma^r) \cdot \dot{\epsilon} \geq 0 \quad in\ \Omega^1 \cup \Omega^2 \\ D = (R_N - R^r_N)\,[\dot{u}_N] + (R_T - R^r_T) \cdot [\dot{u}_T] + G_\beta \dot{\beta} \geq 0 \quad on\ \Gamma_c \,. \end{array} \right. \tag{9}$$

The only dissipative processes are : friction, associated with the tangential velocity $[\dot{u}_T]$, and adhesion, associated with $\dot{\beta}$. By introducing a pseudo-potential $\Phi$, which is positive, convex and equal to zero in (0,0), we can formulate the complementary constitutive laws, in agreement with the inequalities (9) :

$$\sigma = \sigma^r, \quad R_N = R_N^r(= -\sigma_N) \text{ and } (R_T^{ir}, G_\beta) \in \partial\Phi([\dot{u}_T], \dot{\beta}). \tag{10}$$

We adopt the following form for $\Phi$, with $\chi_N = (\sigma_N, [u_N], \beta)$ :

$$\Phi\left([\dot{u}_T], \dot{\beta}; \chi_N\right) = \mu\left|\sigma_N^* + C_N[u_N]\beta^2\right| \,\|[\dot{u}_T]\| + \frac{b}{2}\dot{\beta}^2 + I_{C^-}(\dot{\beta}), \tag{11}$$

where $\mu$ is a friction coefficient,
$\sigma_N^*$ is a regularization of the normal contact force $\sigma_N$,
$C^- = \{\gamma; \gamma \leq 0\}$.

2.4. CONSTITUTIVE LAWS

Finally, by remembering that $-R_T^{ir} = \sigma_T + R_T^r$, we derive from the state laws (7)-(8) and the complementary laws (10), the following relations :

• *In* $\Omega^\alpha$

*Generalized Hooke's law*

$$\sigma^\alpha(u) = A^\alpha : \epsilon(u^\alpha) \quad \alpha = 1, 2. \tag{12}$$

• *On* $\Gamma_c$

*Unilateral contact conditions*

$$\sigma_N + C_N[u_N]\beta^2 \leq 0, \quad [u_N] \leq 0 \text{ and } (\sigma_N + C_N[u_N]\beta^2)\,[u_N] = 0. \tag{13}$$

*Adhesion and friction law*

$$\begin{array}{l} \left|\sigma_T + C_T[u_T]\beta^2\right| \leq \mu\left|\sigma_N^* + C_N[u_N]\beta^2\right| \\ \left|\sigma_T + C_T[u_T]\beta^2\right| < \mu\left|\sigma_N^* + C_N[u_N]\beta^2\right| \Rightarrow [\dot{u}_T] = 0 \\ \left|\sigma_T + C_T[u_T]\beta^2\right| = \mu\left|\sigma_N^* + C_N[u_N]\beta^2\right| \Rightarrow \\ \qquad \exists\lambda \geq 0 \;\; \sigma_T + C_T[u_T]\beta^2 = -\lambda[\dot{u}_T]. \end{array} \tag{14}$$

*Damage evolution equation*

$$b\dot{\beta} = -\left[w - (C_N[u_N]^2 + C_T[u_T]^2)\beta\right]^- . \tag{15}$$

The conditions (13) represent a contact law where no interpenetration between the two bodies occurs. But in a traction situation, i.e. if $[u_N] < 0$, the interface can develop a compliance force which is proportional to the jump between the two boundaries and is characterized by a specific coefficient $C_N$. The friction-adhesion law (14) expresses the coupling between a Coulomb force and an extra compliance force in the tangential direction, with a characteristic coefficient $C_T$. Evolution of the damage coefficient $\beta$ is described by (15), where b characterizes the viscosity of adhesion. When decohesion occurs ($\beta = 0$), the relations (13), (14) are the classical unilateral contact conditions (Signorini's conditions) and friction conditions.

## 3. Variational formulation

### 3.1. THE CONTINUOUS PROBLEM

We denote by $V^\alpha$ the following spaces :

$$V^\alpha = \left\{v^\alpha \in [H^1(\Omega^\alpha)]^d ; v^\alpha = 0 \text{ a.e. on } \Gamma_U^\alpha\right\}, \ \alpha = 1,2, \text{and we set}$$
$$V = V^1 \times V^2, \quad K = \{v = (v^1, v^2) \in V; [v_N] \leq 0 \text{ a.e. on } \Gamma_c\},$$
$$H = L^\infty(\Gamma_c).$$

For all $v \in V$ the norm on $V$ is given by $||v|| = ||v^1||_{V^1} + ||v^2||_{V^2}$ and $\langle .,. \rangle$ shall denote the duality pairing on $H^{\frac{1}{2}}(\Gamma_c) \times H^{-\frac{1}{2}}(\Gamma_c)$ .

We suppose that :

$$f \in W^{1,2}(0,T; [L^2(\Omega^1)]^d \times [L^2(\Omega^2)]^d),$$
$$\phi \in W^{1,2}(0,T; \left[L^2(\Gamma_\phi^1)\right]^d \times \left[L^2(\Gamma_\phi^2)\right]^d),$$

which imply $F \in W^{1,2}(0,T;V)$, where

$$(F,v) = \sum_{\alpha=1,2} \left[\int_{\Omega^\alpha} f^\alpha.v^\alpha dx + \int_{\Gamma_\phi^\alpha} \phi^\alpha.v^\alpha ds\right] \ \forall v \in V.$$

We shall adopt the following notations :

- $a : V \times V \longrightarrow I\!R,$
  $a(u,v) = a^1(u^1,v^1) + a^2(u^2,v^2) \ \forall u = (u^1,u^2), v = (v^1,v^2) \in V,$
  where $a^\alpha(u^\alpha,v^\alpha) = \int_{\Omega^\alpha} A_{ijkl}^\alpha \epsilon_{ij}(u^\alpha)\epsilon_{kl}(v^\alpha)dx, \ \alpha = 1,2.$

- $j : H \times V \times V \longrightarrow I\!R,$
  $j(\beta,u,v) = \int_{\Gamma_c} \mu|\sigma_N^*(Pu^1) + C_N\beta^2[u_N]||[v_T]|ds,$

  where $P$ is the projection of $V^1$ on $V_0^1$ with
  $V_0^1 = \{w^1 \in V^1; div\sigma(w^1) = -f^1\},$
  $\mu \in L^\infty(\Gamma_c)$ and $\mu \geq 0,$
  $(\,.\,)^* : H^{-\frac{1}{2}}(\Gamma_c) \to L^2(\Gamma_c)$ is a linear and compact mapping.

• $b_N, b_T : H \times V \times V \longrightarrow I\!R$,
$b_N(\beta, u, v) = \int_{\Gamma_c} C_N \beta^2 [u_N][v_N] ds$ and $b_T(\beta, u, v) = \int_{\Gamma_c} C_T \beta^2 [u_T][v_T] ds$.

• $g(\beta, u) = -\frac{1}{b}\left[w - (C_N [u_N]^2 + C_T [u_T]^2)\beta\right]^-$.

Then the local problem (1)-(4), (12)-(15) admits the following variational formulation :

**Problem P.**
*Find* $(u, \beta) \in W^{1,2}(0,T;V) \times W^{1,2}(0,T;H)$ *such that* $u(0) = u_0$, $\beta(0) = \beta_0 \in [0, 1[$ *and for almost all* $t \in [0, T]$, $u(t) \in K$ *and* $\forall\, v \in V$

$$a(u, v - \dot{u}) + j(\beta, u, v) - j(\beta, u, \dot{u}) + b_T(\beta, u, v - \dot{u}) \geq (F, v - \dot{u}) + \langle \sigma_N(u^1), [v_N] - [\dot{u}_N] \rangle \tag{16}$$

$$\langle \sigma_N(u^1), [z_N] - [u_N] \rangle + b_N(\beta, u, z - u) \geq 0 \quad \forall\, z \in K, \tag{17}$$

$$\dot{\beta} = g(\beta, u) \quad a.e.\ on\ \Gamma_c, \tag{18}$$

where the initial condition $u_0$ belongs to $K$ and satisfies the following compatibility condition :

$$a(u_0, w - u_0) + j(\beta_0, u_0, w - u_0) + b_T(\beta_0, u_0, w - u_0) \geq (F(0), w - u_0) \quad \forall w \in K. \tag{19}$$

This problem is a generalization of the quasi-static unilateral contact problem with friction considered in (Cocu *et al.*, 1996).

### 3.2. INCREMENTAL FORMULATION

First, let us focus on the properties of the bilinear form $a$ and the mapping $j$ involved in the problem **P** :

$a(.,.)$ is continuous on $V \times V$ and coercive i.e. satisfies

$$\exists M > 0 \quad \forall u \in V \quad \forall v \in V \quad |a(u, v)| \leq M||u||\,||v||, \tag{20}$$

$$\exists m > 0 \quad \forall u \in V \quad a(u, u) \geq m||u||^2, \tag{21}$$

and the mapping $j(.,.,.)$ satisfies the following property :

$$\exists C > 0 \quad \forall (u, \bar{u}, v, \bar{v}) \in V^4$$
$$|j(\beta, u, v) - j(\beta, u, \bar{v}) - j(\beta, \bar{u}, v) + j(\beta, \bar{u}, \bar{v})| \leq \bar{\mu} C ||u - \bar{u}||\,||v - \bar{v}||, \tag{22}$$

with $\bar{\mu} = |\mu|_{L^\infty(\Gamma_c)}$ .

An incremental formulation is obtained by operating a time discretization of problem **P**, taking $n \in \mathbb{N}^*$ and setting $\Delta t = T/n$, $t_i = i \cdot \Delta t$ and $F^i = F(t_i)$ for $i = 1, ..., n$. Using an implicit scheme we obtain the following sequence of problems $(P_i^n)$, $i = 0, ..., n-1$, defined for a given $(u^0, \beta^0) \in K \times [0, 1[$ .

**Problem $\mathbf{P_i^n}$ .**
Find $(u^{i+1}, \beta^{i+1}) \in K \times H$ *such that* :

$$\begin{aligned} a(u^{i+1}, v - u^{i+1}) + j(\beta^{i+1}, u^{i+1}, v - u^i) - j(\beta^{i+1}, u^{i+1}, u^{i+1} - u^i) \\ +b(\beta^{i+1}, u^{i+1}, v - u^{i+1}) \geq (F^{i+1}, v - u^{i+1}) \quad \forall v \in K \end{aligned} \tag{23}$$

$$\beta^{i+1} - \beta^i = \Delta t \, g(\beta^{i+1}, u^{i+1}) \quad a.e. \; on \; \Gamma_c , \tag{24}$$

where $b = b_N + b_T$.

The incremental problem $\mathbf{P_i^n}$ is solved using a fixed point method as follows.

For every $\bar{u} \in K$ let us denote by $s(\bar{u}) = \beta$ the solution of

$$\beta = \Delta t \, g(\beta, \bar{u}) + \beta^i. \tag{25}$$

For every $\beta \in H$ we denote by $u(\beta)$ the solution of

$$\begin{aligned} u \in K \quad a(u, v - u) + j(\beta, u, v - u^i) - j(\beta, u, u - u^i) \\ +b(\beta, u, v - u) \geq (F^{i+1}, v - u) \quad \forall v \in K. \end{aligned} \tag{26}$$

The existence of the solution for the problem (25) is clear, while inequalities such as (26) have a unique solution if the friction coefficient is sufficiently small i.e. $\bar{\mu} < \dfrac{m}{C}$. We suppose from now on that $\mu$ satisfies this condition.

We define the mapping $T : K \longrightarrow K$ by

$$\forall \bar{u} \in K \quad T(\bar{u}) = u(s(\bar{u})).$$

By a straightforward computation, we obtain that $\exists \, k_1 > 0$ such that for all $\bar{u}_1, \bar{u}_2 \in K$

$$|s(\bar{u}_1) - s(\bar{u}_2)| \leq k_1 \Delta t \, ||\bar{u}_1| + |\bar{u}_2|| \; |[\bar{u}_1 - \bar{u}_2]| \quad a.e. \; on \; \Gamma_c. \tag{27}$$

Let us now set $u_1 = u(s(\bar{u}_1)), u_2 = u(s(\bar{u}_2))$. Adding the inequalities (26) with $u = u_1$, $v = u_2$ and $u = u_2$, $v = u_1$ and using the properties of $a, j$ and $b$ one can show that $\exists \, k_2 > 0$ such that for all $\bar{u}_1, \bar{u}_2 \in K$

$$||u_1 - u_2||^2 \leq k_2 \int_{\Gamma_c} |[u_2]| \, |[u_1 - u_2]| \, |s(\bar{u}_1) - s(\bar{u}_2)| ds. \tag{28}$$

From the inequalities (27), (28) and the relation $||u|| \leq k_3$ (with $k_3$ independant of $\bar{u}$), satisfied by $u(s(\bar{u}))$, it follows that $T$ is a contraction mapping for sufficiently small $\Delta t$. Then $T$ has a unique fixed point $u$ and $(u, s(u))$ is the solution of the incremental problem $\mathbf{P}_\mathbf{i}^\mathbf{n}$ .

## 4. Numerical application

A numerical scheme based on the previous fixed point method is used to solve the implicit problem $\mathbf{P}_\mathbf{i}^\mathbf{n}$ . For the frictional contact problem without adhesion, various numerical methods have been presented in (Raous *et al.*, 1988). Here, for each state of adhesion ($\beta \in [0, 1[$) the quasi variational inequality (23) is solved as a sequence of minimization problems under constraints (Cangémi *et al.*, 1996 a).
This model has been used to describe the behaviour of the fiber/matrix interface of a ceramic composite. A numerical simulation of a fiber micro-indentation experiment in a SIC/SIC composite was realized in (Cangémi *et al.*, 1996 b). An excellent agreement between numerical and experimental results has been observed.

## References

Cangémi, L., Cocu, M. and Raous, M. (1996 a) Adhesion and friction model for the fiber/matrix interface of a composite, *Proceedings ESDA'96-ASME, July 1-4, Montpellier.*

Cangémi, L., Cocu, M. and Raous, M. (1996 b) Adhérence et frottement, une nouvelle approche pour les interfaces fibre/matrice, *Actes des 10ème Journées Nationales sur les Composites, Paris 29-31 Octobre, ENSAM.*

Cocu, M., Pratt E. and Raous M. (1996) Formulation and approximation of quasistatic frictional contact, *Int. J. Engng. Sci.*,Vol. **34**, pp. 783–798.

Frémond, M. (1987) Adhérence des solides, *J. Méc. Théor. et Appl.*,Vol. **6**, pp. 383–407.

Klarbring, A. (1990) Derivation and analysis of rate boudary-value problems of frictional contact, *European Journal of Mechanics, A/Solids*, Vol. **9**, no. 1, pp. 53–85.

Raous, M., Chabrand P. and Lebon, F. (1988) Numerical methods for frictional contact problems and applications, *J. Méc. Théor. et Appl.*,special issue, supplement no. 1 toVol. **6** , pp. 111–128.

# A CONTINUOUS MODEL FOR ZIEGLER-NATTA POLYMERIZATION

D. ANDREUCCI
*Dipartimento di Metodi e Modelli Matematici,*
*Università La Sapienza,*
*Roma, Italy.*

A. FASANO
*Dipartimento di Matematica,*
*Università di Firenze,*
*Firenze, Italy.*

AND

R. RICCI
*Dipartimento di Matematica,*
*Università di Milano,*
*Milano, Italy.*

## 1. Introduction

Our purpose here is the construction of a mathematical model for the phenomenon of polymerization in a Ziegler-Natta reactor. In this device a gaseous monomer diffuses through an agglomerate of catalytic particles. The process of polymerization takes place on the surface of each particle, so that a polymeric layer grows around it. The growth of such layers is affected, among other factors, by the local concentration of monomer, so that equations describing the expansion of the polymeric layer must be coupled with a diffusion equation for the monomer. Moreover that polymerization releases heat, so that an energy equation, determining temperature, is to be included in the model too. In fact, this is one of the most interesting features of the phenomenon, because the polymerization rate is affected by temperature; moreover, if the polymer becomes too hot, it may partially melt, thereby preventing the monomeric gas from reaching the catalyst surface (see [4]). Of course one tries to avoid the last occurrence.



*P. Argoul et al. (eds.),*
*IUTAM Symposium on Variations of Domains and Free-Boundary Problems in Solid Mechanics,* 219–226.

Another interesting feature of the process, which we do not deal with here, is the fragmentation of catalytic particles: at the early stages of polymerization, the forces exherted on the pellet by the growing polymeric layer break it into smaller pieces. Instead, we assume pellets to be (small) rigid bodies, from the beginning of the process. More precisely, we consider an aggregate of particles (which may be thought as the product of a previous fragmentation stage), surrounded by the monomer; the gas also diffuses into the aggregate. For more information on the physics of polymerization we refer the reader to [3] and to the references therein.

As in [4] we assume polymerization to take place uniformly on the surface of each catalytic pellet, which from now on will be therefore referred to also as a catalytic site. Moreover, we consider a continuous model, rather than a semidiscrete one as in [4].

## 2. Diffusion in the macroscopic polymeric matrix

Our model is based on a two scale approach: a "macroscale" which describes the diffusion of the gaseous monomer through the matrix formed by the polymer itself, and a "microscale" which describes the process of growth of the polymer around the calatyst particle. This picture is justified by the fact that the systems contains a very large number of catalytic particles whose size is very small.

At each positive time, the polymer forms a porous matrix encompassing the catalytic sites. The monomer molecules diffuse through this matrix, as well as inside the polymeric layer surrounding each pellet, and they eventually attach to the matrix itself when they reach a catalytic site. This induces a growth of the porous matrix, with a growth rate depending on the rate of polymerization.

It is reasonable to assume that the porosity $\varepsilon \in (0,1)$ of the matrix remains constant during this swelling process, according to what happens in an ideal model where the porous matrix is made by closely packed spheres of polymer surrounding each catalyst particle.

The catalytic sites distribution is approximated with a continuous distribution with a density function $\rho(x,t)$ representing the number of sites per unit of volume. Since the total number of catalytic sites is preserved, the sites density will decrease in time, as a consequence of the swelling of the porous matrix.

Conversely, the decrease in the sites density affects the growth rate of the matrix, since it reduces the probability of an encounter between a monomer molecule and a catalytic site.

The polymerization process releases heat, and most of the fundamental characteristics of the process are temperature depending. Moreover, the prediction of the temperature history is crucial for the control of the final product.

The four relevant "macroscopic" variables are the monomer concentration $M$ in the porous channels of the matrix, the sites density $\rho$, the velocity field $\mathbf{v}$ induced by the matrix swelling, and the temperature $T$.

Conservation of the number of catalytic sites gives

$$\frac{\mathrm{d}}{\mathrm{d}t}\rho + \rho\,\mathbf{div}(\mathbf{v}) = \mathbf{0}, \tag{1}$$

where $\frac{\mathrm{d}}{\mathrm{d}t}$ indicates the Lagrangian derivative following the local motion of the porous matrix.

The monomer concentration $M$ in the porous volume obeys the diffusion equation

$$\frac{\mathrm{d}}{\mathrm{d}t}(\varepsilon M) + \varepsilon M\,\mathbf{div}\mathbf{v} - \mathbf{div}(\mathbf{D}\nabla\mathbf{M}) = -\rho\mathbf{Q}, \tag{2}$$

where $Q$ is the rate of monomer absorption into the polymeric layer surrounding a catalytic site. We will give later the precise form of $Q$. $D$ denotes the monomer diffusivity and is in general a function of the temperature.

Since the monomers reacting at the catalytic site attach to the porous matrix, the swelling rate of the matrix is proportional to the rate of polymerization. This implies that the velocity field $\mathbf{v}$ solves

$$\mathbf{div}(\mathbf{v}) = \rho\hat{\mathbf{Q}}(\mathbf{1}-\varepsilon)^{-\mathbf{1}}, \tag{3}$$

where $\hat{Q}$ is the volume increase rate of a single pellet, due to polymerization. The factor $(1-\varepsilon)^{-1}$ is the ratio between the volume of the porous matrix and the total volume of the micropellets.

If we assume that our matrix and any other data are initially, and stay, radially symmetric, the unknowns in the velocity field (we are neglecting any rigid motion effect) reduce to the radial component $v\mathbf{u_r}$, where $\mathbf{u_r}$ denotes the unitary vector in the radial direction and $v$ is the Lagrangian derivative of the radial coordinate of a point in the matrix. Then the system can be rewritten, in Eulerian form, as

$$\frac{\partial}{\partial t}\rho + \left(\frac{\partial}{\partial r} + \frac{2}{r}\right)(\rho v) = 0\,, \quad \text{in } \Sigma, \tag{4}$$

$$\frac{\partial}{\partial t}(\varepsilon M) - \left(\frac{\partial}{\partial r} + \frac{2}{r}\right)\left(D\frac{\partial}{\partial r}M - \varepsilon M v\right) = -\rho Q\,, \quad \text{in } \Sigma, \tag{5}$$

$$\left(\frac{\partial}{\partial r} + \frac{2}{r}\right)v = \rho\hat{Q}(1-\varepsilon)^{-1}\,, \quad \text{in } \Sigma, \tag{6}$$

where $\Sigma$ is the moving domain $\Sigma = \{0 < r < R(t), t > 0\}$. In turn, $R(t)$ must be found by solving the ordinary differential equation

$$\dot{R} = v(R,t). \tag{7}$$

We recall that we assume $\varepsilon$ to be a positive constant; $D$ might depend on temperature.
We must also prescribe initial and boundary data as follows

$$M(r,0) = M_0(r)\,, \quad 0 < r < R_0\,, \tag{8}$$

$$M(R(t),t) = M_1(t)\,, \quad 0 < t\,, \tag{9}$$

$$\rho(r,0) = \rho_0(r)\,, \quad 0 < r < R_0\,, \tag{10}$$

$$R(0) = R_0 > 0\,. \tag{11}$$

The choices $M_0(r) \equiv 0$ and $\rho_0(r) \equiv$ *constant* are natural in view of the physical meaning of the model. See also (28) below. A precise definition of $Q$ and $\hat{Q}$ is given in next section (see (31) and (29)).

We assume that the pellets are so small that the temperature $T$ can be considered uniform across a single pellet, so that heat diffusion is relevant only for the macroscopical variables. Then, the heat equation for $T$ takes the form

$$c\frac{\partial T}{\partial t} - \left(\frac{\partial}{\partial r} + \frac{2}{r}\right)(k\nabla T - cTv) = \mu\hat{Q}\rho\,, \qquad \text{in } \Sigma, \tag{12}$$

where the source $\mu\rho\hat{Q}$ accounts for the heat released by polymerization. In practice, $\hat{Q}$ is replaced with the last term in (29) below. The conversion factor $\mu > 0$ is assumed to be constant.
Of course, (12) must be complemented with suitable initial and boundary data, e.g.,

$$T(r,0) = T_0(r)\,, \quad 0 < r < R_0\,, \tag{13}$$

$$-k\frac{\partial T}{\partial r}T(R(t),t) = h(T(R(t),t) - T_1(t))\,, \qquad 0 < t\,; \tag{14}$$

$h > 0$ is a (constant) heat exchange coefficient. Problem (12), (13), (14) is coupled with the macroscopic gas diffusion problem, and with the microscopical free boundary problem described in next section, through the terms $\hat{Q}$ in (12), and perhaps through $c$ and $k$. We also recall that $R$ is a free boundary to be found as the solution of (7), (11). Conversely, the quantities $\lambda$ in (22), and $D$ in (5), $d$ in (16), depend on $T$.

## 3. Polymerization around the microscopical particles

A more refined model for the monomer absorption-polymer growth process relies on the analysis of the phenomenon on a smaller scale, where we take into account a more precise mechanism for polymerization. As monomer molecules reach catalytic particles, the polymer grows into an aggregate of small elementary spheres, each of them containing a catalytic particle in its interior. Monomers have to diffuse into the growing polymer coating in order to reach the catalyst. Once they reach it they attach to the polymer and "push apart" the polymer itself from the catalyst. (This is the same growth mechanism described before but in a smaller scale and with a single catalytic site.)

Let us denote by $\hat{r}(t)$ the radial coordinate at time $t$ of the center of one of these microspheres. At time $t$ the microsphere has radius $s(t)$ and then its volume is $4/3\pi s^3(t)$. We assume that all the catalytic pellets are equal, so that, in all the microspheres,

$$s(0) = y_0 > 0\,, \tag{15}$$

where $y_0$ is a given constant. We denote by $y$ the radial coordinate in the generic microsphere. Then the growing polymer occupies the region between $y = y_0 > 0$, and $y = s(t)$. Inside this region we indicate by $m(y,t)$ the concentration of monomer per unit of free volume. Let us remark that, to be precise, we should use the notation $m(y,t;\hat{r}(t))$, $s(t;\hat{r}(t))$, which is abbreviated to the one above.

The monomer flux inside the microsphere is given by a diffusive term and a convective term due to the expansion of the microsphere itself,

$$\varepsilon_p \frac{\partial}{\partial t} m = \left(\frac{\partial}{\partial y} + \frac{2}{y}\right)\left(d\frac{\partial}{\partial y} m - \varepsilon m v_p\right)\,, \tag{16}$$

in $S = \{y_0 < y < s(t)\,, 0 < t\}$. We assume again that the porosity $\varepsilon_p \in (0,1)$ is constant; the diffusivity $d$ inside the microsphere might depend on temperature. As the swelling occurs only at $y = y_0$, the convective velocity obeys the equation

$$\left(\frac{\partial}{\partial y} + \frac{2}{y}\right) v_p = 0\,, \qquad \text{in } S, \tag{17}$$

and the radial growth rate $\dot{s}(t)$ equals the velocity $v_p$ at the boundary itself

$$\dot{s}(t) = v_p(s(t), t). \tag{18}$$

Let us stress the fact that $\dot{s}$ is a Lagrangian derivative, or, to be completely explicit,

$$\dot{s}(t) = \lim_{h \to 0} \frac{s(t+h;\hat{r}(t+h)) - s(t;\hat{r}(t))}{h}\,. \tag{19}$$

Boundary conditions have to be given for equations (16) and (17) at $y = y_0$. The polymerization rate is assumed to be proportional to the monomer concentration at the catalyst, i.e.,

$$v_p(y_0, t) = \lambda m(y_0, t). \tag{20}$$

The factor $\lambda$ is strongly dependent on $T$. On the other hand, the increase in volume is proportional to the incoming flux of monomers at the surface of the catalyst,

$$\nu v_p(y_0,t) = dm_y(y_0,t) - \varepsilon_p v_p(y_0,t) m(y_0,t), \tag{21}$$

which can be trasformed, by means of (20), into a non-linear boundary condition for $m$ at $y_0$

$$dm_y(y_0,t) = \lambda \nu m(y_0,t) + \lambda \varepsilon_p m(y_0,t)^2\,; \tag{22}$$

here $\nu > 0$ is a given constant. At the boundary of the microsphere the concentration equals the macroscopic monomer concentration at the macroscopic location of the center of the microsphere, i.e.,

$$m(s(t),t) = KM(\hat{r}(t),t)\,, \qquad 0 < t\,, \tag{23}$$

where $K > 0$ is a constant partition factor.

Equation (17) can be easily integrated to yield

$$y^2 v_p(y,t) = constant\,, \tag{24}$$

and (20), (24) can be used to relate the growth rate of the microsphere to the concentration of the monomer at $y = y_0$

$$\dot{s}(t) = \frac{\lambda m(y_0,t) y_0^2}{s^2(t)}, \tag{25}$$

which can be interpreted as a free boundary condition for the radius of the microsphere.

Once the concentration $M(\hat{r}(t),t)$ is known as a function of time, equations (16) and (25), together with the boundary conditions (23) and (22) give a free boundary problem which has been studied in [1].

We have also to take into account that the microsphere moves into the macrosphere with speed $\dot{\hat{r}}$ given by the differential equation

$$\dot{\hat{r}} = v(\hat{r}(t),t)\,. \tag{26}$$

Finally, we give an explicit form for the quantities $\hat{Q}$, $Q$ appearing in the equations for $v$ and $M$. Assuming that the porosity $\varepsilon$ inside the macrosphere stays constant, we get the following relation between the radius $s(t)$ of the microsphere and the density of the sites $\rho(r,t)$

$$1 - \varepsilon = \frac{4}{3}\pi \rho(\hat{r}(t),t) s^3(t), \tag{27}$$

where $\hat{r}(t)$ denotes the location of the center of the microsphere. Equality (27) determines the evolution of the radius of the microsphere as a function of the local site density. Also note that, the radius of the catalytic particle $y_0 > 0$ and $\rho_0(r) = \rho(r, 0)$ must be connected by

$$1 - \varepsilon = \frac{4}{3}\pi\rho_0 y_0^3 \,. \tag{28}$$

Then

$$\rho\hat{Q}(\hat{r}(t), t) = \rho 4\pi y_0^2 v_p(y_0, t) = \rho 4\pi s^2 \dot{s} = 3(1 - \varepsilon)\dot{s}/s\,; \tag{29}$$

in last equality we have made use of (27). Therefore we replace (6) with

$$\left(\frac{\partial}{\partial r} + \frac{2}{r}\right) v = \frac{3\dot{s}}{s}\,, \qquad \text{in } \Sigma, \tag{30}$$

and we recover (27) a posteriori, as a consequence of (28) and of (4), (30). Let us also say that (30) relates the increase of the macroscopic volume with the corresponding microscopical quantity.

The sink term $-\rho Q$ in the diffusion equation for the monomer in the macrosphere equals the flux of monomer entering the microspheres, i.e., $Q$ is given by

$$Q(\hat{r}(t), t) = d\frac{\partial m}{\partial y}(s(t), t; \hat{r}(t))\,, \tag{31}$$

which has to be substituted into (5). Note that in (31) no convective term is present: that is, the monomer is absorbed into the microsphere only through diffusion.

As a side remark, we note that condition (20) can be replaced by

$$m(y_0, t) = 0\,, \tag{32}$$

corresponding to the formal limit $\lambda \to \infty$, i.e., the fast polymerization limit. Accordingly, (22) and (25) become

$$dm_y(y_0, t) = \nu v_p(y_0, t)\,, \qquad \dot{s}(t) = \frac{dm_y(y_0, t)}{\nu s^2(t)}\,. \tag{33}$$

Finally, the complete model consists in: equations (4), (5), (30), (7), complemented with (8)–(11), (31) (macroscopical diffusion problem); equations (15), (16), (22), (23), (25), (26) (microscopical diffusion problem); equations (12), together with (29), and (13), (14) (heat transfer problem).

## 4. Quasi-steady state approximation of the microproblem

A considerable mathematical simplification can be obtained assuming that the diffusion process inside the microsphere is quasi-stationary, i.e. if equation (16) reduces to

$$\left(\frac{\partial}{\partial y}+\frac{2}{y}\right)\left(d\frac{\partial}{\partial y}m-\varepsilon_p m v_p\right)=0\,, \tag{34}$$

In this case we can integrate twice (34) and evaluate at $y = y_0$ and $y = s(t)$, obtaining

$$m(y_0,t)=\exp\left(\alpha m(y_0,t)\left(\frac{1}{s(t)}-\frac{1}{y_0}\right)\right)\left(m(s(t),t)+\frac{\beta}{\alpha}\right)-\frac{\beta}{\alpha}, \tag{35}$$

where $\alpha = \varepsilon_p y_0^2 \lambda/d$ and $\beta = \lambda \nu y_0^2$.

Finally we can substitute the boundary condition (23) and the relation (25) into (35), ending with a relation between $s(t)$, $\dot{s}(t)$ and the value $M(\hat{r}(t),t)$ of the monomer concentration at the location occupied by the microsphere,

$$\frac{\left(\frac{s^2\dot{s}}{y_0^2\lambda}+\frac{\nu d}{\varepsilon_p}\right)}{\left(KM+\frac{\nu d}{\varepsilon_p}\right)}=\exp\left(\frac{\varepsilon_p}{d}s^2\dot{s}\left(\frac{1}{s}-\frac{1}{y_0}\right)\right)\,. \tag{36}$$

If we look at $M(\hat{r}(t),t)$ as a given function of $t$, we can consider equation (36) as an O.D.E. for $s$ ((36) is not in normal form but it can be easly proved that it can be globally solved in time, as long as $M(\hat{r}(t),t)$ exists).

Finally we can rewrite the sink term in the monomer diffusion equation in terms of the local concentration $M$ ant the corresponding solution $s$ of (36). It is also convenient to reset the problem into lagrangean coordinates. This eliminates the difficulty of "following the microparticle" to relate $M$ and $s$, since the lagrangean coordinate is the "flag" of the microparticle itself (see [2] for datail of the equations in lagrangean coordinates).

## References

1. D. Andreucci, A. Fasano, R. Ricci, *On the growth of a polymer layer around a catalytic particle: a free boundary problem*, to appear in Nonlinear Diff. Eq. Appl..
2. D. Andreucci, A. Fasano, R. Ricci, *A continuous model for the heterogeneos Ziegler-Natta polymerization*, Quad. Dip. Mat. Univ. Milano 29/1996.
3. N.A. Dotson, R. Galvan, R.L. Laurence, M. Tirrell, *Polymerization process modeling*, VCH Publishers, Cambridge, 1996.
4. R.L. Laurence, M.G. Chiovetta *Heat and mass transfer during olefin polymerization from the gas phase*, Polymer Reaction Engng, K.H. Reichert and W. Geiseler eds, Hanser (Munich), 1983, pp. 73–112.

# A STEADY DISPLACEMENT-BASED FORMULATION FOR THE ANALYSIS AND DESIGN OF LARGE DEFORMATION CONTINUOUS ELASTOPLASTIC MANUFACTURING PROCESSES

D. BALAGANGADHAR
*Department of Mechanical and Industrial Engineering*
*University of Illinois at Urbana-Champaign, Urbana, IL 61801, USA*

AND

D.A. TORTORELLI
*Department of Mechanical and Industrial Engineering*
*Also Department of Theoretical and Applied Mechanics*
*University of Illinois at Urbana-Champaign, Urbana, IL 61801, USA*

## 1. Introduction

Many manufacturing processes such as metal forming, continuous casting, laser surface heat treatment, etc. are steady state in nature. However, the materials used in these processes exhibit history dependent response *e.g.* plasticity. Hence, the analyst must consider the deformation history which each material particle experiences, even though the process is steady.

In this work we define the manufacturing process boundary value problem on a reference frame (RF) to take advantage of its steady-state nature. The RF is associated with the undeformed configuration and displacement is the primary response field. The formulation has both Lagrangian and Eulerian characteristics; the Lagrangian characteristic arises from the association of the RF with the undeformed configuration and the Eulerian characteristic arises from the translation of the undeformed configuration relative to the RF. History dependent material state variables (*e.g.* the plastic strain and cumulative plastic strain) also appear as primary variables in this mixed formulation. Another novel aspect of this formulation is the treatment of the die/workpiece contact conditions. We exploit the



*P. Argoul et al. (eds.),*
*IUTAM Symposium on Variations of Domains and Free-Boundary Problems in Solid Mechanics*, 227–236.

kinematic description of the formulation to replace the inequality contact constraint with an equality constraint.

Traditionally steady state processes with history dependent material behavior are analyzed with Lagrangian kinematic descriptions in which the displacement is the primary response field [1, 2, 3]. This method requires a transient analysis to determine the evolution of the history dependent material variables. The transient analysis terminates when 'steady state' behavior is observed. However, oscillations appear in the computed steady state Lagrangian fields due to the interaction between the time and space discretizations [2]. Moreover, these methods are computationally inefficient since they require large meshes, transient analyses, complicated contact algorithms and adaptive meshing strategies.

Steady state processes have also been modeled using Eulerian kinematic descriptions [4, 5]. Most of these formulations use velocity as the primary response field, viscoplastic material behavior and neglect elastic deformation [4]. If necessary, elastic strain effects can be included by an 'elastic reanalysis' [4] or by the 'initial stress-rate method' [6]. The material state variables are evaluated by the integration of the evolution equations along their path lines [7, 8]. However, the path-lines are not known *a priori* in the deformed configuration; consequently iterative techniques are required. In contrast, the material state evolution equations can also be solved simultaneously with the momentum balance equation [9]. However, to date, results obtained from this method exhibit oscillations [10] due to numerical instabilities that arise from the hyperbolic evolution equations. Another issue in Eulerian formulations is the treatment of free surfaces. Since the governing equations are expressed on the deformed configuration, the free surface locations are not known *a priori*. To determine their location, time consuming iterative updating or successive recalculation are used [6].

Arbitrary Lagrangian-Eulerian (ALE) kinematic descriptions, which combine both the Eulerian and Lagrangian descriptions, are also used to solve history-dependent problems [11, 12, 13, 14, 15, 16, 17, 18]. ALE formulations are suitable for problems where the region of interest moves in a fixed domain, such as in crack propagation [16]. The formulation requires a transient analysis for history-dependent problems. Both velocity based [12, 13, 14] and displacement based [15, 16, 17, 18] ALE formulations are reported.

Several computational advantages are achieved in this analysis verses the traditional Lagrangian and Eulerian analyses. When comparing to Lagrangian formulations we note that our analysis is steady so that the notion of a 'consistent tangent operator' [19] is moot; transient adaptive refinement schemes are not required [20, 21, 22]; and the inequality contact constraint is replaced with a simple equality constraint. When comparing to Eulerian formulations we note that the RF is associated the undeformed configuration so that free surface corrections are not necessary; the RF uses displacement so that the elastic response is obtained directly; and the material evolution equations are solved directly via the streamline upwind Petrov-Galerkin method to circumvent the numerical instabilities previously encountered.

To exemplify the analysis, we analyze and design the extrusion process of metal slab. The analysis uses a simplified two-dimensional plane strain assumption, a finite strain elastoplastic material model, and a frictionless die. In the optimization, the die shape is designed to minimize the drawing force. The sensitivities for the nonlinear programming algorithm are evaluated via direct differentiation method [23].

## 2. Analysis and Optimization Summary

The body is in configuration $V$ at time $t = 0$ and its material particles are identified in $V$ via the position vector $\mathbf{X}$, defined with respect to the material coordinate frame $\{\mathbf{0} : \mathbf{e}_1, \mathbf{e}_2, \mathbf{e}_3\}$. Here, $\{\mathbf{e}_1, \mathbf{e}_2, \mathbf{e}_3\}$ comprises an orthonormal basis. The motion of the body is represented by the smooth mapping $\mathbf{f}$ such that at any given time $t \in I$, the location $\mathbf{x}$ (with respect to the material coordinate frame) of a material point which occupied the position $\mathbf{X}$ at time $t = 0$ in $V$ is given by

$$\mathbf{x} = \mathbf{f}(\mathbf{X}, t) \tag{1}$$

Here $I = [0, T]$ is the time interval in which the problem is defined and $T$ is the terminal time in the analysis.

In most steady state processes, the region over which the deformation evolves is primarily restricted to the vicinity of the loading. We represent this region by $V_0$ in the undeformed configuration; $V_0$ corresponds to the configuration $V_t$ in the deformed configuration (see Fig. 1), *i.e.* $V_t = \mathbf{f}(V_0, t)$. One may think of $V_0$ as a control volume, however here it is not associated with the deformed configuration (as in a Eulerian analysis), rather it is associated with the undeformed configuration (as in a Lagrangian analysis).

We now introduce a reference configuration $V_r$. Material particles are identified in $V_r$ via the position vector $\mathbf{r}$, defined with respect to its distinct reference coordinate frame $\{\hat{\mathbf{0}} : \hat{\mathbf{e}}_1, \hat{\mathbf{e}}_2, \hat{\mathbf{e}}_3\}$ , where $\{\hat{\mathbf{e}}_1, \hat{\mathbf{e}}_2, \hat{\mathbf{e}}_3\}$ comprises an orthonormal basis. At any given time $t$, $V_r$ is mapped into the undeformed configuration $V_0$ by the map $\mathbf{g}$ (see Fig. 1). Note that $V_r$ only maps to the control volume portion of the undeformed body and this portion changes in time to follow the loading zone *i.e.*, $V_0 = \mathbf{g}(V_r, t)$. The position vectors $\mathbf{r}$ in the reference configuration and $\mathbf{X}$ in the undeformed configuration are related by

$$\mathbf{X} = \mathbf{g}(\mathbf{r}, t) \tag{2}$$

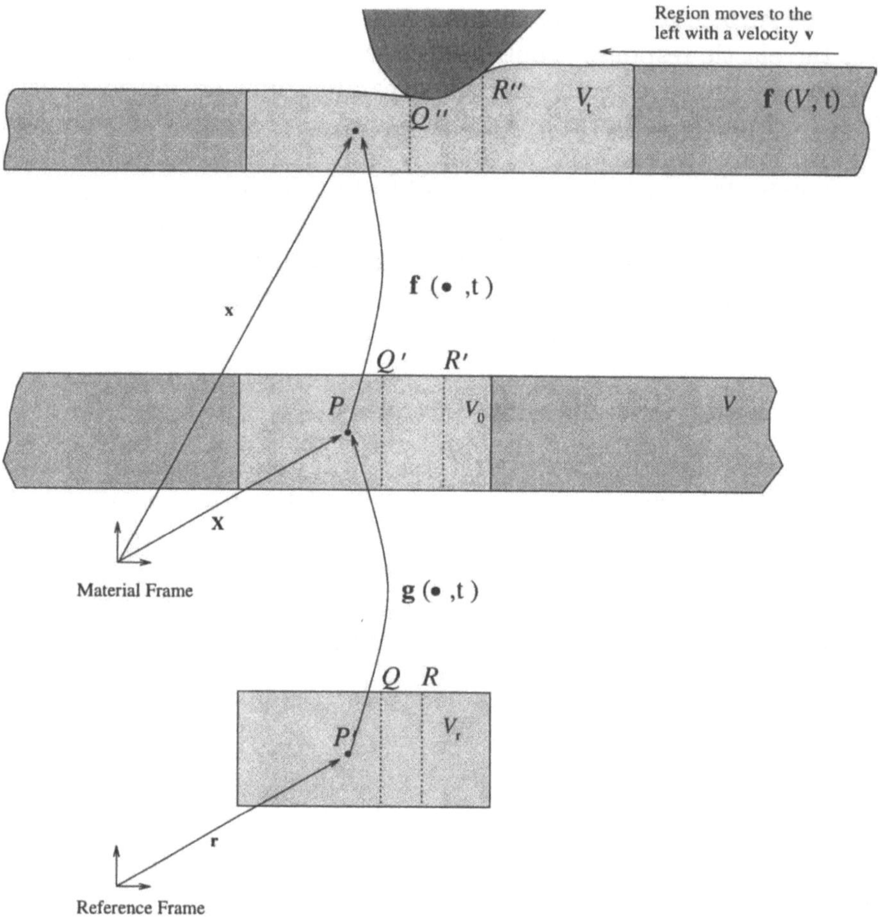


*Figure 1.* Kinematic model

Fields defined on the undeformed configuration $V_0$ are denoted as $()(\mathbf{X}, t)$, and for each of these fields we define the corresponding reference field $\hat{()}(\mathbf{r}, t)$ on $V_r$ such that their values are equal at corresponding points, *i.e.*

$$\hat{()}(\mathbf{r}, t) = ()(\mathbf{g}(\mathbf{r}, t), t) \tag{3}$$

Note that at different times, *e.g.* $t$ and $t'$, the material point associated with $\mathbf{r}$ changes, *e.g.* $\mathbf{X} = \mathbf{g}(\mathbf{r}, t)$ and $\mathbf{X}' = \mathbf{g}(\mathbf{r}, t')$, and hence the reference fields when evaluated at $(\mathbf{r}, t)$ and $(\mathbf{r}, t')$ will generally not be equal. If the function values for the pairs $(\mathbf{r}, t)$ and $(\mathbf{r}, t')$ agree for all $\mathbf{r}$ in $V_r$ and for all $t$ in $I$, then the process is deemed steady and we have *e.g.* $\hat{\mathbf{u}}(\mathbf{r},t) = \hat{\mathbf{u}}(\mathbf{r}, t')$; and consequently $\mathbf{u}(\mathbf{X}, t) = \mathbf{u}(\mathbf{X}', t')$.

In Lagrangian elastoplastic finite deformation formulations, the momentum balance and the material state evolution equations are formulated over the undeformed configuration into a *transient* initial-boundary value problem. For this purpose, we adopt a multiplicative decomposition of the deformation gradient and the associative flow rule based on the principle of maximum plastic dissipation [19]. Here, however these equations are transformed to the reference configuration via the appropriate transformation rules induced by the map **g** and the steady state nature of the process is used to formulate a *steady* boundary value problem.

The objective of the present contact analysis is to determine the contact surfaces $Q''R''$, $Q'R'$ and/or $QR$ in the deformed, undeformed, and RF configurations, respectively. In Eulerian formulations the die/workpiece contact problem is a non-issue since the contact surface in the deformed configuration $Q''R''$ is known *a priori*. In Lagrangian analyses, the surface $Q'R'$ on the undeformed configuration that contacts the die is not known *a priori*; rather it is identified via an iterative procedure in which the *inequality* contact constraints are satisfied [24]. In the RF formulation, the surface $QR$ is identified *a priori*. Displacement boundary conditions are prescribed over this surface to ensure that the surface $Q'R'$ contacts the die surface $Q''R''$. This 'prescribed displacement' and the mapping $\mathbf{g}(\bullet, t)$ are determined so that the traction at $Q$ and $R$ (and hence $Q'R'$) is zero and the tangential traction over the remainder of the surface satisfies the given friction conditions. Thus, the traditional inequality contact constraint is replaced with an equality constraint.

For the optimization problem, design variables that parameterize the die shape as well as cost and constraint functions that quantify the product quality are identified. A nonlinear programming algorithm [25] is then used to solve the optimization problem. This algorithm requires the sensitivities of the cost and constraint functions with respect to the design variables. These sensitivities are accurately and efficiently computed using the analytical direct differentiation method [23].

## 3. Numerical results

The formulations discussed in the previous sections are used to optimize the die shape for the extrusion of the metal slab shown in Fig. 3. (Normalized dimensions and material data are also provided in Fig. 3.) An infinitely long slab of height $h_i = 2.0$ is extruded through a convergent die with an exit height $h_o = 1.5$. The slab is drawn at a constant velocity (however, the magnitude of the velocity does not affect the results due to the rate independent material response). The objective of the optimization is to design the die geometry that will minimize the drawing force.

$$h_i = 2.0 \qquad h_o = 1.5 \qquad L = 4.0$$

$$Q = e_p$$
$$\phi = \left[(\mathbf{CS}\cdot\mathbf{SC}) - \tfrac{1}{3}(\mathbf{S}\cdot\mathbf{C})^2\right]^{\frac{1}{2}} - \kappa(e_p)$$
$$\kappa(e_p) = 0.1 + 0.1e_p$$
$$H = \sqrt{\tfrac{2}{3}}$$

$$\Psi = \tfrac{\mu}{2}\left(J_{\mathbf{C}}^{-\frac{1}{3}}\mathbf{C}_p^{-1}\cdot\mathbf{C} - 3\right) + U(J_{\mathbf{C}})$$
$$\mu = 1.0$$
$$U(J_{\mathbf{C}}) = 5.0\left[\log\left(J_{\mathbf{C}}{}^{0.5}\right)\right]^2$$
$$\rho_0 = 1.0$$

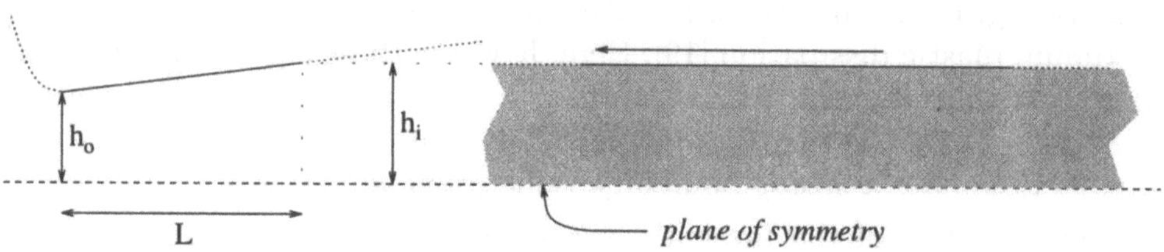


*Figure 2.* **Laser surface treatment: problem definition**

### 3.1. ANALYSIS

The standard Galerkin finite element method is used to discretize the momentum balance equation, whereas to ensure numerical stability, the streamline upwind Petrov-Galerkin finite element method is used to discretize the hyperbolic material state evolution equations. Relaxation is used for the first few Newton iterations and then the standard Newton scheme is employed to solve the nonlinear discretized equations.

The domain $V_r$ (and $V_0$) for the finite element analysis is 20 *units* long; the exit of the die is located 10 *units* from the outflow boundary. The discretization includes local mesh refinement in the region adjacent to the die. A convergence study is performed and it is found that the $25 \times 6$ mesh shown in Fig. 3 (c) is adequate. The analysis for the original die geometry (see Fig. 4) converges in the 11 iterations and the results are shown in Figs. 5 - 8.

For comparison purposes, the same problem is also analyzed via a transient Lagrangian analysis using ABAQUS [26]. The domain for the Lagrangian analysis is of length 45 *units*. A convergence study is performed to consider the effects of the mesh size and the time step. The results converge for a $60 \times 6$ mesh and a time step of $0.1s$. The ABAQUS results predict the same trends that are obtained from our analysis; the numerical values vary by approximately 30%. This variation is attributed to the differences in the material and kinematic models.

The RF and Lagrangian analyses require 200 and 3000 CPU seconds respectively on the same IBM 590 machine, thus establishing the computational efficiency of the RF method. However the unsymmetric stiffness matrix and the increased number of degrees of freedom per element (due the material state variables), increases the memory requirement for the RF method. The RF analysis requires 6 MB of memory to store the stiffness matrix in unsymmetric, banded form versus 2.5 MB required to store the stiffness matrix in the symmetric banded form for the Lagrangian analysis.

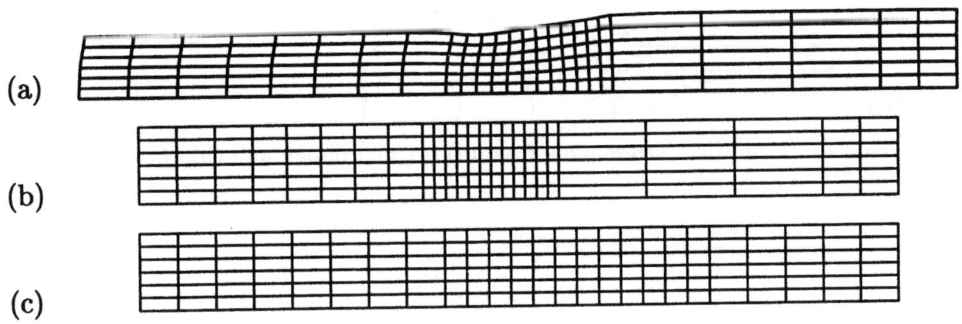


*Figure 3.* The meshes in the (a) deformed, (b) undeformed and (c) reference configurations

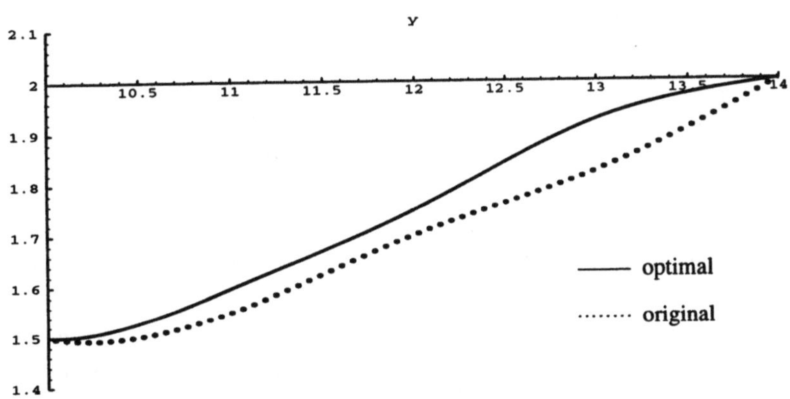


*Figure 4.* Die shapes for the original and optimal designs

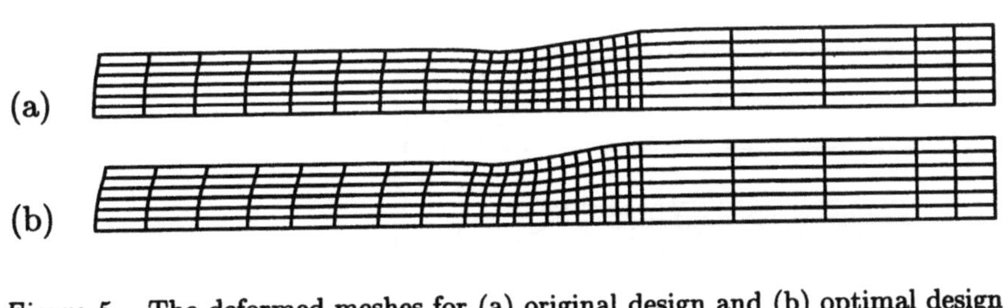


*Figure 5.* The deformed meshes for (a) original design and (b) optimal design

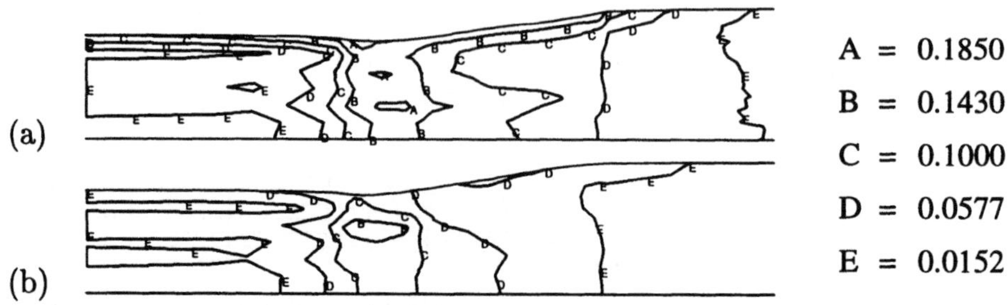


*Figure 6.* Equivalent stress plots from (a) original design and (b) optimal design

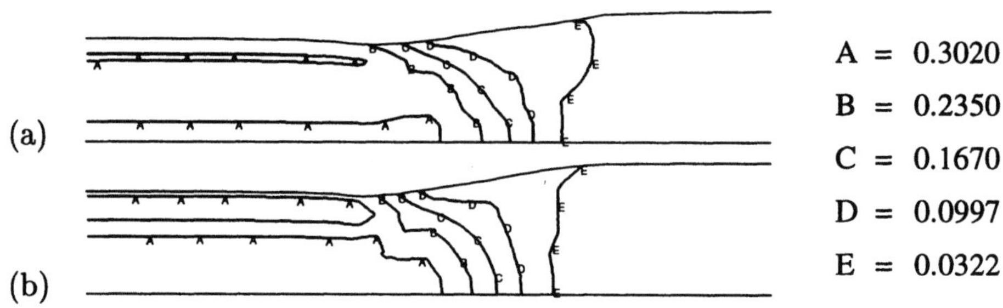


*Figure 7.* $C_p^{11}$ plots from (a) original design and (b) optimal design

### 3.2. OPTIMIZATION

The die shape is represented by cubic splines [24] and their parameters are optimized to reduce the drawing force required to extrude the slab. This force is obtained by integrating the normal tractions across the outflow boundary. Constraint functions are included to ensure the monotonicity of the die shape.

The optimization requires 5 iterations to reduce the drawing force from $6.357 \times 10^{-2}$ to $3.335 \times 10^{-2}$. The original and optimal die shapes appear in Fig. 4. The response for the optimal and original designs are shown in Figs. 5 - 8. It is seen from Fig. 6 that the equivalent stress is reduced at the outlet section in the optimal design. Further, note that the optimal die

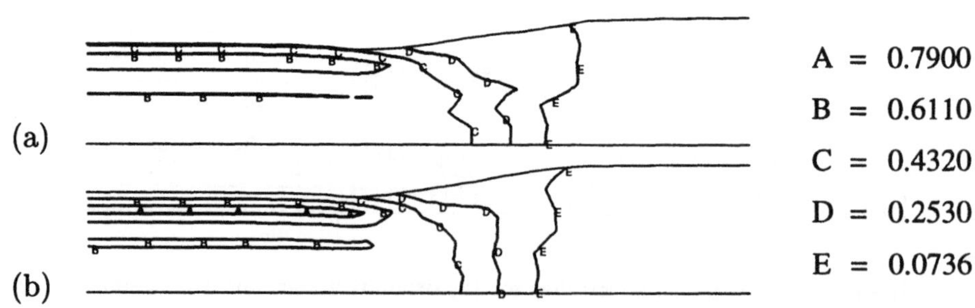


*Figure 8.* $e_p$ plots from (a) original design and (b) optimal design

lowers the equivalent stress in the other regions of the slab. The material state variables are uniformly reduced as seen in Fig. 7 which illustrates the longitudinal plastic strain and Fig. 8 which illustrates the equivalent plastic strain. Elastic spring backs are 14.9% and 12.7% in the original and optimal designs, respectively.

## 4. Conclusions

A moving reference frame formulation is developed for large deformation steady processes that uses displacement and material state variables as the primary response fields. This formulation is then applied to analyze the extrusion of a metal slab through a die and to optimize the die shape.

## 5. Acknowledgments

The research was supported by the National Science Foundation grant NYI-DMI 9358132.

## References

1. E. H. Lee, R. L. Mallet, and W. H. Yang. Stress and deformation analysis of metal extrusion processes. *Sudam report, 76-2*, Stanford University, 1976.
2. E. J. Appleby, C. Y. Lu, R. S. Rao, M. L. Devenpeck, P.K.Wright, and O.Richmond. Strip drawing : A theoretical-experimental comparison. *International Journal of Mechanical Science*, 26-5:351–362, 1984.
3. J. T. Carroll and J. S. Strenkowski. Finite element models of orthogonal cutting with applications to single point diamond turning. *International Journal of Mechanical Science*, 30-12:899–920, 1988.
4. O. C. Zienkiewicz, P. C. Jain, and E. Onate. Flow of solids during forming and extrusion: some aspects of mumerical solution. *International Journal of Solids and Structures*, 14:15–38, 1978.
5. J. S. Strenkowski and K-J. Moon. Finite element prediction of chip geometry and tool/workpiece temperature distributions in orthogonal metal cutting. *Journal of Engineering for Industry*, 112:313–318, 1990.
6. P. R. Dawson and E. G. Thompson. Finite element analysis of steady-state elasto-visco-plastic flow by the initial stress-rate method. *International Journal of Numerical Methods in Engineering*, 12:47–57, 1978.
7. M. Viriyauthakorn and B. Caswell. Finite element simulation of viscoelastic flow. *Journal of Non-Newtonian Fluid Mechanics*, 6:245–267, 1980.
8. P. R. Dawson. Viscoplastic finite element analysis of steady-state forming processes including strain history and stress flux dependence. *Applications of Numerical Methods to Forming Processes*, 28:55–67, 1978.
9. E. G. Thompson, J. F. T. Pittman, and O. C. Zienkiewicz. Some integration techniques for the analysis of viscoelastic flows. *International Journal for Numerical Methods in Fluids*, 3:165–177, 1983.
10. A. Agrawal and P. R. Dawson. A comparison of Galerkin and streamline techniques for integrating strains from an Eulerian flow field. *International Journal of Numerical Methods in Engineering*, 21:853–881, 1985.
11. J. Huetink, P. T. Vreede, and J Van Der Lugt. Progress in mixed Eulerian-Lagrangin finite element simulation of forming proccesses. *International Journal for Numerical*

*Methods in Engineering*, 30:1441–1457, 1990.
12. R. Rakotomalala and P. Joyot. Arbitrary Lagrangin-Eulerian thermomechanical finite-element model of material cutting. *Communications in Numerical Methods in Engineering*, 9:975–987, 1993.
13. Y-K. Hu and W. K. Liu. An ale hydrodynamic lubrication finite element method with application to strip rolling. *International Journal for Numerical Methods in Engineering*, 36:855–880, 1993.
14. W. K. Liu, H. Chang, J-S. Chen, and T. Belytschko. Arbitrary Lagrangin-Eulerian Petrov-Galerkin finite elements for nonlinear continua. *Computer Methods in Applied Mechanics and Engineering*, 68:259–310, 1988.
15. R. B. Haber. A mixed Eulerian-lagrangian displacement model for large deformation analysis in solid mechanics. *Computer Methods in Applied Mechanics and Engineering*, 43:277–292, 1984.
16. H. S. Lee and R. B. Haber. Eulerian-lagrangian methods for crack growth in creeping materials. *Advanced Computational Methods for Material Modeling - ASME*, 268:141–153, 1993.
17. S. Ghosh and N. Kikuchi. An arbitrary Lagrangin-Eulerian finite element method for large deformation analysis os elastic-viscoplastic solids. *Computer Methods in Applied Mechanics and Engineering*, 86:127–188, 1991.
18. S. Ghosh and R. Suresh. R-s adapted arbitrary Lagrangin-Eulerian finite element method for metal forming problems with strain localization. *International Journal for Numerical Methods in Engineering*, 39:3247–3272, 1996.
19. J. C. Simo and R. L. Taylor. Consistent Tangent Operators for Rate-Independent Elasto-Plasticity. *Computer Methods in Applied Mechanics and Engineering*, 48:101–118, 1985.
20. Special issue on adaptive meshing. *Internatioonal Journnal for Numerical methods in Engineering*, 32(4), 1991.
21. Reliability in computational mechanics. *Computer Methods in Applied Mechanics and Engineering*, 10, 1992.
22. N. Palle. An adaptive mesh refinement scheme for solidification problems. *Ph.D Thesis*, University of Illinois at Urbana-Champaign, 1993.
23. D. A. Tortorelli and Z. Wang. A systematic approach to shape sensitivity analysis. *International Journal of Solids and Structures*, 30:1181–1212, 1993.
24. C. O. Spivey and D. A. Tortorelli. Tangent operators, sensitivity expressions, and optimal design of nonlinear elastica in contact with applications to beams. *International Journal for Numerical Methods in Engineering*, 37:49–73, 1994.
25. *DOT 3.0.* Vanderplaats, Miura & Associates, Inc., Goleta, CA.
26. *ABAQUS 5.4.* Hibbitt, Karlsson and Sorensen, Inc., Providence, RI.

# STRONG SOLUTIONS TO EQUATIONS OF VISCO-THERMO-PLASTICITY WITH A TEMPERATURE-DEPENDENT HYSTERETIC STRAIN – STRESS LAW

PAVEL KREJČÍ AND JÜRGEN SPREKELS
*Weierstrass Institute for Applied Analysis and Stochastics*
*Mohrenstr. 39, D-10117 Berlin, Germany*

**Abstract.** We propose a thermodynamically consistent model for the uniaxial behaviour of visco-thermo-elasto-plastic materials where the thermoplastic stress component $\sigma^p$ is characterized by a constitutive law of the form $\sigma^p(x,t) = \mathcal{P}[\varepsilon, \theta(x,t)](x,t)$, where $\varepsilon$, $\theta$ denote the fields of strain and absolute temperature, respectively, and where $\{\mathcal{P}[\cdot,\theta]\}_{\theta>0}$ is a family of (rate-independent) hysteresis operators of Prandtl-Ishlinskii type, parametrized by the absolute temperature. The system of state equations governing the space-time evolution of the material is derived. A survey of existence, uniqueness and continuous dependence results for an initial-boundary value problem for this system is given.

## 1. Introduction

In [7] we initiated the study of systems of equations of thermo-elasto-plasticity with a strain – temperature – stress law which is assumed in the form of a Prandtl-Ishlinskii hysteresis operator with a temperature-dependent density function as a generalization of standard isothermal elasto-plasticity models [11, 2, 3, 14, 1, 6]. This research was motivated by the fact that at each constant temperature the classical Prandtl-Ishlinskii theory provides a good qualitative description of the loading – unloading experiment, while at different temperatures the corresponding hysteresis diagrams are in general different.

The main question consists in incorporating such a model into the general framework of thermodynamics. In other words, we have to define the state functions of internal energy, free energy and entropy in such a way that for every process governed by the system "equation of motion + energy



*P. Argoul et al. (eds.),*
*IUTAM Symposium on Variations of Domains and Free-Boundary Problems in Solid Mechanics*, 237–244.

balance" the absolute temperature is positive and and the second principle of thermodynamics is fulfilled e.g. in the form of the Clausius-Duhem inequality [10, 12].

Since the constitutive law itself is given by an operator which takes into account not only the instantaneous values of the internal variables, but also the memory configuration of the system, it is natural to expect that the internal energy, free energy and entropy will also be expressed in terms of "state operators" invoving memory rather than "state functions". It turns out that explicit formulas for these operators can be derived in a form analogous to Prandtl-Ishlinskii hysteresis potentials [1, 6].

In this paper we consider an initial-boundary value problem in the one-dimensional thermo-elasto-plastic case, where also the effects of viscosity, kinematic hardening and thermic dilation are taken into account. From the mathematical point of view, the viscosity term is substantial for the proof of existence and uniqueness of solutions for every given data. Recall that this is not the case for temperature - independent models, where the dissipation properties of hysteresis operators turn out to be sufficiently strong, see [4, 6].

We only give here a survey of existence, uniqueness and continuous dependence results without proofs. The complete detailed analysis will be performed in [8].

## 2. Hysteresis Constitutive Operators of Elastoplasticity

L. Prandtl's normalized elastic - perfectly plastic model, corresponding to the rheological combination in series of one elastic (with elasticity modulus 1) and one rigid-perfectly plastic element, provides the simplest example for a hysteresis constitutive operator. It can formally be described as follows.

Let $r > 0$ (the yield limit) and $\sigma_r^0 \in [-r, r]$ (the initial stress) be given numbers. For any input function $\varepsilon \in W^{1,1}(0,T)$, we define the output $\sigma_r \in W^{1,1}(0,T)$ as the solution to the variational inequality (the superimposed dot denotes the time derivative)

$$\sigma_r(t) \in [-r, r] \quad \forall\, t \in [0,T], \tag{1}$$

$$(\dot{\varepsilon}(t) - \dot{\sigma}_r(t))\,(\sigma_r(t) - \tilde{\sigma}) \geq 0 \quad \forall\, \tilde{\sigma} \in [-r, r], \quad \text{a.e. in } (0,T), \tag{2}$$

$$\sigma_r(0) = \sigma_r^0. \tag{3}$$

It can easily be proved that the problem (1) – (3) admits a unique solution $\sigma_r \in W^{1,1}(0,T)$ for every $\varepsilon \in W^{1,1}(0,T)$ and $\sigma_r^0 \in [-r, r]$ (even in the multi-dimensional case, see [5, 6, 13, 14]). The solution operator

$$\mathfrak{s}_r : [-r, r] \times W^{1,1}(0,T) \to W^{1,1}(0,T); \quad (\sigma_r^0, \varepsilon) \mapsto \sigma_r \tag{4}$$

is called *stop operator*.

It is immediately seen that for piecewise monotone inputs $\varepsilon$ the output $\mathfrak{s}_r[\sigma_r^0, \varepsilon]$ can be explicitly described in each monotonicity interval $[t_0, t_1] \subset [0,T]$. Indeed, from the definition we can infer that

$$\mathfrak{s}_r[\sigma_r^0, \varepsilon](t) = \begin{cases} \min\{r, \mathfrak{s}_r[\sigma_r^0, \varepsilon](t_0) + \varepsilon(t) - \varepsilon(t_0)\}, & t \in [t_0, t_1], \\ \quad \text{if } \varepsilon \text{ is non-decreasing in } [t_0, t_1], & \\ \max\{-r, \mathfrak{s}_r[\sigma_r^0, \varepsilon](t_0) + \varepsilon(t) - \varepsilon(t_0)\}, & t \in [t_0, t_1], \\ \quad \text{if } \varepsilon \text{ is non-increasing in } [t_0, t_1]. & \end{cases} \tag{5}$$

The stop operator has the following properties (For a proof, see [1, 6]).

**Proposition 1**

(i) *Let* $\sigma_r^0 \in [-r, r]$ *and* $\varepsilon \in W^{1,1}(0,T)$ *be given, and let* $\sigma_r := \mathfrak{s}_r[\sigma_r^0, \varepsilon]$. *Then*

$$\sigma_r(t)\,(\dot\varepsilon(t) - \dot\sigma_r(t)) \geq 0, \quad a.e.\ in\ (0,T), \tag{6}$$

$$(\dot\sigma_r(t))^2 = \dot\varepsilon(t)\,\dot\sigma_r(t), \quad a.e.\ in\ (0,T). \tag{7}$$

(ii) *For every* $\sigma_r^{01}, \sigma_r^{02} \in [-r, r]$, $\varepsilon_1, \varepsilon_2 \in W^{1,1}(0,T)$, *and* $\sigma_r^i := \mathfrak{s}_r[\sigma_r^{0i}, \varepsilon_i]$, $i = 1, 2$, *it holds*

$$\int_0^T |\dot\sigma_r^1(t) - \dot\sigma_r^2(t)|\,dt \leq |\sigma_r^{01} - \sigma_r^{02}| + 2\int_0^T |\dot\varepsilon_1(t) - \dot\varepsilon_2(t)|\,dt, \tag{8}$$

$$|\sigma_r^1(t) - \sigma_r^2(t)| \leq |\sigma_r^{01} - \sigma_r^{02}| + 2\max_{0\leq\tau\leq t} |\varepsilon_1(\tau) - \varepsilon_2(\tau)|, \quad \forall\, t \in [0,T]. \tag{9}$$

Notice that the inequality (8) enables us to extend the domain of definition of the stop operator to the whole space $C[0,T]$ and to consider $\mathfrak{s}_r : [-r, r] \times C[0,T] \to C[0,T]$ as a Lipschitz continuous operator.

For functions $\varepsilon$ of two variables, $\varepsilon : \mathcal{I} \times [0,T] \to \mathbb{R}$, where $\mathcal{I} \subset \mathbb{R}$ is an interval such that $\varepsilon(x, \cdot) \in C[0,T]$ for almost every $x \in \mathcal{I}$, we define the output of the stop operator with initial configuration $|\sigma_r^0(x)| \leq r$, $x \in \mathcal{I}$, through the formula

$$\mathfrak{s}_r[\sigma_r^0, \varepsilon](x,t) = \mathfrak{s}_r[\sigma_r^0(x), \varepsilon(x, \cdot)](t), \quad (x,t) \in \mathcal{I} \times [0,T]. \tag{10}$$

Here, we have used the same symbol $\mathfrak{s}_r$ since there is no risk of confusion.

Following the approach of Prandtl [11] and Ishlinskii [2], we now consider a parallel rheological combination of simple elasto-plastic elements defined by the variational inequality (1) – (3) with a density function $\varphi$ which, in our case, *is assumed to depend also on the temperature* $\theta$; that is, we consider the thermo-plastic stress component $\sigma^p$ in the form

$$\sigma^p(x,t) = \int_0^\infty \varphi(r, \theta(x,t))\,\sigma_r(x,t)\,dr. \tag{11}$$

For the sake of simplicity, we assume that the initial stress configuration $\sigma_r^0 = \sigma_r^0(x)$ is of the form

$$\sigma_r^0(x) = \text{sign}\, \varepsilon(x,0) \min\{r, |\varepsilon(x,0)|\}, \quad x \in \mathcal{I}, \tag{12}$$

which characterizes the state without initial memory.

For this choice of $\sigma_r^0$, we may simply write $\mathfrak{s}_r[\varepsilon]$ instead of $\mathfrak{s}_r[\sigma_r^0, \varepsilon]$, and the operator $\mathfrak{s}_r : C(\bar{\mathcal{I}} \times [0,T]) \to C(\bar{\mathcal{I}} \times [0,T])$ thus defined is still Lipschitz continuous. This enables us to rewrite the constitutive equation (11) in operator form

$$\sigma^p = \mathcal{P}[\varepsilon, \theta] := \int_0^\infty \varphi(r,\theta) \mathfrak{s}_r[\varepsilon]\, dr, \tag{13}$$

with a given non-negative function $\varphi$ whose properties will be specified below in Hypothesis 3.

We remark that for each constant temperature $\theta$ the function $\varphi(\cdot,\theta)$ can be identified from the initial loading curve $\sigma^p = \Phi(\varepsilon,\theta)$ which is obtained by plotting the value of $\sigma^p$ against a monotonically increasing value of $\varepsilon$ from the imperturbed state $\varepsilon(0) = \sigma_r^0 = 0$ for every $r > 0$. We then have (see [1, 4, 6]) $\varphi(r,\theta) = -\Phi_{rr}(r,\theta)$ for constant $\theta$, and secondary branches of the hysteresis loops are given by the functions $C \pm 2\Phi(\frac{1}{2}|\varepsilon - \varepsilon_*|, \theta)$, where $\varepsilon_*$ is a turning point.

Denoting by $u$ the displacement, we consider the usual equation of motion

$$\varrho u_{tt} - \sigma_x = f(\theta, x, t), \tag{14}$$

with a constant mass density $\varrho > 0$ and a given volume force density $f$ which will be allowed to depend also on temperature. The stress $\sigma$ is assumed in the form

$$\sigma = \sigma^p + \gamma(\varepsilon) + \mu\varepsilon_t - \beta\theta, \qquad \varepsilon = u_x, \tag{15}$$

where $\mu > 0$, $\beta \in \mathbb{R}$ are given constants and $\gamma$ is a nondecreasing Lipschitz continuous function. The terms $\gamma(\varepsilon)$, $\mu\varepsilon_t$, $-\beta\theta$ represent the stress components corresponding to the effects of kinematic hardening, viscosity and thermic dilation, respectively.

We need to couple (14) with the balance law for the density of internal energy.

## 3. The Energy Balance

We first construct an *internal energy operator* $\mathcal{U}$ and an *entropy operator* $\mathcal{S}$ that assign to each pair $(\varepsilon, \theta)$ of input functions the densities $U =$

$\mathcal{U}[\varepsilon,\theta]$ of internal energy and $S = \mathcal{S}[\varepsilon,\theta]$ of entropy, respectively. For thermodynamic consistency, the first and second laws of thermodynamics, expressed by the balance of internal energy,

$$U_t = \sigma\,\varepsilon_t + g - q_x\,, \tag{16}$$

and by the Clausius-Duhem inequality,

$$S_t \geq \frac{g}{\theta} - \left(\frac{q}{\theta}\right)_x, \tag{17}$$

respectively, must hold in a suitable sense together with the condition $\theta > 0$. Here, $q$ denotes the heat flux and $g = g(\theta, x, t)$ is a given heat source density. As usual, the subscripts $t$, $x$ denote partial derivatives with respect to these variables.

For the heat flux $q$, we assume Fourier's law

$$q = -\kappa\,\theta_x\,, \tag{18}$$

with a constant heat conductivity $\kappa > 0$; hence, assuming $\theta > 0$ (this will have to be verified later on), we can rewrite (17) as

$$U_t - \theta\,S_t - \sigma\,\varepsilon_t \leq \frac{\kappa}{\theta}\,\theta_x^2\,, \tag{19}$$

or, introducing the *free energy* $F := U - \theta\,S$, as

$$F_t - \sigma\,\varepsilon_t + \theta_t\,S \leq \frac{\kappa}{\theta}\,\theta_x^2\,. \tag{20}$$

In order to find a suitable expression for $F$ in operator form $F = \mathcal{F}[\varepsilon,\theta]$, we now make use of the well-known fact that the inequality (6) can be interpreted as an energy inequality for the individual stop operator $\mathfrak{s}_r$, where the operator $(1/2)\mathfrak{s}_r^2$ plays the role of a *hysteresis potential* for $\mathfrak{s}_r$ (cf. [5] or Section 2.5 in [1]). This observation suggests to define the free energy operator $\mathcal{F}$ in the form

$$\mathcal{F}[\varepsilon,\theta] := F_0(\theta) - \beta\theta\varepsilon + \Gamma(\varepsilon) + \frac{1}{2}\int_0^\infty \varphi(r,\theta)\,\mathfrak{s}_r^2[\varepsilon]\,dr\,, \tag{21}$$

where

$$F_0(\theta) = \varrho\,(-C_V\,\theta\,\ln(\theta/\tilde{\theta}) + C_V\,\theta + C_0) \tag{22}$$

is the purely caloric part of the free energy, with positive physical constants $\tilde{\theta}$, $C_V$, $C_0$, and $\Gamma$ is an antiderivative of $\gamma$. This choice of $\mathcal{F}$ yields (at least formally)

$$\theta_t\left(\mathcal{S}[\varepsilon,\theta] + F_0'(\theta) - \beta\varepsilon + \tfrac{1}{2}\textstyle\int_0^\infty \varphi_\theta(r,\theta)\,\mathfrak{s}_r^2[\varepsilon]\,dr\right)$$
$$\leq \tfrac{\kappa}{\theta}\,\theta_x^2 + \textstyle\int_0^\infty \varphi(r,\theta)\,\mathfrak{s}_r[\varepsilon]\,(\varepsilon_t - (\mathfrak{s}_r[\varepsilon])_t)\,dr\,. \tag{23}$$

Taking (6) into account, we see that (20) is satisfied provided the density $\varphi$ is non-negative and the entropy operator $\mathcal{S}$ is defined as

$$\mathcal{S}[\varepsilon,\theta] := -F_0'(\theta) + \beta\varepsilon - \frac{1}{2}\int_0^\infty \varphi_\theta(r,\theta)\,\mathfrak{s}_r^2[\varepsilon]\,dr\,. \tag{24}$$

In addition,

$$\begin{aligned}\mathcal{U}[\varepsilon,\theta] &= \mathcal{F}[\varepsilon,\theta] + \theta\,\mathcal{S}[\varepsilon,\theta]\\ &= F_0(\theta) - \theta\,F_0'(\theta) + \Gamma(\varepsilon) + \mathcal{V}[\varepsilon,\theta],\end{aligned} \tag{25}$$

where

$$\mathcal{V}[\varepsilon,\theta] := \frac{1}{2}\int_0^\infty (\varphi(r,\theta) - \theta\,\varphi_\theta(r,\theta))\,\mathfrak{s}_r^2[\varepsilon]\,dr\,. \tag{26}$$

Consequently, the balance of internal energy (16) takes the form

$$(\varrho C_V\,\theta + \mathcal{V}[\varepsilon,\theta])_t - \kappa\,\theta_{xx} = (\mathcal{P}[\varepsilon,\theta] + \mu\varepsilon_t - \beta\theta)\,\varepsilon_t + g\,. \tag{27}$$

The analysis of the system (14), (15), (27) is independent of the concrete values of the positive constants $\varrho, \kappa$; we therefore assume in the sequel $\varrho = \kappa = 1$.

## 4. Survey of main results

We consider a model problem for a system of the form (13) – (15), (27), namely

$$u_{tt} - (\mathcal{P}[u_x,\theta] + \gamma(u_x) + \mu u_{xt} - \beta\theta)_x = f(\theta,x,t)\,, \tag{28}$$

$$(C_V\,\theta + \mathcal{V}[u_x,\theta])_t - \theta_{xx} = (\mathcal{P}[u_x,\theta] + \mu u_{xt} - \beta\theta)\,u_{xt} + g(\theta,x,t) \tag{29}$$

in $]0,1[\times]0,T[$ coupled with initial and boundary conditions

$$u(x,0) = u^0(x),\ u_t(x,0) = u^1(x),\ \theta(x,0) = \theta^0(x), \quad x\in]0,1[. \tag{30}$$

$$u(0,t) = u(1,t) = \theta_x(0,t) = \theta_x(1,t) = 0\,, \quad t\in[0,T]\,, \tag{31}$$

under the following hypotheses.

**Hypothesis 2.**

(i) $u^0, u^1 \in W^{2,2}(0,1) \cap \mathring{W}^{1,2}(0,1)$, $\theta^0 \in W^{1,2}(0,1)$ *and there exists a constant* $\delta > 0$ *such that*

$$\theta^0(x) \geq \delta \qquad \forall x \in [0,1]. \tag{32}$$

(ii) $\gamma: \mathbb{R} \to \mathbb{R}$ *is an absolutely continuous function,* $\gamma(0) = 0$ *and there exists a constant* $\gamma_0 > 0$ *such that*

$$0 \leq \frac{d\gamma(\varepsilon)}{d\varepsilon} \leq \gamma_0 \qquad a.e. \tag{33}$$

(iii) *The functions* $f, g$ *are measurable,* $f(\cdot, x, t)$, $g(\cdot, x, t)$ *are absolutely continuous in* $[0, \infty[$ *for a.e.* $(x,t) \in ]0,1[\times]0,T[$. *Moreover, there exists a constant* $K > 0$ *and functions* $f_0, g_0 \in L^2(]0,1[\times]0,T[)$ *such that*

$$g(0,x,t) = g_0(x,t) \;\geq\; 0 \qquad a.e., \tag{34}$$

$$|f(\theta,x,t)| + |f_t(\theta,x,t)| \;\leq\; f_0(x,t) \quad a.e. \tag{35}$$

$$|f_\theta(\theta,x,t)| + |g_\theta(\theta,x,t)| \;\leq\; K \quad a.e. \tag{36}$$

**Hypothesis 3.** *The function* $\varphi: ]0,\infty[^2 \to [0,\infty[$ *is measurable,* $\varphi(r,\cdot)$, $\varphi_\theta(r,\cdot)$ *are absolutely continuous for a.e.* $r > 0$ *and there exists a constant* $L > 0$ *such that for a.e.* $\theta > 0$ *we have*

$$\int_0^\infty \Big(\varphi(r,\theta) + r|\varphi_\theta(r,\theta)| + (1+r^2)\,|\varphi(r,\theta) - \theta\varphi_\theta(r,\theta)|\Big)\,dr \;\leq\; L, \tag{37}$$

$$\int_0^\infty \theta|\varphi_{\theta\theta}(r,\theta)|r^2 dr \leq C_V, \tag{38}$$

*where* $C_V$ *is the purely caloric specific heat in (22).*

**Theorem 4** (Existence). *Let Hypotheses 2, 3 hold. Then there exists a solution* $(u,\theta)$ *to the problem (28) – (31) such that*

$$u_{tt}, u_{xx}, u_{xxt}, \theta_x \;\in\; L^\infty(0,T;L^2(0,1)), \tag{39}$$

$$u_{xtt}, \theta_t, \theta_{xx} \;\in\; L^2(]0,1[\times]0,T[), \tag{40}$$

$$\theta, u, u_x, u_t, u_{xt} \;\in\; C([0,1]\times[0,T]), \tag{41}$$

*there exists a constant* $c_0 > 0$ *depending only on the given data such that for all* $t \in [0,T]$ *and* $x \in [0,1]$ *we have*

$$\theta(x,t) \geq \delta e^{-c_0 t} > 0 \tag{42}$$

*and (28) – (29) are satisfied almost everywhere.*

**Theorem 5** (Uniqueness and continuous dependence). *Let Hypotheses 3, 2(ii) hold, let* $(u^0, u^1, \theta^0, f, g)$, $(u'^0, u'^1, \theta'^0, f', g')$ *be two sets of functions satisfying Hypothesis 2, and let* $(u, \theta)$, $(u', \theta')$ *be solutions of (28) – (31) corresponding to these data, respectively, which satisfy (39) – (42). Assume moreover that there exist a constant* $\tilde{K} > 0$ *and functions* $d_f, d_g \in L^2(]0,1[\times]0,T[)$ *such that*

$$|f(\theta_1, x, t) - f'(\theta_2, x, t)| \leq \tilde{K}|\theta_1 - \theta_2| + d_f(x,t), \tag{43}$$

$$|g(\theta_1, x, t) - g'(\theta_2, x, t)| \leq \tilde{K}|\theta_1 - \theta_2| + d_g(x,t) \tag{44}$$

*holds for all* $\theta_1, \theta_2 \in \mathbb{R}^+$ *and a.e.* $(x,t) \in ]0,1[\times]0,T[$. *Then there exists a constant* $C$ *depending only on the size of the data in their respective spaces such that for all* $t \in [0,T]$ *the differences* $\bar{u} = u - u'$, $\bar{\theta} = \theta - \theta'$ *satisfy*

$$\begin{aligned} &||\bar{u}_t(t)||^2 + \int_0^t \left(||\bar{\theta}||^2 + ||\bar{u}_{xt}||^2\right)(\tau)\, d\tau \\ &\quad \leq C\left(||\bar{u}_t(0)||^2 + ||\bar{u}_x(0)||^2 + ||\bar{\theta}(0)||^2 + \int_0^t \int_0^1 \left(d_f^2 + d_g^2\right) dx\, dt\right), \end{aligned} \tag{45}$$

*where* $||\cdot||$ *denotes the norm of* $L^2(0,1)$.

## References


1. M. Brokate, J. Sprekels (1996): *Hysteresis and Phase Transitions.* Heidelberg, Springer-Verlag.
2. A. Yu. Ishlinskii (1944): Some applications of statistical methods to describing deformations of bodies. *Izv. AN SSSR, Techn. Ser.* **9**, 583–590.
3. M.A. Krasnosel'skii, A.V. Pokrovskii (1989): *Systems with hysteresis*, Springer-Verlag, Berlin. Russian edition: Nauka, Moscow 1983.
4. P. Krejčí (1986): Hysteresis and periodic solutions of semilinear and quasilinear wave equations. *Math. Z.* **193**, 247–264.
5. P. Krejčí (1991): Vector hysteresis models. *European J. Appl. Math.* **2**, 281–292.
6. P. Krejčí (1996): *Hysteresis, convexity and dissipation in hyperbolic equations*, Gakkōtosho, Tokyo.
7. P. Krejčí, J. Sprekels *(to appear)*: On a system of nonlinear PDE's with temperature-dependent hysteresis in one-dimensional thermoplasticity.
8. P. Krejčí, J. Sprekels *(in preparation)*: Global solutions to a system of one-dimensional thermo-visco-elasto-plasticity.
9. G.A. Maugin (1992): *The thermomechanics of plasticity and fracture*, Cambridge University Press, Cambridge .
10. I. Müller (1985): *Thermodynamics.* Pitman, New York.
11. L. Prandtl (1928): Ein Gedankenmodell zur kinetischen Theorie der festen Körper. *Z. Ang. Math. Mech.* **8**, 85–106.
12. M. Šilhavý (1996): *The mechanics and thermodynamics of continuous media.* Springer, Berlin – Heidelberg.
13. A. Visintin (1987): Rheological models and hysteresis effects. *Rend. Sem. Mat. Univ. Padova* **77**, 213–241.
14. A. Visintin (1994): *Differential models of hysteresis*, Springer - Verlag, Berlin.

# THERMOMECHANICAL APPROACH OF RUNNING DISCONTINUITIES

C. STOLZ
*Laboratoire de Mécanique des Solides, CNRS URA 317*
*Ecole Polytechnique- F-91128 Palaiseau Cedex*

AND

R.M. PRADEILLES-DUVAL
*DGA-DCE-CREA*
*Ecole Polytechnique- F-91128 Palaiseau Cedex*

The propagation of moving surface inside a body is analyzed within the framework of thermodynamical couplings, when the moving surface is associated with an irreversible change of mechanical properties. The moving surface is a surface of heat sources and of entropy production, intensities of which are related to particular energy release rates defined in terms of Hamiltonian gradients. As examples, we analyze the dynamical evolution of partial damage in a bar, the problem of stability of an equilibrium state and the bifurcation of the quasistatic evolution.

## 1. Introduction

In the recent past, the propagation of damage has been studied in connection with fracture mechanics. Different approaches based on macroscopic or microscopic description of mechanical degradation of properties have been proposed. In the framework of thermomechanical coupling as in fracture mechanics, the analysis defines two different energy release rate associated with heat production and entropy production. This paper is concerned mostly with a description of damage involved on the evolution of a moving interface along which mechanical transformation occurs.

The dynamical evolution is analysed in an Hamiltonian formulation. In connection with an energetic analysis, the evolution of the surface is then characterized.



*P. Argoul et al. (eds.),*
*IUTAM Symposium on Variations of Domains and Free-Boundary Problems in Solid Mechanics*, 245–251.

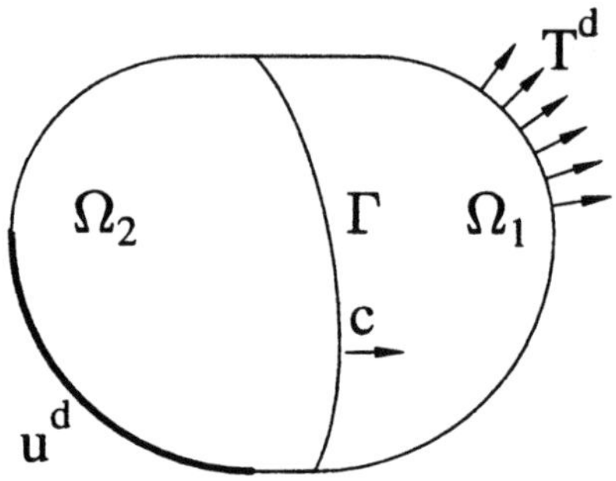


*Figure 1.* The boundary value problem

The moving surface $\Gamma$ is a surface of heat sources and of entropy production, the intensities of which are defined in terms of Hamiltonian gradients.

## 2. General features

The paper deals with partial damage. At each time, two materials coexist in the structure, of volume $\Omega$. The domain is composed of two materials with different mechanical charateristics. The interface between them is perfect and denoted by $\Gamma$. Material 1 changes into material 2 along $\Gamma$ by an irreversible process. Hence $\Gamma$ moves with a normal celerity $c = \phi N$,($N$ is the normal outward $\Omega_2, \phi$ is then positive along $\Gamma$). When this surface propagates, all the mechanical quantities such $f$ can have a jump denoted by $[f] = f_1 - f_2$.

Then all volume average has a rate defined by :

$$\frac{d}{dt}\int_\Omega f \, d\omega = \int_\Omega \dot{f} \, d\omega - \int_\Gamma [f] \, c.N \, da$$

Assuming that the two materials are thermoelastic. The actual state is characterized by the displacement $u$, from which the strain $\varepsilon$ is defined. The free energy of each materials $w_i$ are function of the strain and of the temperature $\theta$. The equations of state are

$$\sigma = \frac{\partial w_i}{\partial \varepsilon}, \quad s = -\frac{\partial w_i}{\partial \theta},$$

where $\sigma$ is the stress, and $s$ the entropy, the internal energy $(w + s\theta)$ is denoted by $e$.

Assuming that the external loading is prescribed on the boundary $\partial\Omega_T$, and the displacement on $\partial\Omega_u$. The problem is to characterize the evolution of $\Gamma$, that is to determine the normal propagation $\phi$, during the time.

## 3. Dissipation analysis

The mass conservation leads to the continuity of the mass flux $m = \rho\phi$, where $\rho$ denotes the mass density. The first law and the second law of thermodynamics give rise to local equations inside the volume and along the moving surface $\Gamma$ :

$$\rho\dot{e} = \sigma : \dot{\varepsilon} - divq, \text{in } \Omega,$$

$$m[e + \frac{v^2}{2}] + N.[\sigma.v] - N.[q] = 0, \text{ along } \Gamma.$$

The displacememt is continuous and the momentum is conserved along the moving interface :

$$[u] = 0, \ [\sigma].N = m[v].$$

Then the heat power supply is given in terms of a release rate of internal energy :

$$N.[q] = G_{th}\phi,$$

$$G_{th} = \rho[e] - N.\bar{\sigma}.[\nabla u].N$$

where $\bar{\sigma} = \frac{1}{2}(\sigma_1 + \sigma_2)$.

*Entropy production* The entropy production is given by :

$$\int_\Omega (\rho\dot{s} + \frac{divq}{\theta} - q.\frac{\nabla\theta}{\theta^2})d\omega + \int_\Gamma (-m[s] + N.[\frac{q}{\theta}])da \geq 0.$$

The interface is perfect at each time, under the assumption of separability of the two dissipation, the term inside the volume is reduced to the conduction, and the term along the surface is then :

$$D_\Gamma = \frac{\rho[w] - N.\bar{\sigma}.[\nabla u].N}{\theta}\phi = \frac{G_s}{\theta}\phi$$

where $G_s$ has also the form of a release rate of energy.

## 4. Hamiltonian formalism

The total Hamiltonian of the structure is

$$H = \int_\Omega \frac{1}{2}\rho v^2 d\omega + \mathcal{P} + \int_\Omega \rho s\theta d\omega$$

where $\mathcal{P}$ is the total potential energy :

$$\mathcal{P}(u, T^d) = \int_\Omega \rho w \ d\omega - \int_{\partial\Omega_T} T^d.u \ da$$

The equations of the momentum conservation are then :

$$\frac{\partial H}{\partial p}.\delta p = \int_\Omega v.\delta p \; d\omega,$$

$$\frac{\partial H}{\partial u}.\delta u = -\frac{d}{dt}\int_\Omega p.\delta u \; d\omega,$$

where $p$ is the impulsion, these equations lead to the classical equation of movement.

The first law of thermodynamics can be written as follows :

$$\frac{dH}{dt} - \frac{\partial H}{\partial T^d}.\dot{T}^d = \int_{\partial\Omega} -q.n \; da,$$

and taking into account of the momentum conservation, we have

$$\frac{\partial H}{\partial \Gamma}.\dot{\Gamma} = \int_\Gamma [q].N \; da = -\int_\Gamma G_{th}\phi \; da$$

the second law has the same form as previously.

In a thermomechanical coupling, two different release rate must be distinguish, one defined in term of variation of the Hamiltonian give rise the heat source associated with the moving surface, the second one give rise the production of entropy.

*Isothermal evolution* In that case, we can define another Hamiltonian

$$H = \int_\Omega \frac{1}{2}\rho v^2 d\omega + \mathcal{P}$$

and the total dissipation is then given by :

$$\frac{dH}{dt} - \frac{\partial H}{\partial T^d}.\dot{T}^d = \int_{\partial\Omega} -q.n \; da = -\int_\Gamma G_{dyn}\phi da,$$

where $G_{dyn} = \rho[w] - N.\bar{\sigma}.[\nabla u].N$. In this case, the driving traction of Abeyratne and œknowles is recovered [1].

These relations generalize the case of quasistatic evolution, where the potential energy plays the rule of the Hamiltonian. A similar analysis has been given for the case of running cracks,[7],[6].

As an example, a bar composed by two material is considered, one of the material is transformed into the second one when the thermodynamical force $G_{dyn}$ reaches a critical value $G_c$. Such a description can modelize the impact test on quasi-brittle material with Hopkinson bars (See figures 2,3).

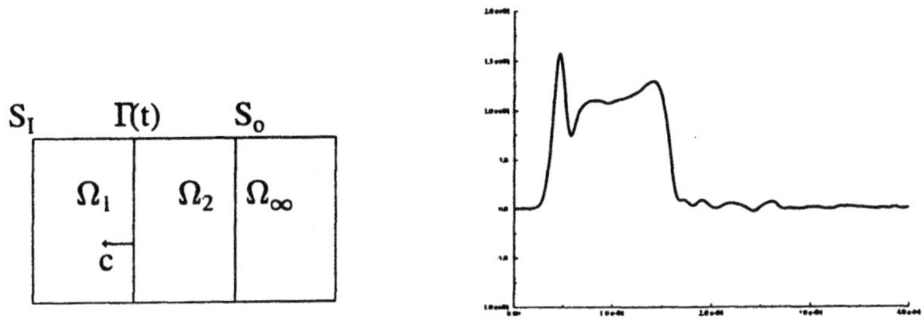


*Figure 2.* The impact test and the rate of the input displacement on $S_I$

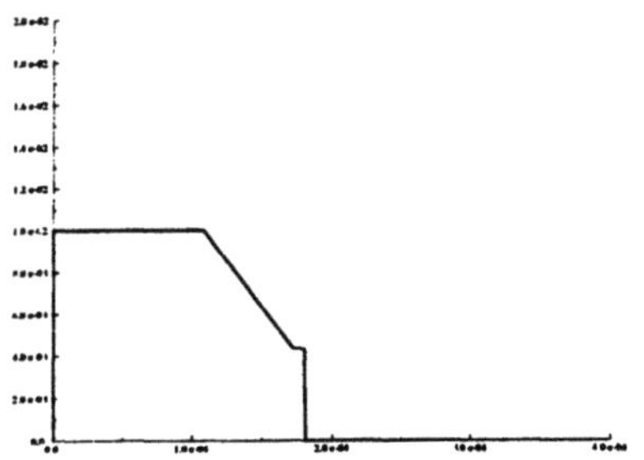

*Figure 3.* The position of the interface versus the time

The results shown that the rate of propagation of $\Gamma$ is constant, about the order of ten percent of the celerities of waves.

## 5. Quasistatic Evolution

In isothermal evolution, the same criterium of evolution can be assumed :

$$\phi \geq 0, \text{ if } G = G_c \text{on } \Gamma.$$

The evolution of the interface is governed by the consistency condition, during the evolution of $\Gamma$, at the geometrical point $x_\Gamma(t)$ the criterion is reached :

$$G(x_\Gamma(t), t) = G_c$$

then the derivative of $G$ following the moving surface is null :$D_\phi G = 0$. Thus leads to the consistency condition:

$$(\phi - \phi^*) D_\phi G \geq 0, \forall \phi^* \geq 0, \text{ over } \Gamma^+.$$

$\Gamma^+$ is the subset of $\Gamma$ where the critical value $G_c$ is reached, on the complementary part $\phi = 0$.

$$\begin{aligned} D_\phi G &= [t.\sigma].\nabla v_1.t - N.\dot{\sigma}_2.[\nabla u].N - \phi K \\ K &= t.div_\Gamma \sigma_2.[\nabla u].N - t.\sigma_2.\nabla([\nabla u].N) \\ &\quad +[\sigma : \nabla \varepsilon.N] - N.\sigma.(\nabla\nabla u.N).N] \end{aligned}$$

In that case, the evolution is determined by the functional

$$\begin{aligned} F(v, \phi, \dot{T}^d) &= \int_\Omega \frac{1}{2}\varepsilon(v) : \frac{\partial^2 w}{\partial\varepsilon\partial\varepsilon} : \varepsilon(v)\, d\omega - \int_{\partial\Omega_T} \dot{T}^d.v\, da \\ &\quad + \int_\Gamma (\phi[t.\sigma].\nabla v_1.t + \frac{\phi^2}{2}K)\, da \end{aligned}$$

Then the evolution is given by :

$$\frac{\partial F}{\partial v}.(v - v^*) + \frac{\partial F}{\partial \phi}(\phi - \phi^*) \geq 0$$

for all v* kinematically admissible field, and $\phi^*(s) \geq 0$ along $\Gamma^+$.

The discussion of the stability and bifurcation along an evolution process can be investigate as presented in the paper, [2].

*Stability and no Bifurcation* Let $W$ be the value of $F$ at the solution in the rate of displacement $v$ of the local equations for a given velocity of propagation $\phi$ :

$$div\dot{\sigma} = 0, \ \dot{\sigma} = \frac{\partial^2 w}{\partial\varepsilon\partial\varepsilon} : \varepsilon(v)$$

$$D_\phi(\sigma.N) = 0, \ D_\phi[u] = 0.$$

Then

$$W(\phi, \dot{T}^d) = F(v(\phi, \dot{T}^d), \phi, \dot{T}^d)$$

The stability of the actual state is determined by the condition of the existence of a solution

$$\delta\phi \frac{\partial^2 W}{\partial\phi\partial\phi}\delta\phi \geq 0, \ \text{for } \delta\phi \geq 0 \text{ on } \Gamma^+, \delta\phi \neq 0$$

and the uniqueness amd non bifurcation is characterized by

$$\delta\phi \frac{\partial^2 W}{\partial\phi\partial\phi}\delta\phi \geq 0, \ \text{for } \delta\phi \neq 0 \text{ on } \Gamma^+.$$

In paper [2] some analytical results are presented for composite spheres or cylinders.

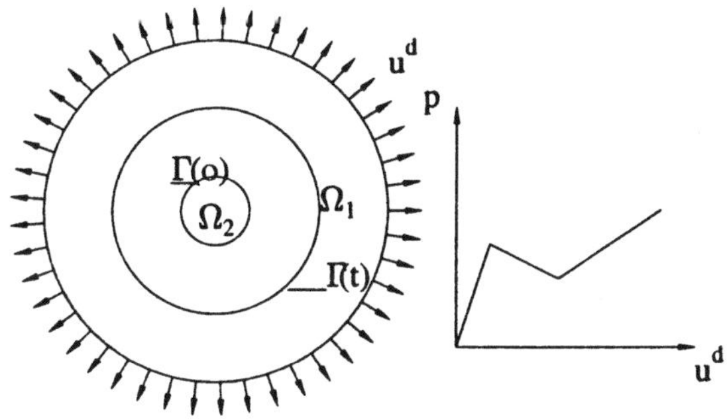


*Figure 4.* The case of composite sphere, typical answer.

## 6. Conclusion

In dynamics, the definition of the Hamiltonian is only defined by a potential energy and a kinetic energy, in the same spirit the definition of the functional $F$ is governed only by the existence of a potential energy, and by the definition of the thermodynamical forces associated to the propagation. Then all the definition of the thermodynamical forces in terms of release rate of energy are valid for all cases, for which a potential energy exist. Some applications has been performed in generalized media, such as beams and plates, to study the delamination of laminate [5].

## References


1. Abeyratne, R., Knowles, J.,(1990) On the driving traction acting on a surface of strain discontinuity in a continuum, *J. Mech. Phys. of Solids*, **38, 3**, pp.345-360
2. Pradeilles-Duval, R.M., Stolz, C., (1995) Mechanical transformation and discontinuities along a moving surface, *J. Mech. Phys. of Solids*, **43, 1**,pp.91-121
3. Stolz, C., (1988) Sur les équations de la dynamique des milieux continus anélastiques, *C. R. Acad. Sci. Paris*, **t 307, Série II,b**, pp.1997-2000
4. Pradeilles-Duval, R.M., Stolz, C., (1991) Sur les problèmes d'évolution des solides avec changement de phase irréversible, *C.R. Acad. Sci. Paris*, **t 313, Série II,b**, pp.297-302
5. Pradeilles-Duval, R.M., (1992) Evolution de système avec surfaces de discontinuités mobiles, application au délaminage. Doctorat Ecole Polytechnique, Décembre 1992.
6. Stolz, C., Pradeilles-Duval, R.M., (1996) Approches énergétiques des discontinuités mobiles en dynamique non linéaire, *C.R. Acad. Sci. Paris*, **t 322, Série II,b**, p.525-532
7. Stolz, C., (1995) Functional Approach in non linear dynamics, *Arch. Mech*, **47, 3**, pp.421-435

# GLOBAL SOLUTIONS WITH CONTACT DISCONTINUITIES FOR QUASILINEAR HYPERBOLIC SYSTEMS OF CONSERVATION LAWS

LI TA-TSIEN (LI DA-QIAN)
*Department of Mathematics, Fudan University*
*Shanghai 200433, China.*

## 1. Introduction and main result

We consider the quasilinear system of conservation laws

$$\frac{\partial u}{\partial t} + \frac{\partial f(u)}{\partial x} = 0, \tag{1}$$

where $u = (u_1, ..., u_n)^T$ is an unknown vector function of $(t, x)$, $f : \mathbb{R}^n \to \mathbb{R}^n$ is a given smooth vector function of $u$. Suppose that on the domain under consideration, system (1) is strictly hyperbolic, i.e., the Jacobi matrix $A(u) = \nabla f(u)$ possesses n distinct real eigenvalues:

$$\lambda_1(u) < \lambda_2(u) < ... < \lambda_n(u). \tag{2}$$

Let $l_i(u) = (l_{i1}(u), ...l_{in}(u))$ and $r_i(u) = (r_{i1}(u), ...r_{in}(u))^T$ be a left eigenvector and a right eigenvector corresponding to $\lambda_i(u)$ respectively for $i = 1, ..., n$.

Suppose furthermore that on the domain under consideration, system (1) is linearly degenerate in the sense of P.D. Lax:

$$\nabla\lambda_i(u)r_i(u) \equiv 0 \quad (i = 1, ..., n). \tag{3}$$

Prescribing the following piecewise constant initial condition

$$t = 0: \ u = \begin{cases} \hat{u}_l, & x \le 0, \\ \hat{u}_r, & x \ge 0, \end{cases} \tag{4}$$

where $\hat{u}_l$ and $\hat{u}_r$ are constant vectors with

$$\hat{u}_l \neq \hat{u}_r, \tag{5}$$

we get the Riemann problem (1) and (4).



*P. Argoul et al. (eds.),*
*IUTAM Symposium on Variations of Domains and Free-Boundary Problems in Solid Mechanics,* 253–260.

**Lemma 1 (P.D. Lax[1])** *Suppose that on the domain under consideration, system (1) is strictly hyperbolic and linearly degenerate. If the initial amplitude* $|\hat{u}_r - \hat{u}_l|$ *is suitably small, then the Riemann problem (1) and (4) admits a unique similarity solution* $u = U(\frac{x}{t})$ *composed of* $n+1$ *constant states* $\hat{u}_0 = \hat{u}_l$, $\hat{u}_1, \ldots, \hat{u}_{n-1}$ *and* $\hat{u}_n = \hat{u}_r$ *and* $n$ *contact discontinuities with small amplitude* $x = \hat{s}_i t$ $(i = 1, 2, \ldots, n)$ *passing through the origin, the states* $\hat{u}_{i-1}$ *and* $\hat{u}_i$ *being connected by the i-th contact discontinuity* $x = \hat{s}_i t$.

We now consider the generalized Riemann problem for system (1) with the following piecewise smooth initial condition

$$t = 0: \ u = \begin{cases} \bar{u}_l(x), & x \leq 0, \\ \bar{u}_r(x), & x \geq 0, \end{cases} \tag{6}$$

where $\bar{u}_l(x)$ and $\bar{u}_r(x)$ are given $C^1$ vector functions on $x \leq 0$ and on $x \geq 0$, respectively, with

$$\bar{u}_l(0) = \hat{u}_l \qquad \bar{u}_r(0) = \hat{u}_r. \tag{7}$$

The generalized Riemann problem (1) and (6) can be regarded as a perturbation of the corresponding Riemann problem (1) and (4).

**Lemma 2 (Li & Yu[2])** *Under the same assumptions of Lemma 1, the generalized Riemann problem (1) and (6) admits a unique local piecewise* $C^1$ *solution* $u = u(t,x)$ *which possesses a structure similar to the similarity solution* $u = U(\frac{x}{t})$ *of the Riemann problem (1) and (4) in a neighborhood of the origin. Here the similar structure means that the solution* $u = u(t,x)$ *still consists of* $n$ *contact discontinuities passing through the origin; moreover, at the origin the i-th contact discontinuity still connects the known constant states* $\hat{u}_{i-1}$ *and* $\hat{u}_i$*; and coincides with the i-th contact discontinuity* $x = \hat{s}_i t$ *in* $u = U(\frac{x}{t})$.

Lemma 2 shows that the similarity solution $u = U(\frac{x}{t})$ to the Riemann problem (1) and (4) under consideration possesses a local nonlinear structural stability and then the piecewise smooth solution to the Riemann problem (1) and (4), consisting of the preceding $n$ contact discontinuities with small amplitude, must be the similarity solution $u = U(\frac{x}{t})$ at least in the local sense (i.e., for small $t > 0$).

Thus, it is natural to propose the following problem: Under what conditions does the generalized Riemann problem (1) and (6) admit a unique global piecewise $C^1$ solution $u = u(t,x)$ on $t \geq 0$, which still consists of $n$

contact discontinuities passing through the origin and has a global structure similar to the similarity solution $u = U(\frac{x}{t})$ of Riemann problem (1) and (4)?

The main result of this paper, Theorem 1, will give an affirmative answer to this problem, which shows that the similarity solution $u = U(\frac{x}{t})$ of Riemann problem (1) and (4) possesses a global nonlinear structural stability and then the piecewise smooth solution to Riemann problem (1) and (4), consisting of the preceding $n$ contact discontinuities, must be the similarity solution $u = U(\frac{x}{t})$.

**Theorem 1 (Li & Kong[3])** *Suppose that system (1) is strictly hyperbolic and linearly degenerate. Suppose furthermore that there exists a constant $\mu > 0$ such that*

$$\begin{aligned}\theta \ \triangleq\ & \sup_{x\le 0}\left\{(1+|x|)^{1+\mu}(|\bar{u}_l(x)|+|\bar{u}_l'(x)|)\right\}+\\ & \sup_{x\ge 0}\left\{(1+|x|)^{1+\mu}(|\bar{u}_r(x)|+|\bar{u}_r'(x)|)\right\}<\infty.\end{aligned} \tag{8}$$

*Then there exists $\theta_0 > 0$ so small that for any given $\theta \in (0, \theta_0]$, the generalized Riemann problem (1) and (6) admits a unique global piecewise $C^1$ solution $u = u(t,x)$ containing only $n$ contact discontinuities with small amplitude $x = x_i(t)$ $(i = 1, ..., n)$ passing through the origin on $t \geq 0$. This solution possesses a global structure similar to that of the similarity solution $u = U(\frac{x}{t})$ to the corresponding Riemann problem (1) and (4), hence, $u = U(\frac{x}{t})$ has a global nonlinear structural stability.*

Precisely speaking, $u = u(t,x)$ is given by

$$u = u(t,x) = \begin{cases} u^{(0)}(t,x), & x \le x_1(t),\\ u^{(1)}(t,x), & x_1(t) \le x \le x_2(t),\\ \cdots\cdots & \\ u^{(n-1)}(t,x), & x_{n-1}(t) \le x \le x_n(t),\\ u^{(n)}(t,x), & x_n(t) \le x, \end{cases} \tag{9}$$

where $u^{(i)}(t,x)$ $(i = 0, 1, ..., n)$ are all $C^1$ solutions to system (1) on each corresponding angular domain respectively, and for $i = 1, ..., n$, $u^{(i-1)}(t,x)$ and $u^{(i)}(t,x)$ are connected by the i-th contact discontinuity $x = x_i(t)$ $(x_i(0) = 0)$ with

$$x_i'(0) = \hat{s}_i \qquad (i = 1, ..., n) \tag{10}$$

and

$$u^{(i)}(0,0) = \hat{u}^i \qquad (i = 0, 1, ..., n). \tag{11}$$

Here, $u^{(i-1)}(t,x)$ and $u^{(i)}(t,x)$ are connected by the i-th contact discontinuity $x = x_i(t)$ means that the Rankine-Hugoniot condition

$$f(u^+) - f(u^-) = s(u^+ - u^-) \tag{12}$$

and the condition that each i-th contact discontinuity must be an i-th characteristic on both sides:

$$s = \lambda_i(u^+) = \lambda_i(u^-) \tag{13}$$

hold on $x = x_i(t)$, where

$$u^+ = u^{(i)}(t, x_i(t)), \qquad u^- = u^{(i-1)}(t, x_i(t)) \tag{14}$$

and

$$s = x_i'(t). \tag{15}$$

**Remark 1** *In Theorem 1, a part of contact discontinuities may degenerate into weak discontinuities, on which the first order derivatives of the solution are discontinuous, while the solution itself is continuous.*

**Remark 2** *Similar results hold in the case that system (1) might be non-strictly hyperbolic but each characteristic has a constant multiplicity, say, on the domain under consideration,*

$$\lambda_1(u) \equiv \ldots \equiv \lambda_p(u) < \lambda_{p+1}(u) < \ldots < \lambda_n(u) \qquad (1 \le p \le n). \tag{16}$$

*In this case, we get a global piecewise $C^1$ solution with $n-p+1$ contact discontinuities.*

**Remark 3** *When $\hat{u}_l = \hat{u}_r$, Theorem 1 gives the global existence and uniqueness of weak discontinuous solution ($C^0$ and piecewise $C^1$ solution) $u = u(t,x)$ containing only $n$ weak discontinuities $x = x_i(t)$ $(i = 1, \ldots, n)$ passing through the origin on $t \ge 0$. The same result still holds for general quasilinear hyperbolic systems*

$$\frac{\partial u}{\partial t} + A(u)\frac{\partial u}{\partial x} = 0, \tag{17}$$

*in which all the characteristics are only supposed to be weakly linearly degenerate. Thus, the theory on the global existence and uniqueness of $C^1$ solution to the Cauchy problem with $C^1$ initial data, established by Li, Zhou & Kong [4]-[6], can be generalized to the global weak discontinuous solution for the Cauchy problem with $C^0$ and piecewise $C^1$ initial data.*

## 2. Sketch of the proof

Since all the contact discontinuities are unknown, in order to prove Theorem 1 we have to solve the corresponding free boundary problem, however, by Lemma 2, it suffices to establish a uniform a priori estimate for the piecewise $C^1$ norm of the solution.

Taking the decomposition of $u$ and $u_x$ with respect to the right eigenvectors:

$$u = \sum_{k=1}^{n} v_k r_k(u) \tag{18}$$

and

$$u_x = \sum_{k=1}^{n} w_k r_k(u), \tag{19}$$

it is only necessary to establish a uniform a priori estimate for the piecewise $C^0$ norm of $v = (v_1, ..., v_n)$ and $w = (w_1, ..., w_n)$.

It is easy to see that

$$\frac{dv_i}{d_i t} = \sum_{j,k=1}^{n} \beta_{ijk}(u) v_j w_k \qquad (i = 1, ..., n) \tag{20}$$

and

$$\frac{dw_i}{d_i t} = \sum_{j,k=1}^{n} \gamma_{ijk}(u) w_j w_k \qquad (i = 1, ..., n), \tag{21}$$

where

$$\frac{d}{d_i t} = \frac{\partial}{\partial t} + \lambda_i(u) \frac{\partial}{\partial x} \tag{22}$$

denotes the directional derivative along the i-th characteristic and

$$\beta_{ijk}(u) = (\lambda_k(u) - \lambda_i(u)) l_i(u) \nabla r_j(u) r_k(u) \tag{23}$$

and

$$\gamma_{ijk}(u) = \frac{1}{2} \left\{ (\lambda_j(u) - \lambda_k(u)) l_i(u) \nabla r_k(u) r_j(u) - \nabla \lambda_k(u) r_j(u) \delta_{ik} + (j|k) \right\}, \tag{24}$$

where $(j|k)$ stands for all the terms obtained by changing $j$ and $k$ in the previous terms. Noting that system (1) is linearly degenerate, we have

$$\gamma_{ijj}(u) \equiv 0 \qquad \forall i, j. \tag{25}$$

The whole proof is similar to the proof of the global existence and uniqueness of $C^1$ solution for the Cauchy problem with $C^1$ initial data in [5], however, noting that in the present situation each characteristic passing through the origin is a contact discontinuity, we have to pay more attention to the relationship between the values of $v$ and $w$ on both sides of a contact discontinuity. The following lemmas are specially presented for this purpose.

**Lemma 3** *The Rankine-Hugoniot condition (12) on the i-th contact discontinuity $x = x_i(t)$ can be rewritten as*

$$v_j^+ = v_j^- + O\left(|v^\pm|^2\right) \qquad (j = 1, ..., i-1, i+1, ..., n), \tag{26}$$

*where $v^\pm = v^\pm(t, x_i(t)) \triangleq v(t, x_i(t) \pm 0)$.*

**Lemma 4** *On the i-th contact discontinuity $x = x_i(t)$, for $j = 1, ..., i-1$, $i+1, ..., n$, we have*

$$w_j^- - w_j^+ = \left(\lambda_i(u^-) - \lambda_j(u^-)\right)^{-2} \Phi, \tag{27}$$

*where*

$$\begin{aligned}\Phi = {} & \left[(\lambda_i(u^+) - \lambda_j(u^+))^2 - (\lambda_i(u^-) - \lambda_j(u^-))^2\right] w_j^+ \\ & + \textstyle\sum_{k\neq i} \left(\lambda_i(u^+) - \lambda_k(u^+)\right)^2 w_k^+ \left(l_j(u^-) - l_j(u^+)\right) r_k(u^+) \\ & + \textstyle\sum_{k\neq i} \left(\lambda_i(u^\pm) - \lambda_k(u^\pm)\right) w_k^\pm \left(\nabla\lambda_i(u^\pm) r_k(u^\pm)\right) l_j(u^-) \left(u^+ - u^-\right)\end{aligned} \tag{28}$$

*and $w^\pm = w^\pm(t, x_i(t)) \triangleq w(t, x_i(t) \pm 0)$.*

**Lemma 5** *There exist positive constants $C_1$ and $C_2$ independent of $\theta$, such that*

$$|\hat{v}^{(i)} - \hat{v}^{(i-1)}| \leq C_1\theta \qquad (i = 1, ..., n) \tag{29}$$

*and*

$$|\hat{w}^{(i)}| \leq C_2\theta \qquad (i = 1, ..., n). \tag{30}$$

By means of Lemmas 3-5, in a manner similar to the proof of global existence of $C^1$ solution to the Cauchy problem in [5], we can get a uniform a priori estimate for the piecewise $C^1$ norm of the solution $u = u(t,x)$ containing $n$ contact discontinuities, then finish the proof of Theorem 1.

## 3. Application - System of the motion of an elastic string

Consider the following generalized Riemann problem for the system of the motion of an elastic string:

$$\begin{cases} u_t - v_x = 0, \\ v_t - \left(\frac{r-1}{r}u\right)_x = 0, \end{cases} \tag{31}$$

$$t = 0 \,:\, (u, v) = \begin{cases} \left(\tilde{u}^0 + \bar{u}_l(x), \tilde{v}^0 + \bar{v}_l(x)\right), & x \leq 0, \\ \left(\tilde{u}^0 + \bar{u}_r(x), \tilde{v}^0 + \bar{v}_r(x)\right), & x \geq 0, \end{cases} \tag{32}$$

where $u = (u_1, ..., u_n)^T$, $v = (v_1, ..., v_n)^T$, $r = |u| = \sqrt{u_1^2 + ... + u_n^2}$ and $\tilde{u}^{(0)} = (\tilde{u}_1^{(0)}, ..., \tilde{u}_n^{(0)})^T$ and $\tilde{v}^0 = (\tilde{v}_1^0, ..., \tilde{v}_n^0)^T$ are constant vectors with $\tilde{r}^{(0)} = |\tilde{u}^{(0)}| > 1$, $(\tilde{u}_l(x), \tilde{v}_l(x))$ and $(\tilde{u}_r(x), \tilde{v}_r(x))$ are $C^1$ vector functions on $x \leq 0$ and on $x \geq 0$ respectively, satisfying the decay property as shown in (8). Let

$$U = (u_1, ..., u_n, v_1, ..., v_n)^T = \begin{pmatrix} u \\ v \end{pmatrix}. \tag{33}$$

In a neighborhood of $U = \begin{pmatrix} \tilde{u}^0 \\ \tilde{v}^0 \end{pmatrix}$, (31) is a hyperbolic system with the following real eigenvalues:

$$\begin{aligned} \lambda_1 &\triangleq -1 < \lambda_2 \equiv ... \equiv \lambda_n \triangleq -\sqrt{\tfrac{r-1}{r}} \\ &< \lambda_{n+1} \equiv ... \equiv \lambda_{2n-1} \triangleq \sqrt{\tfrac{r-1}{r}} < \lambda_{2n} \triangleq 1. \end{aligned} \tag{34}$$

When $n = 2$, (31) is a strictly hyperbolic system; while, when $n > 2$, (31) is a quasilinear hyperbolic system of conservation laws with characteristics with constant multiplicity. Moreover, It is easy to see that system (31) is linearly degenerate in the sense of P.D. Lax.

By Theorem 1 and Remarks 2 and 3, we have

**Theorem 2** *(1) Suppose that*

$$\left(\bar{u}_l(0), \bar{v}_l(0)\right) \neq \left(\bar{u}_r(0), \bar{v}_r(0)\right), \tag{35}$$

*there exists $\theta_0 > 0$ so small that for any given $\theta \in (0, \theta_0]$ the generalized Riemann problem (31)-(32) admits a unique global piecewise $C^1$ solution containing only 4 contact discontinuities with small amplitude on $t \geq 0$. This solution possesses a global structure similar to that of the similarity solution to the corresponding Riemann problem (31) and*

$$t = 0 : (u, v) = \begin{cases} \left(\tilde{u}^0 + \bar{u}_l(0), \tilde{v}^0 + \bar{v}_l(0)\right), & x \leq 0, \\ \left(\tilde{u}^0 + \bar{u}_r(0), \tilde{v}^0 + \bar{v}_r(0)\right), & x \geq 0. \end{cases} \tag{36}$$

*(2) Suppose that*

$$(\bar{u}_l(0), \bar{v}_l(0)) = (\bar{u}_r(0), \bar{v}_r(0)), \tag{37}$$

*there exists a unique global $C^0$ and piecewise $C^1$ solution for all $t \in \mathbb{R}$, to the Cauchy problem (31)-(32) with $C^0$ and piecewise $C^1$ initial data, provided that $\theta$ is suitably small.*

## References


1. Lax, P.D., (1957) Hyperbolic systems of conservation laws II, *Comm. Pure Appl. Math.*, **10**, pp. 537-556.
2. Li Ta-tsien and Yu Wen-ci, (1985) Boundary Value Problems for Quasilinear Hyperbolic Systems, *Duke University Mathematics* Series V.
3. Li Ta-tsien and Kong De-xing, (1997) Solutions globales avec discontinuités de contact pour les systèmes hyperboliques quasi linéaires de lois de conservation, *C.R. Acad. Sci. Paris, t. 324, Série* **1**, pp. 621-626.
4. Li Ta-tsien, Zhou Yi and Kong De-xing, (1994) Weak linear degeneracy and global classical solutions for general quasilinear hyperbolic systems, *Comm. in Partial Differential Equations*, **19**, pp. 1263-1317.
5. Li Ta-tsien, Zhou Yi and Kong De-xing, (1997) Global classical solutions for general quasilinear hyperbolic systems with decay initial data, *Nonlinear Analysis, Theory, Methods & Applications*, **28**, pp. 1299-1332.
6. Li Ta-tsien, Kong De-xing and Zhou Yi, (1996) Global classical solutions for quasilinear non-strictly hyperbolic systems, *Nonlinear Studies*, **3**, pp. 203-229.

# STABILITY OF A SYSTEM WHOSE BOUNDARY EVOLUTIONS ARE GOVERNED BY A STANDARD CONSTITUTIVE LAW

PIERRE BEREST
*Laboratoire de Mécanique des Solides*
*Centre commun Polytechnique, Mines, Ponts, URA 317 CNRS*
*91128 Palaiseau cedex*

## Introduction

Many studies have been devoted to the problems of peeling or unsticking of adhesive tape, for instance Wu (1982), Ericksen (1991), Jensen and Thouless (1992).
In this paper we suggest a theoretical framework inspired by the works of Ehrlacher (1985) and Nguyen Q.S. (1984), that allows to discuss a large family of such problems.

## The Euler condition

We consider here the quasi-static isothermal evolutions of a variable system whose free energy can be described as a functional of

(i) the displacement $u = u(\underline{x})$, $u \in \mathbb{R}$, $\underline{x} \in \Omega \subset \mathbb{R}^n$,
(ii) the domain $\Omega$ on which the displacement is defined,
(iii) the boundary conditions (i.e., the value $u_d$ of $u(\underline{x})$ on the boundary $\partial\Omega$ of the domain $\Omega$) and
(iv) a loading parameter $\lambda$.

In other words, we consider free energies of the following form :

$$E(u;\Omega,u_d,\lambda) = \int_\Omega L(u,\nabla u,\lambda)d\Omega\,;\; u = u_d \text{ on } \partial\Omega$$

The first variation of $E$ (with respect to $u$) can be written :

$$\delta E = E'_u \cdot \delta u = \int_\Omega \{L'_u - \operatorname{div}(L'_{\nabla u})\}\delta u\, d\Omega$$



*P. Argoul et al. (eds.),*
*IUTAM Symposium on Variations of Domains and Free-Boundary Problems in Solid Mechanics*, 261–267.

In order that $u^* = u^*(\underline{x})$ realize a minimum of $E = E(u)$, its first variation must be zero. If system evolutions can be described by the free energy (i.e., no other forces are considered), the relation $\delta E = 0$ stands for the principle of virtual work. This relation is the Euler-Lagrange condition, and a solution of this equation is called an extremal.

### The Legendre and Jacobi conditions

The second variation of $E$ (with respect to $u$) can be written as follows

$$\begin{aligned}\delta^2 E^* &= \delta u \cdot E''_{uu} \cdot \delta u \\ &= \int_\Omega \{L''^*_{uu}\delta u + L''^*_{u\nabla u}\delta\nabla u - \operatorname{div}(L''^*_{\nabla u u}\delta u + L''^*_{\nabla u\nabla u}\delta\nabla u)\}\delta u\, d\Omega \\ &= \int_\Omega J^*(\delta u)\delta u\, d\Omega\end{aligned}$$

(Where * means that the quantity is calculated for a solution of the Euler-Lagrange equation). The second order operator $J^*(\delta u)$ is called the Jacobi operator. If $L''^*_{\nabla u\nabla u}(\underline{x}) > 0$ for any $\underline{x} \in \Omega$ (Legendre condition) and if $J^*(\delta u) = 0$, $\delta u = 0$ on $\partial\Omega$ has no solution except $\delta u = 0$ (Jacobi condition), then $\delta^2 E^* > 0$ and $E^*$ is a (local) minimum of $E$. (For details, see Berest 1997). In the following, we assume that these two conditions are satisfied.

### Varying boundary

In this paragraph, the domain $\Omega$ and the boundary conditions $u = u_d$ on $\partial\Omega$ are considered to be variables. We introduce a new variation, $\delta\underline{x}$, defined on every point of the domain $\Omega$, and the variations $\delta u$ are now $\delta u = \overline{\delta} u + \underline{\nabla u} \cdot \underline{\delta x}$ where $\overline{\delta} u$ is the variation "at $\underline{x}$ constant". In particular, $\delta u_d = \overline{\delta} u + \underline{\nabla u} \cdot \underline{\delta x}$ on the boundary $\partial\Omega$.
Now, the first variation of $E = E(u; \Omega, u_d)$ can be written as :

$$\delta E = E'_u \cdot \overline{\delta} u + E'_\Omega \cdot \delta\Omega + E'_{u_d} \cdot \delta u_d = \int_\Omega \overline{\delta} L\, d\Omega + \int_{\partial\Omega} L(\underline{\delta x} \cdot \underline{n}) da$$

or, after re-arranging,

$$\delta E = \int_\Omega \{L'_u - \operatorname{div}(\underline{L}'_{\nabla u})\}\overline{\delta} u\, d\Omega + \int_{\partial\Omega} \{(\underline{L}'_{\nabla u} \cdot \underline{n})\delta u_d + L(\underline{\delta x} \cdot \underline{n}) - (\underline{\nabla u} \cdot \underline{\delta x})(\underline{L}'_{\nabla u} \cdot \underline{n})\} da$$

Let $u^* = u^*(\underline{x}; \Omega, u_d, \lambda)$ be an extremal such that $u^*(\underline{x}) = u_d$ on $\partial\Omega$ ; then the first integral is zero and the first variation of $E$, when calculated for the extremal, can be written as follows

$$\boxed{\delta^* E = E'^*_\Omega \cdot \delta\Omega + E'^*_{u_d} \cdot \delta u_d = \int_{\partial\Omega} -\underline{\delta x} \cdot \underline{\underline{H}}^* \cdot \underline{n}\, da + \int_{\partial\Omega} \delta u_d\, \underline{p}^* \cdot \underline{n}\, da}$$

where $\underline{p}^* = L'_{\nabla u}$, $\underline{\underline{H}}^* = {}^T\underline{\nabla u}^* \otimes \underline{L}'^*_{\nabla u} - L\underline{\underline{1}}^*$ and the principle of virtual work can be expressed as :

$$\delta^* E + A \cdot \delta\Omega + B \cdot \delta u_d = 0$$

where $A$ and $B$ are the generalized forces conjugated to the variables $\Omega$ and $u_d$, respectively. For solving a particular problem, a constitutive relation linking the boundary $\partial\Omega$ and boundary condition $u_d$ to the generalized forces $A$ and $B$ is needed.

**Elastic Systems**

In some cases, the generalized forces $A$ and $B$ derive from a Lagrangian potential $D = D(\Omega, u_d)$ –i.e., $A \cdot \delta\Omega + B \cdot \delta u_d = \delta D(\Omega, u_d)$, which means that $D$ can be simply added to the initial free energy of the system (which in now called $E_0$, $E^* = E_0^* + D$) and we can state the principle of virtual work in its simplest form :

$$\delta E^* = E'^*_\Omega(\Omega, u_d, \lambda) \cdot \delta\Omega + E'^*_{u_d}(\Omega, u_d, \lambda) \cdot \delta u_d = 0$$

In the following, we discuss the stability of the equilibrium positions defined by the following constitutive relation for the boundary :

$$\boxed{u_d = 0 \quad E'^*_\Omega \cdot \delta\Omega = 0 (\forall \delta\Omega) \quad \text{on} \quad \partial\Omega}$$

The second relation can also be written as $\underline{\underline{H}}^* \cdot \underline{n} = (-L^* \underline{\underline{1}} + {}^T\underline{\nabla u}^* \otimes \underline{L}'^*_{\nabla u}) \cdot \underline{n} = 0$ on $\partial\Omega$, and the second variation will be

$$\delta^2 E^* = \delta\Omega \cdot E''^*_{\Omega\Omega} \cdot \delta\Omega = \delta \int_{\partial\Omega} -\underline{\delta x} \cdot \underline{\underline{H}}^* \cdot \underline{n} da$$

When calculating such an expression, it must be noted that quantities such as $\delta^* u$, which will appear during the calculation, are variations of extremals and will satisfy the equation

$$\delta^* u = \overline{\delta}^* u + \underline{\nabla u}^* \cdot \underline{\delta x} = 0 \quad \text{on} \quad \partial\Omega, \quad J^*(\overline{\delta}^* u) = 0 \quad \text{in} \quad \Omega$$

Then re-arranging leads to

$$^* = \int_{\partial\Omega} -(-\underline{\delta x} \cdot \underline{\underline{H}}^* \cdot {}^T\underline{\underline{\text{grad}\delta x}} - (L'^*_x \cdot \underline{\delta x})\underline{\delta x} + \delta\underline{\nabla u} \cdot (\underline{\delta x} \otimes \underline{L}'^*_{\nabla u} - \underline{L}'^*_{\nabla u} \otimes \underline{\delta x}) + (\underline{\delta x} \cdot \underline{\nabla u})\delta \underline{L}^*_{\nabla u}) \cdot \underline{n}\, d$$

When $\Omega$ is a plane surface, $\underline{\delta x} \cdot \underline{\underline{H}}^* \cdot {}^T\underline{\underline{\text{grad}\delta x}} = (\underline{\delta x} \cdot \underline{\underline{H}}^* \cdot \underline{t})(\underline{n} \cdot \frac{d}{ds}\underline{\delta x})$ ; furthermore, if $L$ is an isotropic function of $\underline{\nabla u}$, $\underline{L}^*_{\nabla u}$ is colinear to $\underline{n}$, and if we take $\underline{\delta x} = \delta x \underline{n}$ on $\partial\Omega$, then

$$\boxed{\delta^2 E^* = -\int_{\partial\Omega} \{-(\underline{L}'^*_x \cdot \underline{\delta x})(\underline{\delta x} \cdot \underline{n}) + (\underline{\delta x} \cdot \underline{\nabla u}^*)(\delta \underline{L}'^*_{\nabla u} \cdot \underline{n})\} ds}$$

where $\delta \underline{L}'_{\nabla u} = L''_{\underline{\nabla u}\,\underline{x}} \cdot \underline{\delta x} + L''_{\nabla u \nabla u} \cdot \underline{\delta \nabla u}$, and $\underline{\delta \nabla u} = \overline{\delta}\underline{\nabla u} + \underline{\underline{\nabla \nabla}} u \cdot \underline{\delta x}$. In the particular case when $L(u, \underline{\nabla u})$ is independent of $\underline{x}$, then $\delta^2 E^*$ reduces to

$$\delta^2 E^* = \int_{\partial\Omega} -(\underline{\delta x} \cdot \underline{\nabla^* u})\underline{\delta \nabla^* u} \cdot L''^*_{\nabla u \nabla u} \cdot \underline{n}\, ds$$

and the system is unstable, for it is invariant by translation ($\underline{\delta x} = \underline{\delta x}_0 (\forall \underline{x})$ and $\delta u = 0$ leads to $\underline{\delta \nabla^* u} = \underline{0}$ and $\delta^2 E^* = 0$).

## Simple Dissipative Systems

The previous analysis can be enlarged to the case of "simple dissipative systems" (Ehrlacher, 1985). In such systems, the generalized force $A$ can be derived from a potential, $A\delta\Omega = \delta D(\Omega)$, but it must belong to a convex set $C$ defined by the inequality $f(A) \leq 0$. If $E = E_0(u, \Omega, \lambda) + D(\Omega)$ then the boundary constitutive law is

$$\boxed{u_d = 0\ ,\ E'_\Omega \cdot \dot{\Omega} = 0\ ,\ (\forall A^* \in C)(A - A^*) \cdot \dot{\Omega} \geq 0 \quad \text{on} \quad \partial\Omega}$$

In other words, the actual rate $\dot{\Omega}$ must belong to the convex cone of the normals to $C$, or $\dot{\Omega} \in N$. In many cases, this simply means that the boundary remains fixed except when the Hamiltonian reaches some criterion $|H_0^*| = C$, where $C$ is a positive constant. If $|H_0^*| < C$, then $\dot{x} = 0$ on the corresponding point of the boundary ; if $|H_0^*| = 0$, then $\underline{\dot{x}} \cdot \underline{n} \geq 0$. In the homogeneous and isotropic case, $H = H(u, \nabla u)$ is a function of $|\nabla u|$ only on the boundary $\partial\Omega$, and the criterion can also be written $|\nabla u| \leq \tilde{C}$, where $\tilde{C}$ is another constant.

The stability of an equilibrium state can be discussed in a similar way to the elastic case, as noticed by Nguyen Quoc Son (1984) : the second variation $\delta^2 E = \delta\Omega \cdot (E''_{\Omega\Omega} + D''_{\Omega\Omega}) \cdot \delta\Omega$ must be definite positive, but only for $\delta\Omega \in N$. This means that even if the equilibrium is stable, the solution of the evolution rate problem can be non-unique, because if $\underline{X}$ is an eigenvector of $\delta^2 E$, which does not belong to $N$, and $\dot{Y}$ is a solution of the rate-problem which belongs to $N$, then $\underline{X} + \mu\underline{\dot{Y}}$ can belong to $N$, at least for small values of $\mu$.

## Example

A perfectly flexible membrane is stuck on a rigid plane frame, except for a domain $\Omega$. A volume $v$ of incompressible fluid is encapsulated between the membrane and the plane surface $\Omega$. The surface tension constant is $A$. Only small diplacements, $u$, are considered. The energy of the system is

$$E = \int_\Omega \left( \frac{1}{2} A\, \underline{\nabla u}^2 - pu \right) d\Omega \ ; \ u = 0 \text{ on } \partial\Omega \ ; \ v = \int_\Omega u\, d\Omega$$

The system is controlled by the volume $v$ ; $p$ is a Lagrange multiplier. The Euler-Lagrange equation is $A\Delta u^* + p = 0$ and the Jacobi equation is $A\Delta\delta u + q = 0$, $q$ is a second Lagrange multiplier ; it has no solution such that $\delta u = 0$ on $\partial\Omega$ and $\delta v = 0$, except for $\delta u = 0$ in $\Omega$, which proves that the equilibrium is stable.

On the boundary $\partial\Omega$, we get $E'_\Omega \cdot \delta\Omega = -\frac{1}{2}A\,\underline{\nabla u}^2\delta x$. Unsticking now becomes possible, $\dot{x} \geq 0$, if $\underline{\nabla u}\cdot\underline{n} + tg\alpha = 0$. If we set $D(\Omega) = \Omega\, A\, tg^2\alpha/2$, then

$$\delta\Omega\cdot(E''^*_{\Omega\Omega} + D''_{\Omega\Omega})\cdot\delta\Omega = \int_{\partial\Omega} -A(\underline{\nabla u}\cdot\underline{\delta x})(\underline{\nabla\bar{\delta}u} + \underline{\underline{\nabla\nabla^* u^*}}\cdot\underline{\delta x})\cdot\underline{n}\,ds$$

where $\bar{\delta}u$ is such that $A\Delta\bar{\delta}u + q = 0$, $\delta v = 0$, and $\delta u = \delta x\, tg\,\alpha$ on $\partial\Omega$ if $\delta x \geq 0$. It is clear that the second variation is zero if $\underline{\delta x} = \underline{\delta x}_0$, $\delta u = 0$ but this solution is not such that $\delta x = \underline{\delta x}\cdot\underline{n} \geq 0$ for any $\underline{x}$ of $\partial\Omega$. The equilibrium is stable but, in general, the solution of the evolution rate problem is non-unique (For example, if $\Omega$ is a circle and $v$ increases, $\Omega$ will become any larger circle that contains $\Omega$).

If the membrane is changed into a soap bubble containing some incompressible fluid, we have the additional condition $\partial u/\partial n = -tg\,\alpha$ on $\partial\Omega$, where $\alpha$ is the contact angle ; the Lagrangian will simply be

$$E = \int_\Omega \frac{1}{2}A(\underline{\nabla u}^2 + tg^2\alpha) - pu\,d\Omega\ ;\ u = 0 \text{ on } \partial\Omega\ ; v = \int_\Omega u\,d\Omega$$

which will give the same solution as before ; however this time, the equilibrium is unstable for any global translation does not change the solution.

## One dimensional examples

The one-dimensional case provides us with simple examples. Here $n = 1$, $\Omega$ is a segment $\Omega = [e, b]$ and $b$ is fixed, but $e$ can decrease ; for $x \leq e$, we simply have $u(x) = 0$. For simplicity we assume that the density of free energy is not explicitely dependent on $x$ :

$$E = \int_e^b L(u, u', \lambda)dx\ ;\ u(e) = 0\ ;\ u(b) = \beta$$

We have the first integral, $h^* = u'^*\cdot L'^*_{u'} - L^* =$ constant, then, $h'^*(x) = 0$. The first variation of $E$ (with respect to the boundary $e$) is

$$E'_e\cdot\delta e = \delta e\, h^*$$

The second variation of $E$ can be easily calculated from $\delta h^*(e) = \bar{\delta}h^*(e) + h'(e)\delta e$ and

$$\bar{\delta}h^* = L''^*_{u'u'}(x)\{u'^*(x)\bar{\delta}u'(x) - u''^*(x)\bar{\delta}u(x)\}$$
$$\bar{\delta}u^*(b) = 0\ ;\ \bar{\delta}u^*(e) + u'^*(e)\delta e = 0$$

from which we get :

$$\overline{\delta}u^*(x) = -\overline{\delta}h^*\, u'^*(x) \int_x^b d\xi/\{u'^2(\xi)L''_{u'u'}(\xi)\}$$

$$\delta e\, \delta h^* = \delta e^2 / \int_e^b d\xi/\{u'^2(\xi)L''_{u'u'}(\xi)\}$$

which proves that, for small values of $(b-e)$, $\delta e\, \delta h^* > 0$ and $\delta u^*(x)$ is not zero for $x \in [e, b[$. The second variation, $\delta e \cdot E''_{ee} \cdot \delta e$, changes its sign when $u'^*(e) = 0$ ; it is noted that, in such a circumstance, $\overline{\delta}^* x$ has a constant sign in $[e, b[$ which means that the second variation, $\delta u \cdot E''_{uu} \cdot \delta u$, is still positive : the instability due to the movement of the boundary always pre-exists to the instability of the system with the fixed boundary.
This conclusion remains when we discuss "unsticking" problems. We assume that $e$ can move if and only if the force associated to the boundary rate reaches some threshold :

$$|h^*(e, \lambda)| < c\,, \quad \dot{e} = 0$$
$$|h^*(e, \lambda)| = c\,, \quad \dot{e} \leq 0$$

Consider a beam stuck on a rigid substratum for any $x \leq e$ ; $y = y(x)$ is the displacement normal to the beam axis and $\alpha(x) = dy/dx$. If the angle $\alpha(b) = \beta$ is fixed, then

$$\tilde{E} = \int_e^b \frac{EI}{2} \alpha'^2(x) dx$$

from which it is easy to deduce $\alpha = \beta(x-e)/(b-e)$, $h = EI\, \beta^2/[2(b-e)^2]$ and

$$\delta^2 F^* = \delta^2(\tilde{E}^* - c\,e) = EI\, \beta^2\, \delta e^2/(b-e)^3 > 0$$

Unsticking will occur when $h = c$ and will be stable.
If now we take $\alpha(b) = 0$ and apply a force $\lambda$ perpendicular to the beam on the point $B$, then

$$\tilde{E} = \int_e^b \left( \frac{EI}{2} \alpha'^2 - \lambda\alpha \right) dx$$

from which $\alpha = \lambda(x-e)(x-b)/(2EI)$ and $h = \lambda^2(e-b)^2/(8EI)$ ; then

$$\delta^2 F^* = \delta^2(\tilde{E}^* - c\,e) = \lambda^2(e-b)\delta e^2/(4EI) < 0$$

Unsticking will occur when $h = c$ and will be unstable.

## References

1. Berest, P. , (1997), *Calcul des variations*, 256 p., Ellipses, Paris.

2. Fedelich, B. , Berest, P. , (1988), Torsion d'un cylindre élasto-fragile : stabilité de l'équilibre. *Arch. Mech.* **40, 5-6**, pp. 641-652.
3. Ehrlacher, A. , (1985), Principe d'extremum et équation en vitesse pour les systèmes à dissipation simple 7[e] *Cong. Fr. Méca.*, **A. 4**, pp. 20-23.
4. Nguyen, Q.S. , (1984), Bifurcation et stabilité des systèmes irréversibles obéissant au principe de dissipation maximale, *J.M.T.A.*, **vol. 3, n°1**, pp. 41–61.
5. Ericksen, J.-L. , (1991), *Introduction to the Thermodynamics of Solids*, 204 p., Chapman and Hall, London.
6. Wu, S. , (1982), *Polymer Interface and Adhesion*, M. Dekker, New York.
7. Jensen, H.M. , Thouless, M.D. , (1992), *Effect of residual stresses in the blister test.* Report 439, The Technical University of Denmark.

# STABILITY OF MOVING FRONTS UNDER GRIFFITH CRITERION: A COMPUTATIONAL APPROACH USING INTEGRAL EQUATIONS AND DOMAIN DERIVATIVES

MARC BONNET
*Laboratoire de Mécanique des Solides*
*Ecole Polytechnique, 91128 Palaiseau cedex,* FRANCE

PRESENTATION. Consider a linearly elastic body $\Omega \subset \mathbf{R}^3$. Its external boundary is divided into two complementary parts $S_u$ (supporting prescribed displacements: $\boldsymbol{u} = \bar{\boldsymbol{u}}$) and $S_t$ (supporting prescribed tractions: $\boldsymbol{\sigma}.\boldsymbol{n} = \bar{\boldsymbol{t}}$). Besides, a crack (described by an open surface $S$ across which the displacement is discontinuous: $\boldsymbol{\varphi} = \boldsymbol{u}^+ - \boldsymbol{u}^-$ denotes the crack opening displacement (COD)) is embedded in $\Omega$.

In the classical Griffith approach, crack propagation may occur at points of the crack front $\partial S = \Gamma$ where the energy release rate $G(s)$ ($s$: arc length along the crack front $\Gamma$, for three-dimensional problems) reaches a certain threshold. $G$ is mathematically defined (Eq. (5) below) as (minus) the kernel associated with the domain derivative of the equilibrium value $W(\bar{\boldsymbol{u}}, \bar{\boldsymbol{t}}; S)$ of the potential energy for the current crack configuration. In the present context (linear elasticity, infinitesimal strains, no body forces), the latter is known in terms of the elastic variables on the external boundary:

$$W \equiv W(\bar{\boldsymbol{u}}, \bar{\boldsymbol{t}}; S) = \frac{1}{2}\int_{S_u} \boldsymbol{t}.\bar{\boldsymbol{u}}\,\mathrm{d}S - \frac{1}{2}\int_{S_t} \bar{\boldsymbol{t}}.\boldsymbol{u}\,\mathrm{d}S \tag{1}$$

Then, the determination of the actual infinitesimal crack front perturbation involves the second-order domain derivative of $W$ (Eqs. 7–9).

This paper investigates a computational approach which (i) uses analytical expressions for the first- and second-order domain derivatives of $W$, with no recourse to finite-difference evaluations, and (ii) is based on a Galerkin symmetric boundary integral equation (SGBIE)[8, 3] of the elastic equilibrium, in view of the major role played by geometry (the crack surface) and geometry perturbations in the problem. The latter governs the boundary unknowns $\boldsymbol{t}\,|_{S_u}\ \boldsymbol{t}\,|_{S_t}, \boldsymbol{\varphi}\,|_S$ and has the general form:

$$\forall(\tilde{\boldsymbol{u}}, \tilde{\boldsymbol{t}}, \tilde{\boldsymbol{\varphi}}) \qquad \left\{ \begin{array}{lcl} \mathcal{B}_{tt}(\tilde{\boldsymbol{t}}, \boldsymbol{t}) + \mathcal{B}_{tu}(\tilde{\boldsymbol{t}}, \boldsymbol{u}) + \mathcal{B}_{t\varphi}(\tilde{\boldsymbol{t}}, \boldsymbol{\varphi}) & = & \mathcal{L}_t(\tilde{\boldsymbol{t}}) \\ \mathcal{B}_{ut}(\tilde{\boldsymbol{u}}, \boldsymbol{t}) + \mathcal{B}_{uu}(\tilde{\boldsymbol{u}}, \boldsymbol{u}) + \mathcal{B}_{u\varphi}(\tilde{\boldsymbol{u}}, \boldsymbol{\varphi}) & = & \mathcal{L}_u(\tilde{\boldsymbol{u}}) \\ \mathcal{B}_{\varphi t}(\tilde{\boldsymbol{\varphi}}, \boldsymbol{t}) + \mathcal{B}_{\varphi u}(\tilde{\boldsymbol{\varphi}}, \boldsymbol{u}) + \mathcal{B}_{\varphi\varphi}(\tilde{\boldsymbol{\varphi}}, \boldsymbol{\varphi}) & = & \mathcal{L}_\varphi(\tilde{\boldsymbol{\varphi}}) \end{array} \right. \tag{2}$$



*P. Argoul et al. (eds.),*
*IUTAM Symposium on Variations of Domains and Free-Boundary Problems in Solid Mechanics,* 269–276.

where the $\mathcal{B}_{ij}$ are known in terms of double surface integrals (see [2] or [6] for detailed expressions) and such that the above formulation is symmetric, and the $\tilde{\boldsymbol{u}}, \tilde{\boldsymbol{t}}, \tilde{\varphi}$ are trial functions of support $S_u, S_t, S$, respectively. After boundary element discretization, the above set of equations lead to a symmetric linear system of equations for the unknowns boundary DOFs.

As a result, a boundary-only formulation, based solely on a boundary element discretisation (i.e. no part of the domain needs discretization) is obtained for the instantaneous crack perturbation problem. Our approach is in fact an adaptation of the so-called $\theta$-method [5, 11], originally defined in the framework of weak formulations and implemented using finite elements, to boundary integral equation formulations. More specifically, Galerkin symmetric BIEs are used because their symmetry is allows (like in the FEM context) to formulate $W_{,\Omega\Omega}$ in terms of the elastic field variables and their first-order domain derivatives (i.e. no second-order field variable derivatives are needed). Also, like in the classical $\theta$-method, Lagrangian-type domain differentiation formulas are used, in order not to increase the crack front singular behaviour of the field variables through the differentiation process.

ENERGY FORMULATION FOR THE CRACK EXTENSION PROBLEM A quasistatic in-plane crack extension process, induced by a load increment, is considered. Using a small (non-physical) time $t$, a perturbed configuration of $S$ is described by $S(t) = S+\boldsymbol{\theta}t$, where $S \equiv S(0)$ is the initial crack surface and $\boldsymbol{\theta}$, the "initial velocity" of extension, must satisfy the requirement

$$\boldsymbol{\theta} \in \Theta = \{\boldsymbol{\theta} \in C^0(S) \mid \theta_n = 0 \text{ on } S,\, \theta_\nu = 0 \text{ on } \Gamma\} \tag{3}$$

where $\boldsymbol{\nu}$ denotes the unit normal to $\Gamma$ which lies in the tangent plane to $S$ and points outwards to $S$. The variations of any field quantity induced by this domain perturbation are described in a Lagrangian manner, using $S$ as the initial configuration. Denoting by $\overset{\star}{f} = f_{,t} + \boldsymbol{\nabla} f.\boldsymbol{\theta}$ the lagrangian derivative of a field quantity $f$, then the material derivative of a surface integral at $t = 0$ is given by:

$$I(f, S; t) = \int_{S(t)} f \,\mathrm{d}S \qquad \overset{\star}{I}(f, S) = \frac{dI}{dt} = \int_S \overset{\star}{f} \,\mathrm{d}S + \int_S f \mathrm{div}_S \boldsymbol{\theta} \,\mathrm{d}S \tag{4}$$

where $\mathrm{div}_S(\cdot) \equiv \mathrm{div}\,(\cdot) - \boldsymbol{n}.\boldsymbol{\nabla}(\cdot).\boldsymbol{n}$ is the surface divergence of a vector field. Moreover, the crack front singularity of $\boldsymbol{\nabla}\overset{\star}{\varphi}$ equals that of $\boldsymbol{\nabla}\varphi$ instead of increasing it; this is the main motivation for our choice of the lagrangian framework, reminiscent of the '$\theta$-method' (Mialon [7], Suo [11]).

The energy release rate $G(s)$ associated to (virtual) tangent crack extensions is such that:

$$\int_\Gamma G(s)\theta_\nu(s) \,\mathrm{d}s + \overset{\star}{W} = 0 \qquad (\forall \boldsymbol{\theta} \in \Theta) \tag{5}$$

where $W$ is given by Eq. (1) and $(\overset{\star}{\cdot})$ denotes the lagrangian derivative with *fixed* load. Then, under the Griffith propagation criterion, the actual crack extension velocity $\boldsymbol{\mu} \in \Theta$ verifies at any point $s$ of the crack front $\Gamma$:

$$\mu \in \Theta_c = \{\boldsymbol{\mu} \in \Theta \mid G(s) < G_c \Rightarrow \mu_\nu(s) = 0\,,\ G(s) = G_c \Rightarrow \mu_\nu(s) \geq 0\}$$

Following Nguyen [9], Pradeilles-Duval [10], define

$$Q(\boldsymbol{\theta}, \boldsymbol{\mu}) = \overset{\star\vee}{W} + \int_\Gamma G_c \theta_\nu \mu_\nu \kappa(s)\,\mathrm{d}s \tag{6}$$

where $(\overset{\vee}{\cdot})$ denotes the lagrangian derivative in the transformation velocity $\boldsymbol{\mu}$ under *fixed* load; $\overset{\star\vee}{W}$ is the second-order domain derivative of $W$, i.e. $\overset{\star\vee}{W} \equiv (\overset{\star}{W})^\vee$, evaluated under the assumption $\overset{\vee}{\boldsymbol{\theta}} = \mathbf{0}$; $\kappa(s)$ is the algebraic curvature on $G$). Then the three following statements hold [9, 10]: first, $\boldsymbol{\mu}$ is governed by the variational inequality:

$$\text{Find } \mu \in \Theta_c,\ \forall \theta \in \Theta_c \quad Q(\boldsymbol{\theta} - \boldsymbol{\mu}, \boldsymbol{\mu}) + \overset{\star}{W}{}' \geq 0 \tag{7}$$

where $()'$ denotes the derivative with respect to load variations and with crack fixed. Second, a solution $\boldsymbol{\mu}$ to (7) exists, i.e. the infinitesimal crack propagation is stable, if

$$\forall \theta \in \Theta_c \quad (\boldsymbol{\theta} \neq \mathbf{0}) \Rightarrow Q(\boldsymbol{\theta}, \boldsymbol{\theta}) > 0 \tag{8}$$

Third, the solution $\boldsymbol{\mu}$ to (7) is unique if (non-bifurcation criterion):

$$\forall (\boldsymbol{\theta}, \boldsymbol{\mu}) \in \Theta \quad (\boldsymbol{\theta} \neq \boldsymbol{\mu}) \Rightarrow Q(\boldsymbol{\theta} - \boldsymbol{\mu}, \boldsymbol{\theta} - \boldsymbol{\mu}) > 0 \tag{9}$$

Our objective is to propose a treatment of the statements (7–9) based on boundary-only expressions of the energy derivatives $\overset{\star}{W}, \overset{\star\vee}{W}, \overset{\star}{W}{}'$.

FIRST-ORDER DERIVATIVE OF $W$. Define the Lagrangian $L$:

$$L = W + \mathcal{B}(\tilde{\boldsymbol{u}}, \boldsymbol{u}) - \mathcal{L}(\tilde{\boldsymbol{u}}) \tag{10}$$

where $\boldsymbol{u}, \tilde{\boldsymbol{u}}, \mathcal{B}, LL$ are compact notations for the set of unknowns, trial functions, bilinear forms and linear forms appearing in Eq. (2). Upon Lagrangian differentiation with $(\tilde{\boldsymbol{u}})^\star = \mathbf{0}$, the derivative $\overset{\star}{\mathcal{L}}$ takes the form

$$\overset{\star}{\mathcal{L}} = \frac{1}{2}\int_{S_u} \overset{\star}{\boldsymbol{t}}.\bar{\boldsymbol{u}}\,\mathrm{d}S - \frac{1}{2}\int_{S_t} \bar{\boldsymbol{t}}.\,\overset{\star}{\boldsymbol{u}}\,\mathrm{d}S + \mathcal{B}(\tilde{\boldsymbol{u}}, \overset{\star}{\boldsymbol{u}}) + \mathcal{B}^1(\tilde{\boldsymbol{u}}, \boldsymbol{u}; \boldsymbol{\theta}) - \mathcal{L}^1(\tilde{\boldsymbol{u}}; \boldsymbol{\theta}) \tag{11}$$

where $\mathcal{B}^1, \mathcal{L}^1$, respectively the Lagrangian derivatives of $\mathcal{B}$ with $\overset{\star}{\boldsymbol{u}} = \mathbf{0}$ and of $\mathcal{L}$ with fixed load, are linear expressions of $\boldsymbol{\theta}$. Now define the adjoint state $\boldsymbol{u}^A$ as the solution to the variational equation:

$$(\forall\, \overset{\star}{\boldsymbol{u}}) \quad \mathcal{B}(\boldsymbol{u}^A, \overset{\star}{\boldsymbol{u}}) = \frac{1}{2}\int_{S_t} \bar{\boldsymbol{t}}.\,\overset{\star}{\boldsymbol{u}}\ \mathrm{d}S \frac{1}{2}\int_{S_u} \overset{\star}{\boldsymbol{t}}.\bar{\boldsymbol{u}}\,\mathrm{d}S \tag{12}$$

Then, using the symmetry of $\mathcal{B}$, one can eliminate $\overset{\star}{\boldsymbol{u}}$ between Eqs. (11,12), thus obtaining the following expression for $\overset{\star}{W}$:

$$\overset{\star}{W} = \mathcal{B}^1(\boldsymbol{u}^A, \boldsymbol{u}; \boldsymbol{\theta}) - \mathcal{L}^1(\boldsymbol{u}^A; \boldsymbol{\theta}) \tag{13}$$

SECOND-ORDER DERIVATIVES OF $W$. First, one has:

$$\overset{\star\vee}{W} = \mathcal{B}^1(\boldsymbol{u}^A, \overset{\vee}{\boldsymbol{u}}; \boldsymbol{\theta}) + \mathcal{B}^1(\overset{\vee}{\boldsymbol{u}}{}^A, \boldsymbol{u}; \boldsymbol{\theta}) + \mathcal{B}^2(\boldsymbol{u}^A, \boldsymbol{u}; \boldsymbol{\theta}, \boldsymbol{\mu}) - \mathcal{L}^2(\boldsymbol{u}^A; \boldsymbol{\theta}, \boldsymbol{\mu}) \tag{14}$$

whereas the first-order domain derivative states are governed by:

$$\mathcal{B}(\tilde{\boldsymbol{u}}, \overset{\star}{\boldsymbol{u}}) = \mathcal{L}^1(\tilde{\boldsymbol{u}}; \boldsymbol{\theta}) - \mathcal{B}^1(\tilde{\boldsymbol{u}}, \boldsymbol{u}; \boldsymbol{\theta}) \quad \mathcal{B}(\tilde{\boldsymbol{u}}, \overset{\star}{\boldsymbol{u}}{}^A) = -\mathcal{B}^1(\tilde{\boldsymbol{u}}, \boldsymbol{u}^A; \boldsymbol{\theta}) \quad (\forall \tilde{\boldsymbol{u}}) \tag{15}$$

upon choosing $\tilde{\boldsymbol{u}} = \overset{\vee}{\boldsymbol{u}}{}^A$ in (15a) and $\tilde{\boldsymbol{u}} = \overset{\vee}{\boldsymbol{u}}$ in (15b), one then obtains an expression for the second-order domain derivative of $W$:

$$\overset{\star\vee}{W} = \mathcal{B}^2(\boldsymbol{u}^A, \boldsymbol{u}; \boldsymbol{\theta}, \boldsymbol{\mu}) - \mathcal{L}^2(\boldsymbol{u}^A; \boldsymbol{\theta}, \boldsymbol{\mu}) - \mathcal{B}(\boldsymbol{u}^A, \overset{\vee}{\boldsymbol{u}}) - \mathcal{B}(\overset{\star}{\boldsymbol{u}}, \overset{\vee}{\boldsymbol{u}}{}^A) \tag{16}$$

Next, the second-order mixed derivative $\overset{\star}{W}{}'$ is given by:

$$\overset{\star}{W}{}' = \mathcal{B}^1(\boldsymbol{u}^A, \boldsymbol{u}'; \boldsymbol{\theta}) + \mathcal{B}^1(\boldsymbol{u}^{A\prime}, \boldsymbol{u}; \boldsymbol{\theta}) - \frac{\partial}{\partial t}\mathcal{L}^1(\boldsymbol{u}^A; \boldsymbol{\theta}) \tag{17}$$

whereas the first-order load derivative states are governed by:

$$\mathcal{B}(\tilde{\boldsymbol{u}}, \boldsymbol{u}') = \frac{\partial}{\partial t}\mathcal{L}(\tilde{\boldsymbol{u}}) \quad (\forall \tilde{\boldsymbol{u}}) \tag{18}$$

$$\mathcal{B}(\tilde{\boldsymbol{u}}, (\boldsymbol{u}^A)') = \frac{1}{2}\int_{S_u} \boldsymbol{t}.\bar{\boldsymbol{u}}' \, dS - \frac{1}{2}\int_{S_t} \bar{\boldsymbol{t}}'.\boldsymbol{u} \, dS \tag{19}$$

SPECIAL CASE: CRACK IN AN INFINITE BODY. For this particular configuration, assuming symmetrical loading of the crack faces (i.e. $\boldsymbol{\sigma}.\boldsymbol{n} = \pm\bar{\boldsymbol{t}}$ on $S^{\pm}$), the only nonzero terms in the formulation (2) are (see [8], and also [4, 1] for plane cracks):

$$\mathcal{B}_{\varphi,\varphi}(\tilde{\varphi}, \varphi) = \int_S \int_S B_{ikqs}(\boldsymbol{y} - \boldsymbol{x}) R_q \tilde{\varphi}_i(\boldsymbol{x}) R_s \varphi_k(\boldsymbol{y}) \, dS_x \, dS_y \qquad \mathcal{L}_\varphi = \int_S \bar{\boldsymbol{t}}.\tilde{\boldsymbol{\varphi}} \, dS$$

with $R_a f \equiv e_{bca} n_b f_{,c}$ (a tangential differential operator) and:

$$B_{ikqs}(\boldsymbol{y} - \boldsymbol{x}) = -\frac{\mu}{8\pi}\left(\delta_{pr}\delta_{qs} + \delta_{ps}\delta_{qr} + \frac{2\nu}{1-\nu}\delta_{pq}\delta_{rs}\right) e_{iep} e_{kgr} r_{,eg}$$

$$r = \| \boldsymbol{y} - \boldsymbol{x} \| \qquad r_{,i} = (y_i - x_i)/r \qquad r_{,ij} = (\delta_{ij} - r_{,i} r_{,j})/r$$

This in turn leads to the following expressions:

$$\mathcal{B}^1(\tilde{\varphi}, \varphi; \boldsymbol{\theta}) = \int_S \int_S B^1_{ikqs}(\boldsymbol{x}, \boldsymbol{y}) R_q \tilde{\varphi}_i(\boldsymbol{x}) R_s \varphi_k(\boldsymbol{y}) \, dS_x \, dS_y \tag{20}$$

$$\mathcal{B}^2(\tilde{\varphi}, \varphi; \boldsymbol{\theta}, \boldsymbol{\mu}) = \int_S \int_S B^2_{ikqs}(\boldsymbol{x}, \boldsymbol{y}) R_q \tilde{\varphi}_i(\boldsymbol{x}) R_s \varphi_k(\boldsymbol{y}) \, dS_x \, dS_y \tag{21}$$

$$\mathcal{L}^1(\tilde{\varphi};\boldsymbol{\theta}) = -\int_S \bar{\boldsymbol{t}}.\boldsymbol{\nabla}\varphi.\boldsymbol{\theta}\,\mathrm{d}S \tag{22}$$

$$\mathcal{L}^2(\tilde{\varphi};\boldsymbol{\theta}) = \int_S \left[\bar{\boldsymbol{t}}.(\boldsymbol{\nabla}\phi.\boldsymbol{\nabla}\mu.\boldsymbol{\theta}) + (\boldsymbol{\nabla}\bar{\boldsymbol{t}}.\boldsymbol{\mu} + (\mathrm{div}_S\boldsymbol{\mu})\bar{\boldsymbol{t}}).\boldsymbol{\nabla}\varphi.\boldsymbol{\theta}\right]\mathrm{d}S \tag{23}$$

with

$$\begin{aligned}
B^1(\boldsymbol{x},\boldsymbol{y}) &= [\theta_m(y)-\theta_m(x)]B_{ikqs,m}(\boldsymbol{y}-\boldsymbol{x}) \\
&\quad + B_{ikqv}(\boldsymbol{y}-\boldsymbol{x})\theta_{v,s}(\boldsymbol{y}) + B_{ikvs}(\boldsymbol{y}-\boldsymbol{x})\theta_{v,q}(\boldsymbol{x}) \\
B^2(\boldsymbol{x},\boldsymbol{y}) &= [\theta_m(y)-\theta_m(x)][\mu_n(y)-\mu_n(x)]B_{ikqs,mn}(\boldsymbol{y}-\boldsymbol{x}) \\
&\quad + [\mu_n(y)-\mu_n(x)]\{\theta_{m,s}(\boldsymbol{y})B_{ikqm,n}(\boldsymbol{y}-\boldsymbol{x}) + \theta_{m,q}(\boldsymbol{y})B_{ikms,n}(\boldsymbol{y}-\boldsymbol{x})\} \\
&\quad + [\theta_n(y)-\theta_n(x)]\{\mu_{m,s}(\boldsymbol{y})B_{ikqm,n}(\boldsymbol{y}-\boldsymbol{x}) + \mu_{m,q}(\boldsymbol{y})B_{ikms,n}(\boldsymbol{y}-\boldsymbol{x})\} \\
&\quad + B_{ikmn}(\boldsymbol{y}-\boldsymbol{x})\{\mu_{m,s}(\boldsymbol{y})\theta_{n,q}(\boldsymbol{x}) + \mu_{m,q}(\boldsymbol{x})\theta_{n,s}(\boldsymbol{y})\}
\end{aligned}$$

NUMERICAL IMPLEMENTATION. The present approach has been numerically implemented for plane cracks of arbitrary shape and mixed-mode loading. The crack surface $S$ and field variables $\varphi, \overset{\star}{\varphi}, \varphi'$ are discretized using 9-noded quadrilateral boundary elements. It is worth noting that the bilinear forms $\mathcal{B}, \mathcal{B}^1, \mathcal{B}^2$ are actually formulated in terms of *tangential* derivatives of boundary variables. Indeed, using a mapping $(\xi_1,\xi_2) \in \Delta \to \boldsymbol{y}(\xi_1,\xi_2) \in S$ (where e.g. $\Delta$ is a reference element) one can show that

$$e_{abc}n_a(\boldsymbol{y})\phi_{b,c}(\boldsymbol{y})\boldsymbol{e}_c\, dS_y = (\phi_{,\xi_1}\boldsymbol{y}_{,\xi_2} - \phi_{,\xi_2}\boldsymbol{y}_{,\xi_1})d\xi_1 d\xi_2 \tag{24}$$

This formula is also applied to the numerical evaluation of the element integrals which arise in the discretization process of $\mathcal{B}, \mathcal{B}^1, \mathcal{B}^2$.

Due to the present use of material-type differentiations, the known fact that ultimately the derivatives of $P$ depend on the crack front velocities only through its normal component $\theta_\nu\mid_{\partial\Gamma}, \mu_\nu\mid_{\partial\Gamma}$ [7], [10] is not apparent in Eqs. (16,17). It is then necessary to construct the velocity fields $\boldsymbol{\theta}, \boldsymbol{\mu}$ such as to be entirely determined by $\theta_\nu\mid_{\partial\Gamma}, \mu_\nu\mid_{\partial\Gamma}$. At the BE level, this is achieved by introducing interpolations of the form:

$$\boldsymbol{\theta}(\boldsymbol{y}) = \boldsymbol{B}^k(\boldsymbol{y})\theta_k \qquad \text{pour } \boldsymbol{y} \in E(\partial\Gamma) \tag{25}$$

where $\theta_k$ $(1 \leq k \leq \text{NC})$ are the values *at the* NC *crack front nodes* $\boldsymbol{A}_k$ of the normal velocity $\theta_\nu$. The local numbering of the elements adjacent to $\Gamma$ is made so that the line $\xi_2 = -1$ is on the approximate crack front. The vector shape functions $\boldsymbol{B}^k(1 \leq k \leq \text{NC})$, which must realize:

$$\boldsymbol{B}^k\mid_{\xi_2=1} = 0 \qquad \boldsymbol{B}^k(\boldsymbol{A}_l) = \delta_{kl}\boldsymbol{\nu}(\boldsymbol{A}_k) \qquad \boldsymbol{B}^k.\boldsymbol{n} = 0 \tag{26}$$

are given (in local numbering) by

$$\boldsymbol{B}^j(\boldsymbol{\xi}) = \tfrac{1}{2}(1-\xi_2)S_j(\xi_1)\boldsymbol{a}\wedge\boldsymbol{n} \qquad (j=1,2,3 \quad (\boldsymbol{a} = (\|\boldsymbol{y}_{,\xi_1}\|)^{-1}\boldsymbol{y}_{,\xi_1}) \tag{27}$$

with $S_1(\xi) = \xi(\xi - 1)/2$, $S_2(\xi) = 1 - \xi^2$, $S_3(\xi) = \xi(\xi + 1)/2$.

NUMERICAL EXAMPLE A circular crack of radius $a$ located in the $(Oy_1y_2)$ plane is loaded by two symmetrical point-forces $\pm F\boldsymbol{e}_3$ applied at points $(0, 0, \pm h)$. This example has an analytical solution for $G$ [12]:

$$G = \frac{1-\nu^2}{E}\frac{P^2}{(\pi h)^3}\frac{\alpha(\kappa+\alpha^2)}{(1+\alpha^2)^4} \qquad \left(\alpha = \frac{a}{h} \ , \ \kappa = \frac{2-\nu}{1-\nu}\right) \tag{28}$$

Three crack meshes $\mathcal{M}(8,2)$, $\mathcal{M}(12,3)$ and $\mathcal{M}(16,5)$ have been used, with $\mathcal{M}(n,p)$ being made of $p$ rings of $n$ 9-noded quadrilateral elements each; hence the mesh $\mathcal{M}(n,p)$ has are $2n$ nodes on the crack front, associated to the same number of nodal normal velocities. Likewise, the discretized $Q(\boldsymbol{\theta}, \boldsymbol{\mu})$ is associated with a $2n \times 2n$ symmetric matrix.

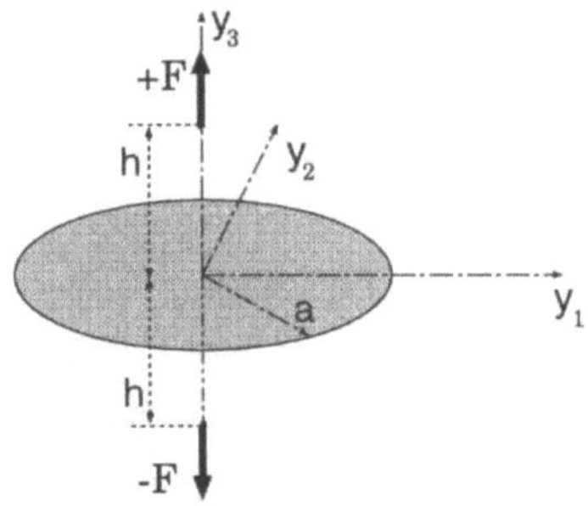


• First, Fig. 1 displays the relative error made on the values of $G$, computed by means of either extrapolation of $\varphi$ or the discretized variational equation (5-13), against $h/a \in [2, 2.5]$, using the mesh $\mathcal{M}(12,3)$.
• Next, an easy consequence of (28) is that:

$$\frac{dG}{d\alpha} > 0 \quad (\alpha > \alpha_m) \quad \text{and} \quad \frac{dG}{d\alpha} < 0 \quad (0 < \alpha < \alpha_m)$$
$$\text{with} \quad \alpha_m^2 = \frac{\sqrt{16\nu^2 - 72\nu + 105} - 2\nu + 9}{2(2-\nu)}$$

thus the circular growth is stable for $\alpha > \alpha_m$ and unstable otherwise. The critical value $\alpha = \alpha_m$ corresponds to a loss of positive definiteness of the quadratic form $Q(\boldsymbol{\theta}, \boldsymbol{\theta})$ (6). The latter has been numerically computed for $h/a = \alpha^{-1} \in [2., 2.5]$ with a 0.005 step and for the mesh $\mathcal{M}(12,3)$; we found $\alpha \approx 2.27$. Then, in order to refine the search, $Q$ has been computed for $\alpha^{-1} \in [2.26, 2.28]$ with a 0.0005 step and for the three meshes $\mathcal{M}(8,2), \mathcal{M}(12,3), \mathcal{M}(16,5)$. The numerical values so obtained for $\alpha_m$ are given in table 1 As can be seen, they approximate very closely the analytical $\alpha_m \approx 2.27589$ (with the present choice $\nu = 0.3$). When $\alpha > \alpha_m$, the

| Mesh | Brackets for $\alpha$ | relative error |
|---|---|---|
| $\mathcal{M}(8,2)$ | $2.267 \leq \alpha_m^{-1} \leq 2.2675$ | $\leq 3.910^{-3}$ |
| $\mathcal{M}(12,3)$ | $2.268 \leq \alpha_m^{-1} \leq 2.2685$ | $\leq 3.510^{-3}$ |
| $\mathcal{M}(16,5)$ | $2.277 \leq \alpha_m^{-1} \leq 2.275$ | $\leq 7.010^{-4}$ |

TABLE 1. Numerical values obtained for $\alpha_m^{-1}$

radius increment $da$ is given in terms of the load increment $dF$ by:

$$da = 2\frac{dF}{F}\frac{\alpha(1+\alpha^2)(\kappa+\alpha^2)}{3\alpha^4+(7\kappa-5)\alpha^2-\kappa} \tag{29}$$

The relative error between the computed values (using mesh $\mathcal{M}(16,5)$) of the extension velocity and the exact solution defined by (29) are shown in Fig. 2. The deterioration of the error as $h/a$ approaches the degeneracy value $\alpha_m^{-1}$, caused by progressive ill-conditioning of $Q$, is clearly visible.

• Finally, table 2 displays relative errors between computed and exact values (with $\alpha = 0.5$) of $W = W(\alpha, F)$, eqn. (1); $G(s)$ (extrapolation of $\varphi$); $G(s)$ (using the formulation (5); the extension velocity $\mu$.

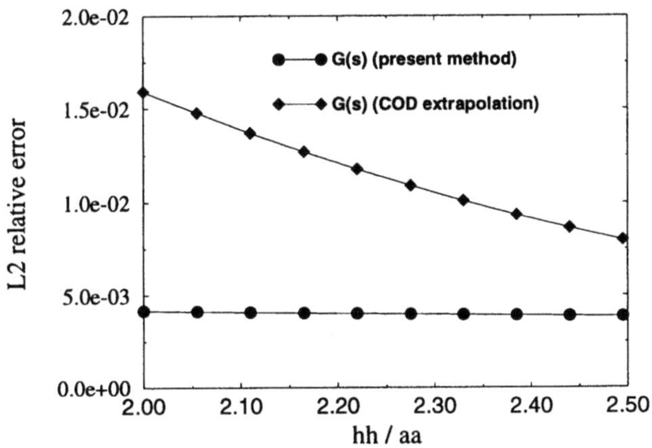


*Figure 1.* Relative $L2$ error on $G$ by (a) the present approach (b) extrapolation of $\phi_3$

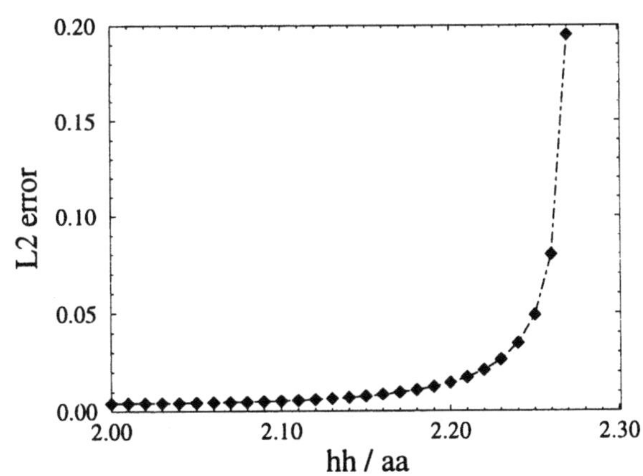


*Figure 2.* Relative $L^2$ error on the extension velocity: mesh $\mathcal{M}(16,5)$, $h/a \in [2, 2.3]$.

| Mesh | $\Delta P/P$ | $\frac{\Vert \Delta G_{present} \Vert_{L^2}}{\Vert G \Vert_{L^2}}$ | $\frac{\Vert \Delta G_{extrapolation} \Vert_{L^2}}{\Vert G \Vert_{L^2}}$ | $\frac{\Vert \Delta \mu \Vert_{L^2}}{\Vert \mu \Vert_{L^2}}$ |
|---|---|---|---|---|
| $\mathcal{M}(8,2)$ | 2.905E-03 | 1.975E-02 | 5.463E-02 | 2.677E-02 |
| $\mathcal{M}(12,3)$ | 3.823E-05 | 7.897E-03 | 2.910E-02 | 1.965E-02 |
| $\mathcal{M}(16,5)$ | -7.176E-04 | 3.789E-03 | 2.142E-02 | 4.466E-03 |

TABLE 2. Relative errors on $P, G, \mu$ for $\alpha = 0.5$

Note that the axisymmetric character of this particular example is not taken into account by the actual computer implementation, but is very well reproduced by the nodal values of $G, \mu_\nu$ along $\partial\Gamma$ numerically obtained.

CONCLUSION. The idea of combining boundary integral equations and analytical domain derivatives is new, to our best knowledge. The numerical example presented shows that in practice an excellent accuracy can be achieved for $G(s)$, $\overset{\star\vee}{W}$ and stability or instability of crack growth. Although the Griffith criterion alone is insufficient for mixed-mode three-dimensional crack propagation simulation, our approach can be combined with other techniques in more complex propagation criterions.

## References


1. P. BONNEMAY. *Equations intégrales pour l'élasticité plane.* PhD thesis, Université Paris VI, France, 1979.
2. M. BONNET. Regularized direct and indirect symmetric variational BIE formulations for three-dimensional elasticity. *Engng. Anal. Bound. Elem.*, 15:93–102, 1995.
3. M. BONNET, G. MAIER, C. POLIZZOTTO. On symmetric galerkin boundary element method. *Appl. Mech. Rev.*, 1997. (in preparation).
4. H. D. BUI. An integral equation method for solving the problem of a plane crack of arbitrary shape. *J. Mech. Phys. Solids*, 25:29–39, 1977.
5. PH. DESTUYNDER, M. DJAOUA, S LESCURE. Quelques remarques sur la mécanique de la rupture élastique. *J. Mécan. Théor. Appl.*, 2:113–135, 1983.
6. S. LI, M.E. MEAR, L. XIAO. Symmetric weak-form integral equation method for three-dimensional fracture analysis. *Comp. Meth. Appl. Mech. Engng.*, 1997. (to appear).
7. P. MIALON. Calcul de la dérivée d'une grandeur par rapport à un fond de fissure par la méthode $\theta$. Bulletin EDF/DER série c vol. 3, Electricité de France, 1987.
8. J. C. NEDELEC. Integral equations with non integrable kernels. *Integral equations and operator theory*, 5:562–572, 1982.
9. Q. S. NGUYEN, R. M. PRADEILLES-DUVAL, C. STOLZ. Sur une loi régularisante en rupture et endommagement fragile. *C.R. Acad. Sci. Paris,* II–309:1515–1520, 1989.
10. R. M. PRADEILLES-DUVAL. *Evolution de systèmes avec surfaces de discontinuité mobiles : application au délaminage.* PhD thesis, Ecole Polytechnique, Palaiseau, France, 1992.
11. X. Z. SUO AND A COMBESCURE. Sur une formulation mathématique de la dérivée de l'énergie potentielle en théorie de la rupture fragile. *C.R. Acad. Sci. Paris, série II,*, 308:1119, 1122–1989.
12. H. TADA, P. PARIS, G. IRWIN. The stress analysis of cracks handbook. Technical report, Del. Research Corporation, Hellertown, Pennsylvania, USA, 1973.

# INSTABILITIES PRODUCED BY EDGES IN THIN SHELLS

J.L. LIONS
*Collège de France*
*3 rue d'Ulm, 75231 Paris Cedex 05*

AND

E. SANCHEZ-PALENCIA
*Laboratoire de Modélisation en Mécanique*
*Université Pierre et Marie Curie, 4 place Jussieu, 75252 Paris Cedex 05*

**Abstract.** It is known that thin shells of elliptic type (i.e. having everywhere principal curvatures of the same sign) which are fixed by a part of the boundary and free by another part of it are sensitive. This means that the limit behaviour as the thickness tends to zero is highly unstable, in the sense that very small and smooth (= belonging to the space $\mathcal{D}$ of test functions for distributions) given forces produces very large perturbations going even out of the distribution space and of the energy space. This instability desappears when the shell is fixed by its whole boundary. We prove here that shells fixed by the whole boundary but having an edge are sensitive.

## 1. Introduction

This communication is concerned with the asymptotic behaviour of thin shells as the thickness $\varepsilon$ tends to zero in the "inhibited" or "membrane dominated" case [3, 12], i.e. when the middle surface $S$ with the kinematical boundary conditions is geometrically rigid, i.e. it does not admit displacements keeping invariant its intrinsic metrics.
The sensitivity phenomenon [8-11] appears in variational problems where the energy space $V^a$, obtained by completion of a space of smooth functions with the energy norm is very "large", going out of the space of distributions. Nevertheless, the duality product is associated with classical integration by parts, and this implies identification of $L^2$ with its dual. As a consequence,



*P. Argoul et al. (eds.),*
***IUTAM Symposium on Variations of Domains and Free-Boundary Problems in Solid Mechanics*, 277–284.**

the dual of the energy space, noted $V^{a\prime}$ does not contain the space $\mathcal{D}'$ of test functions of distributions. Moreover, as the resolvent of the variational problem is an isomorphism of $V^{a\prime}$ on $V^a$, there are "data" $f$ belonging to $\mathcal{D}$ for which there is no solution with finite energy. By the linearity of the problem, this property may be interpreted as an instability property : there exist very small and smooth perturbation of the loading such that the corresponding perturbation of the solution is out of $V^a$ (that is to say, it does not exist in the above framework ; if it exists, it is necessarily a function with infinite energy.

Sensitivity appears in [9, 10] in shells which are uniformly elliptic (i.e. the principal curvatures are everywhere of the same sign and bounded away from zero), fixed (or clamped) by a part $\Gamma_0$ of the boundary, the remainder part $\Gamma_1$ being free. In this communication we prove, using the general ideas of [9], that elliptic shells fixed by their whole boundary but having an edge, are sensitive. A weaker result a weaker result ($V^a$ is not contained in $L^2$) was proved in [5]. Another result of sensitivity for shells with edges in the case of small curvature was given in [11]. It should be pointed out that in this paper we shall make a hypothesis of local symmetry of the shell in a neighbourhood of a point of the edge which is used in the proof for technical reasons, but the result probably holds true without that hypothesis.

Concerning the numerical computation of solutions of shell problems, the sensivity phenomenon leads to a practical impossibility of computing the solutions [6], as the finite element scheme induces a perturbation of the data which is not allowable. Clearly this phenomenon only occurs for the limit behavior as the thickness $\varepsilon$ tends to zero (i.e. for the membrane approximation). Real shells with $\varepsilon > 0$ lead to problems which are "calculable", usually in spaces $H^1 \times H^1 \times H^2$. The convergence of the finite element schemes as $h$ tends to zero holds then true for $\varepsilon > 0$, but the convergence is uniform with respect to $\varepsilon$ only in the "large" space $V^{a\prime}$, in any "usual" space the convergence is not uniform. This is a "locking" phenomenon as defined in [1]. It is known that locking appears mainly in "non-inhibited" or "bending dominated" shells [2], but sensitivity results show that it is also present in inhibited shells.

The notations are the usual ones in shell problems. The convention of summation of repeated indices is used throughout, it goes from 1 to 2 for greek indices and from 1 to 3 for latin indices. Entities which are vectors in the physical 3-dimensional space are usually underlined, as well as the corresponding spaces, for instance

$$\mathbf{u} = (u_1, u_2, u_3) \in \mathbf{L}^2 = (L^2)^3.$$

Some functions will indiferently be considered as defined on the surface $S$ or on the domain $\Omega$ of the parameter plane. Generally speaking, $S$ and its boundary are supposed to be uniformly smooth (unless on the edge !) so that classical uniqueness and unique continuation properties for elliptic systems hold true.

## 2. Description of the shell

Let us first explain some generalities on surfaces. Let $\Omega$ be a bounded and connected domain of $\mathbb{R}^2$ (the parameter plane) with smooth boundary $\Gamma_0$. The middle surface of the shell is a surface $S$ of the Euclidean space $\mathbb{R}^3$ defined by the function $\mathbf{r}$ :

$$(y^1, y^2) \longrightarrow \mathbf{r}(y^1, y^2) \tag{1}$$

from $\Omega$ into $\mathbb{R}^3$. At the points of $S$ we define the tangent vectors

$$\mathbf{a}_\alpha = \mathbf{r}_{,\alpha} \equiv \partial \mathbf{r}/\partial y^\alpha \equiv \partial_\alpha \mathbf{r} \tag{2}$$

which form a local basis (of the tangent plane). We also consider the contravariant basis $\mathbf{a}^\alpha(\mathbf{a}^\alpha.\mathbf{a}_\beta = \delta^\alpha_\beta)$, as well as the unit normal vector $\mathbf{a}_3 = \mathbf{a}^3$. Let

$$\Gamma^\alpha_{\beta\lambda} = \Gamma^\alpha_{\lambda\beta} = \mathbf{a}^\alpha.\mathbf{a}_{\beta,\lambda} = \mathbf{a}^\alpha.\mathbf{a}_{\lambda,\beta} \tag{3}$$

be the Christoffel symbols of the surface and let

$$b_{\alpha\beta} = b_{\beta\alpha} = -\mathbf{a}_\alpha.\mathbf{a}_{3,\beta} = \mathbf{a}_\alpha.\mathbf{a}_{\alpha,\beta} \tag{4}$$

be the second fundamental form, describing the curvatures. We shall assume throughout this paper that $S$ is everywhere elliptic, i.e. the two principal curvatures are of the same sign ; i.e.

$$b_{11}b_{22} - b_{12}^2 > 0. \tag{5}$$

The covariant differentiation is denoted by $D$ (in contrast with ordinary differentiation, denoted by $\partial$, see (2)). For any vector $\mathbf{v}$ we have

$$D_\alpha v_\beta = \partial_\alpha v_\beta - \Gamma^\lambda_{\alpha\beta} v_\lambda \tag{6}$$

Let $\mathbf{u}$ be the displacement vector of $S$ for a deformation of the surface, i.e., we consider that $\mathbf{r}$ changes becoming $\mathbf{r} + \mathbf{u}$, and we linearize for small $\mathbf{u}$. Then the strain tensor is

$$\gamma_{\alpha\beta}(\mathbf{u}) = \frac{1}{2}(D_\alpha u_\beta + D_\beta u_\alpha) - b_{\alpha\beta} u_3 \tag{7}$$

which describes the variation produced by $\mathbf{u}$ in the first-fundamental form (i.e. on the intrinsic metrics of $S$).
Moreover, we consider now a surface $S$ containing an edge. Let the origin $O$ of the parameter plane belong to $\Omega$. Then, the domain of the parameters, $\Omega$ is devided into $\Omega_+$ and $\Omega_-$ by the $y^1 = 0$ axis :

$$\Omega_\pm = \left\{ (y^1, y^2) \in \Omega \; ; \; \pm y^1 \gtrless 0 \right\} \tag{8}$$

the axis $y^1 = 0$ will also be called $\Gamma_1$, which is the "interface". The function $\mathbf{r}$ is supposed to be smooth on $\Omega_+$ and $\Omega_-$, and continuous on $\Gamma_1$, (i.e. the limit values of $\mathbf{r}$ on both sides coincide, and are in their turn smooth with respect to $y^2$. The curve $\mathbf{r}(\Gamma_1)$ is then well defined, and is an edge, i.e. the tangent planes on both sides of this curve form an angle $2\psi$ which is a smooth function of $y^2$ and is supposed to be everywhere different from 0 and $\pi$. In order to describe easily the interface conditions we shall choose the parametrization in such a way that $(\mathbf{a}_1^+, \mathbf{a}_2)$ and $(\mathbf{a}_1^-, \mathbf{a}_2)$ are orthonormal (note that $\mathbf{a}_2 = \mathbf{a}_2^+ = \mathbf{a}_2^-$ is tangent to the edge).
We shall also make a hypothesis of local symmetry. There is a neigbourhood $\mathcal{O}$ of the origin in $\Omega$ such that taking the restriction to $\mathcal{O}$, the edge is plane and the surface $S$ is symmetric with respect to this plane, so that for instance, $b_{11}$ and $b_{22}$ are even functions of $y^1$, where as $b_{12}$ is odd.
The displacement vector $\mathbf{u}$ is defined on $\Omega$ (i.e. on $\Omega_+$ and $\Omega_-$) and must satisfy the boundary conditions on $\Gamma_0$ :

$$u_1 = u_2 = 0 \quad \text{on } \Gamma_0 \text{ (both } \mathbf{u}^+ \text{ and } \mathbf{u}^-\text{)} \tag{9}$$

$$u_3 = 0 \quad \text{on } \Gamma_0 \text{ (both } \mathbf{u}^+ \text{ and } \mathbf{u}^-\text{)} \tag{10}$$

Moreover, the interface conditions on $\Gamma_1$ are

$$u_2^+ = u_2^- \quad \text{on } \Gamma_1 \tag{11}$$

$$u_3^+ = \frac{cos^2\psi}{sin 2\psi}(u_1^+ - u_1^-) - \frac{sin^2\psi}{sin 2\psi}(u_1^+ + u_1^-) \quad \text{on } \Gamma_1 \tag{12}$$

$$u_3^- = \frac{cos^2\psi}{sin 2\psi}(u_1^+ - u_1^-) + \frac{sin^2\psi}{sin 2\psi}(u_1^+ + u_1^-) \quad \text{on } \Gamma_1 \tag{13}$$

**Remark 2.1.** - The boundary conditions (9), (10) show that $S$ is fixed by its boundary $\Gamma_0$. We may consider as well the case when $S$ is clamped, and the normal derivative of $u_3$ also vanishes. Analogously, (11)-(13) show that $\mathbf{u}^+$ and $\mathbf{u}^-$ take the same value on $\Gamma_1$ (note that these two vectors are referred to different frames $\mathbf{a}_i^+$ and $\mathbf{a}_i^-$ on both sides). It may also be imposed the condition that the angle $2\psi$ of the edge be invariant by $u$ ; this condition also involves derivatives of $\mathbf{u}$, see [12] sect. IV.7.4. for all these questions. In any case, it will be seen later (Remark 3.3) that only the conditions (9) and (11) are relevant in our problem. This is the reason why we separated them from the others.■

## 3. Sensitivity property

Let $\mathcal{V}$ be the space of the vectors $\mathbf{u} = (\mathbf{u}^+, \mathbf{u}^-)$ which are smooth on $\Omega_+$ and $\Omega_-$ and satisfy the boundary and interface conditions (9)-(13).

We then define the membrane strain norm

$$\| \mathbf{u} \|_a = \left[ \int_\Omega \sum_{\alpha,\beta} \| \gamma_{\alpha\beta}(\mathbf{u}) \|_0^2 \right]^{1/2} \tag{14}$$

(or roughly speaking, the $L^2$−norm of strain). We give explicitly the expressions of the $\gamma_{\alpha\beta}$ (7) :

$$\begin{cases} \gamma_{11}(\mathbf{u}) &= D_1 u_1 - b_{11} u_3 \\ \gamma_{22}(\mathbf{u}) &= D_2 u_2 - b_{22} u_3 \\ \gamma_{12}(u) &= \gamma_{21}(\mathbf{u}) = \frac{1}{2}(D_2 u_1 + D_1 u_2) - b_{12} u_3 \end{cases} \tag{15}$$

**Lemma 3.1.** - *Under the hypotheses of section 2 (in particular the ellipticity (5)), the expression (14) defines a norm on $\mathcal{V}$.*

**Proof** - We must prove that $\gamma_{\alpha\beta}(\mathbf{u}) = 0$ $(\alpha, \beta = 1, 2)$ implies $\mathbf{u} = 0$ (see (15). From (5), $b_{11}$ and $b_{22}$ are different from zero. We eliminate $u_3$ from $\gamma_{11} = 0$, and we obtain

$$\begin{aligned} D_2 u_2 - (b_{22}/b_{11}) D_1 u_1 &= 0 \\ D_2 u_1 + D_1 u_2 - (2b_{12}/b_{11}) D_1 u_1 &= 0 \end{aligned}$$

which is easily checked to be an elliptic system of first order for $u_1, u_2$. Then (9) are Cauchy boundary conditions for it, and classical uniqueness for such systems in $\Omega_+$ and $\Omega_-$ gives $u_1 = u_2 = 0$ everywhere. Coming back to $\gamma_{11}(\mathbf{u}) = 0$, we also have $u_3 = 0$.■

**Definition 3.2.** - Let $V^a$ be the Hilbert space obtained by completion of $\mathcal{V}$ with the norm (14).

Clearly this means that the elements of $V^a$ are defined as limits of sequences of elements of $\mathcal{V}$ which are Cauchy sequences for the norm (14).

**Remark 3.3.** - It is classical that traces do not make sense in $L^2$. Correspondingly, when handling smooth functions, we may modify the traces by adding variations of the functions such that the $L^2$ norm of the variation is as small as desired. We apply this property to the component $u_3$ of the functions of $\mathcal{V}$. Using variations of $\mathbf{u}$ of the form $(0, 0, \delta u_3)$ with norm (14) as small as desired, we see that the boundary conditions (10), (12), (13) which involve $u_3$ are irrelevant : we obtain the same completion $V^a$ by discarding them from the definition of $\mathcal{V}$. We shall consider in the sequel $\mathcal{V}$ to be the space of smooth vectors on $\Omega_+$ and $\Omega_-$ satisfying (9) and (11).■

The main result of this paper is :

**Theorem 3.4.** - *The space $V^a$ is not a space of distributions, i.e. :*

$$V^a \not\subset \mathcal{D}'(\Omega_+)^3 \times \mathcal{D}'(\Omega_-)^3 \tag{16}$$

**Remark 3.5.** - Passing to the dual spaces, (17) is equivalent to

$$V^{a\prime} \not\supset \mathcal{D}(\Omega_+)^3 \times \mathcal{D}(\Omega_-)^3 \tag{17}$$

(Here duality is understood in the sense of the duality of $\mathcal{D}'^3$ and $\mathcal{D}^3$ in distribution theory or equivalently the identification of $\mathbf{L}^2$ with its own dual, i.e. the classical duality associated with integration by parts in boundary value problems.

**Remark 3.6.** - When the edge $\Gamma_1$ does not exist (or $2\psi = 0$) the situation is very different. In that case we obtain ([4] or [12], chap. VII) :

$$V^a = H_0^1(\Omega) \times H_0^1(\Omega) \times L^2(\Omega).■ \tag{18}$$

**Proof of theorem 3.4.** - We shall prove (17) by contradiction. Let us admit that for any $\varphi = (\varphi^+, \varphi^-) \in \mathcal{D}(S^+)^3 \times \mathcal{D}(S^-)^3$, the linear form

$$\mathbf{v} \rightarrow (\varphi, \mathbf{v}) = \int_S \varphi^i v_i, dS \tag{19}$$

defined on $\mathcal{V}$ continuous for the norm (14), and we shall find a contradiction. Let us define the application

$$\mathbf{v} \rightarrow \gamma(\mathbf{v}) \equiv \{\gamma_{11}(\mathbf{v}), \gamma_{22}(\mathbf{v}), \gamma_{12}(\mathbf{v}) = \gamma_{21}(\mathbf{v})\} \tag{20}$$

from $\mathcal{V}$ into $(L^2(S))^3$. Here according to (20) this space is considered as the space of the symmetric 2×2 matrices with entries in $L^2(S)$. By definition of $V^a$, this application extends by continuity to an isomorphism of $V^a$ on a closed subspace $X$ of $(L^2(\Omega))^3$. It then follows from this isomorphism and from the continuity of (20) that the form

$$\xi \rightarrow (\varphi, \gamma^{-1}(\xi))$$

is continuous on $X$. By the Hahn-Banach theorem, it may be extended to $(L^2(\Omega))^3$. It then follows that there exist $T \in (L^2(\Omega))^3$ (associated with $\varphi$) such that

$$(\varphi, \gamma^{-1}(\xi)) = (T, \xi)_{(L^2)^3}. \tag{21}$$

Denoting $\gamma^{-1}(\xi)$ by $\mathbf{v}$, we see that

$$\left| \begin{array}{l} \text{there exists } T \in (L^2(\Omega))^3 \text{ such that} \\ (T^{\alpha\beta}, \gamma_{\alpha\beta}(\mathbf{v}))_{L^2(\Omega)} = \int_S \varphi^i v_i dS \quad \forall\, \mathbf{v} \in V^a \end{array} \right. \tag{22}$$

Taking in particular $\mathbf{v} \in \mathcal{D}(S^1) \times \mathcal{D}(S^-)$ we obtain (we recall that classical formulas of integration by parts hold true with covariant differentiation) :

$$- D_\beta T^{\alpha\beta} = \varphi^\beta \quad \text{in } \Omega^\pm \ (\alpha = 1, 2) \tag{23}$$

$$- b_{\alpha\beta} T^{\alpha\beta} = \varphi^3 \quad \text{in } \Omega^\pm \tag{24}$$

where we recognize the membrane stress system for the shell.

Moreover as $T^{\alpha\beta} \in L^2$ as well as the left hand sides of (23), the traces of $n_\beta T^{\alpha\beta}$ make sense in $H^{-1/2}$ (cf. [7]) where $\mathbf{n}$ denotes the unit normal vector to $\partial S$ tangent to $S$. Then, taking $\mathbf{v} \in \mathcal{V}$ in (22) integration by parts makes sense and we obtain

$$\int_{\Gamma_1} \left( T^{11+} v_1^+ + T^{12+} v_2^+ - T^{11-} v_1^- - T^{12-} v_2^- \right) dy^2 = 0 \tag{25}$$

for any smooth $\mathbf{v}$ satisfying (11), so that :

$$T^{11+} = T^{11-} = 0 \quad \text{on } \Gamma_1 \tag{26}$$

$$T^{12+} - T^{12-} = 0 \quad \text{on } \Gamma_1 \tag{27}$$

so that for any $\varphi \in \mathcal{D}(S^+) \times \mathcal{D}(S^-)$ there exists a solution $T \in (L^2(S))^3$ of (23), (24), (26), (27). Now we use the hypothesis of symmetry on a neighbourhood $\mathcal{O}$ of the origin in order to deduce a Cauchy problem for an elliptic system. Let us take the restriction of (23), (24), (26), (27) to $\mathcal{O}$, and let us take $\varphi$ symmetric, (i.e. $\varphi^1$ odd and $\varphi^2$, $\varphi^3$ even in $y^1$). If there is a solution $T$, its symmetric is a solution too, and taking half the sum of them we see that there exist a solution which is symmetric, i.e. such that $T^{11}$, $T^{22}$ are even and $T^{12}$ is odd in $y^1$. It them follows from (27) that $T^{12+}$ and $T^{12-}$ vanish on $\Gamma_1$. We now consider the right part (noted $\mathcal{O}^+$) of the neighbourhood ; discarding the superscript +, we see that the system (23)-(24) has a solution in $\mathcal{O}^+$ satisfying

$$T^{11} = 0 \quad , \quad T^{12} = 0 \quad \text{on } \Gamma_1 \tag{28}$$

for any $\varphi \in \mathcal{D}(\mathcal{O}^+)^3$.
Let us take $\varphi^3 = 0$. According to (5) we may solve (24) with respect to $T^{22}$ and replace into (26), which becomes :

$$\left\{ \begin{array}{l} -\partial_1 T^{11} - \partial_2 T^{12} - L^1 = \varphi^1 \\ (b_{11}/b_{22})\partial_1 T^{11} + [-\partial_1 + 2(b_{12}/b_{22})\partial_2] T^{12} + L^2 = \varphi^2 \end{array} \right. \tag{29}$$

where $L^\alpha$ are zero-order terms in $T^{11}$, $T^{12}$. From (5) it follows that (29) is an elliptic first order system for $T^{11}$, $T^{12}$, which is satisfied in $\mathcal{O}^+$, and (28) are Cauchy conditions for it on $\Gamma_1$. We then have a contradiction as it is classically known that such a problem has not solution for any $\varphi^1$, $\varphi^2$ in $\mathcal{D}(\mathcal{O}^+)$. Theorem 3.4 is proved.

## 4. References


1. I. Babuska and M. Suri, On locking and robusting in the finite element method, *SIAM Journ. Num. Anal.*, **29**, p. 1261-1293 (1992).
2. D. Choï, P.J. Palma, E. Sanchez-Palencia and M.A. Vilarino, Membrane Locking in the finite element computation of very thin elastic shells, *Math. Modelling and Numer. Anal.*, to appear.
3. P.G. Ciarlet, *Mathematical elasticity ; vol. III : Shells theory*, North Holland, (to appear).
4. P.G. Ciarlet and E. Sanchez-Palencia, An existence and uniqueness theorem for the two-dimensional linear membrane shell equations, *Journ. Math. Pures Appl.*, **75**, p. 51-67 (1996).
5. G. Geymonat and E. Sanchez-Palencia, On the rigidity of certain surfaces with edges and application to shell theory, *Arch. Rat. Mech. Anal.*, **129**, p. 11-45 (1995).
6. D. Leguillon et E. Sanchez-Palencia, Expériences de non calculabilité dans les problèmes sensitifs. Coques minces chancelantes. Actes 2ème Colloque Nat. Calcul Structures, Hermès, Paris p. 275-280 (1995).
7. J.L. Lions et E. Magenès, *Problèmes aux limites non homogènes et application, vol. I*, Dunod, Paris (1967).
8. J.L. Lions et E. Sanchez-Palencia, Problèmes aux limites sensitifs, *Compt. Rend. Acad. Sci. Paris, série I* **319**, p. 1024-1026 (1994).
9. J.L. Lions et E. Sanchez-Palencia, Problèmes sensitifs et coques élastiques minces, in *Partial Differential Equations and Functional Analysis* in memory of Pierre Grisvard, eds. J.Cea, D. Chenais, G. Geymonat, J.L. lions, Birkhauser, p. 206-220 (1996).
10. J.L. Lions et E. Sanchez-Palencia, Sur quelques espaces de la théorie des coques et de la sensitivité, in *Homogenization and Application to Material Sciences*, ed. D. Cioranescu, A. Damlamian, P. Donato, Gakkotosho, Tokyo (1997), to appear.
11. J.L. Lions and E. Sanchez-Palencia, Sensitivity problems for some shells with edges,*Topol. Meth. in Nonlinear Anal.* (vol. dedicated to O.A. Ledyzhenskaya) to appear.
12. J. Sanchez-Hubert et E. Sanchez-Palencia, *Coques élastiques minces : propriétés asymptotiques*, Masson, Paris (1997).

# ON THE INVERSION OF SUBSURFACE RESIDUAL STRESSES FROM SURFACE STRESS MEASUREMENTS

P. BALLARD AND A. CONSTANTINESCU
*Laboratoire de Mécanique des Solides*
*Ecole Polytechnique*
*91128 Palaiseau Cédex FRANCE*

**Abstract.** X-ray diffraction provides an efficient method to measure the stress tensor at a surface point of a polycrystalline body. When the stress at an interior point of the body is to be measured, it is common to remove matter in order to reach the considered point. However, this matter removal induces, in general, a stress redistribution so that the measured stress is different from the original residual stress. This paper addresses the problem of reconstructing the original stress field from the measurements. The problem is solved by a general method which leads to explicit inversion formulae for the case of simple geometries. Closed-form expressions are obtained for the two-dimensional and three-dimensional half-space. For bodies of arbitrary shape, an existence and uniqueness theorem is proved. The existence proof is constructive and a numerical algorithm may be derived.

## 1. Introduction

A field of solid mechanics where variation of domains plays a central role is experimental measurement of subsurface residual stress fields. The most frequently used experimental method to measure residual stresses is X-ray diffraction. This technique allows to measure an average of all components of the stress tensor at *the surface* of a polycrystalline metal. The average is taken over a small material volume. However, X-ray diffraction does not allow direct measurements at *interior points* of a solid. In practice, layers at the surface of the solid are removed to reach the considered points. The point is, such a variation of domain induces a redistribution of the residual stress field. Thus, the measured stress tensor after removal is generally



*P. Argoul et al. (eds.),*
*IUTAM Symposium on Variations of Domains and Free-Boundary Problems in Solid Mechanics,* 285–292.

different from the original residual stress tensor at the corresponding point. Therefore, the main problem in residual stress analysis by means of X-ray diffraction is: how can the original stress (before matter removal) be related to the surface measured stress (after matter removal) ?

In order to solve this problem, one obviously needs a constitutive assumption. A quite natural assumption is that the stress redistribution due to the matter removal is *elastic*. Experimentally, to minimize the risk of inducing plastic strains, chemical etching is the preferred removal technique.

To our knowledge, the main result in the literature concerning the stated problem is to be found in a paper by Moore and Evans (1958). They give explicit reconstruction formulae in case of special geometries (infinite plate, infinite cylindrical bar) under the assumption that the stress field depends only on one space coordinate.

This paper addresses the general problem of reconstructing the original stresses in a solid of arbitrary shape on the basis of the surface-measured stresses after an arbitrary matter removal.

## 2. Statement of problem

Let $\Omega_0$ be a regular solid body with a residual stress field $\boldsymbol{\sigma}$ (called in the sequel the original stress) in equilibrium with no body forces (this assumption is not essential and body forces could readily be incorporated to the following analysis). Therefore:

$$\text{div}\,\boldsymbol{\sigma} = 0 \quad \text{in } \Omega \tag{1}$$

The boundary $\partial\Omega_0$ is split into two parts $S_0$ and $\partial\Omega^0 \setminus S_0$. $S_0$ is supposed stress-free:

$$\boldsymbol{\sigma} \cdot \boldsymbol{n}_0 = 0 \quad \text{on } S_0 \tag{2}$$

where $\boldsymbol{n}_0$ denotes, as usual, the unit outward normal. No matter is made on the origin of this residual stress field (plastic strain field, surface forces on $\partial\Omega_0 \setminus S_0$, etc...), but to simplify the analysis, it is assumed (but it is absolutely not essential) that $\partial\Omega^0 \setminus S_0$ is stress-free.

A matter removal process is considered. It is assumed to be a smooth process depending on a fictitious time parameter $t$ ($0 \le t \le T$), generating a sequence of regular bodies $\Omega_t$ of boundary $\partial\Omega_t$. More precisely, this process is such that:

$$\forall t \in [0, T] \quad \partial\Omega_0 \setminus S_0 \subset \partial\Omega_t \tag{3}$$

and such that $\partial\Omega_t \setminus (\partial\Omega^0 \setminus S_0)$, which is denoted by $S_t$, is a $C^1$ homotopy of $S_0$ (see fig. 1). $(\alpha, \beta, \gamma)$ is a curvilinear coordinate system on $\Omega_0$ such that:

– $\gamma$ is constant and equal to $t$ on each surface $S_t$.

- $\alpha$ and $\beta$ are arbitrarily choosen on $S_0$.
- $\alpha$ and $\beta$ are 'transported' on the other surfaces $S_t$ following the envelopes of the normals to these surfaces. Therefore, the coordinate $\gamma$ is perpendicular to $\alpha$ and $\beta$ in $\Omega_0 \setminus \Omega_T$.

This matter removal process induces *a priori* a redistribution of the stress field in the whole body. It is assumed that the stress field $\boldsymbol{\sigma}^m$ (the measured stress) is known on the whole surface $S_t$ for all $t$ in $[0, T]$. Therefore $\boldsymbol{\sigma}^m$ is a function of $t$ but also of the space variables $\alpha$ and $\beta$ on the surface $S_t$. In order to assure a free surface condition, $\boldsymbol{\sigma}^m$ verifies:

$$\forall t \in [0, T] \qquad \boldsymbol{\sigma}^m(\alpha, \beta, t) \cdot \boldsymbol{n}(\alpha, \beta, t) = 0 \tag{4}$$

where $\boldsymbol{n}(\alpha, \beta, t)$ denotes the outward unit normal to $\partial\Omega_t$.

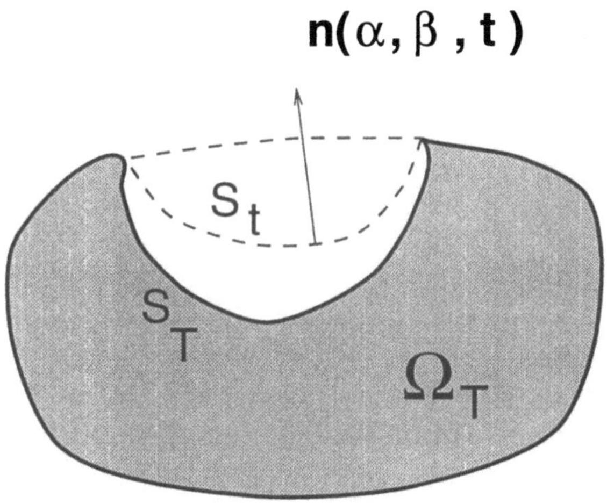


*Figure 1.* Geometry and notations.

The aim of this study is the reconstruction of the original stress field $\boldsymbol{\sigma}$ in the removed part of the body $\Omega_0 \setminus \Omega_T$ from the measured stress data $\boldsymbol{\sigma}^m$. To achieve this purpose, one obviously needs a constitutive assumption. The following is made: the variation of stress due to matter removal is *elastic*.

In this frame, well-posedness of the problem will be proved:

- Uniqueness: for a given measured stress field $\boldsymbol{\sigma}^m$, there corresponds at most one initial stress field $\boldsymbol{\sigma}$ in $\Omega_0 \setminus \Omega_T$.
- Existence: provided some conditions on $\boldsymbol{\sigma}^m$ are fulfilled, every $\boldsymbol{\sigma}^m$ field in $\Omega_0 \setminus \Omega_T$ is derived from an initial stress field $\underline{\underline{\sigma}}$.

## 3. Proof of well-posedness

In this section, enough regularity is assumed on all the fields in order that everything makes sense.

At instant $t$, three elastic problems are considered (see fig. 2):

- Problem 1: equilibrium of $\Omega_0$. The stress field is $\boldsymbol{\sigma}(\alpha, \beta, \gamma)$.
- Problem 2: equilibrium of $\Omega_t$. The stress field on the surface $S_t$ is $\boldsymbol{\sigma}^m(\alpha, \beta, t)$.
- Problem 3: auxiliary problem. One considers an elastic solid at rest of geometry $\Omega_t$. On the part $S_t$ of its boundary, tractions equal to $\boldsymbol{\sigma}(\alpha, \beta, t) \cdot \boldsymbol{n}(\alpha, \beta, t)$ are imposed. Note that this traction distribution is necessary self equilibrated. The resulting stress field is denoted by $\boldsymbol{\sigma}^t(\alpha, \beta, \gamma)$ $(\gamma \geq t)$.

Problem 3 is purely elastic whereas problems 1 and 2 involve the inelastic strain field (identical for the two problems because of the assumption of elastic redistribution of the stress field due to matter removal) at the origin of the residual stress field.

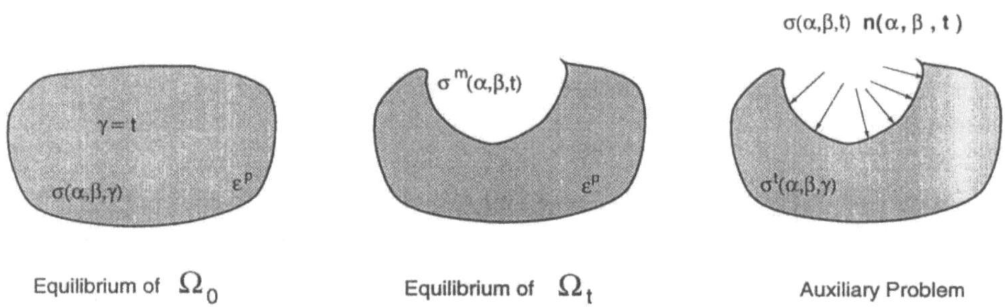


*Figure 2.* **Description of the three elastic problems.**

From the superposition principle, one gets on each surface $S_t$ $(0 \leq t \leq T)$:

$$\boldsymbol{\sigma}(\alpha, \beta, t) = \boldsymbol{\sigma}^m(\alpha, \beta, t) + \boldsymbol{\sigma}^t(\alpha, \beta, t) \tag{5}$$

The objective of what follows is to prove that $\boldsymbol{\sigma}^t(\alpha, \beta, t)$ is fully determined by the data of $\boldsymbol{\sigma}^m$. From equation (5), it will result that the field $\boldsymbol{\sigma}$ in $\Omega_0 \setminus \Omega_T$ is fully determined by the data of the field $\boldsymbol{\sigma}^m$.

As usual, we denote by $\boldsymbol{e}_\alpha$, $\boldsymbol{e}_\beta$, $\boldsymbol{e}_\gamma$ the vector basis associated with the curvilinear coordinate system and $\boldsymbol{e}^\alpha$, $\boldsymbol{e}^\beta$, $\boldsymbol{e}^\gamma$ the vectors of the dual basis for the canonical scalar product of $\mathbb{R}^3$. As a consequence of the construction of the coordinate system $(\alpha, \beta, \gamma)$, one has:

$$\boldsymbol{e}_\gamma = h\boldsymbol{n} \quad \text{and} \quad \boldsymbol{e}^\gamma = \boldsymbol{n}/h \tag{6}$$

where $h$ is a scalar function of $\alpha$, $\beta$ et $\gamma$.

Using standard notations, one writes:

$$\text{div } \boldsymbol{\sigma} = (\nabla_{\boldsymbol{e}_\alpha} \boldsymbol{\sigma}) . \boldsymbol{e}^\alpha + \left(\nabla_{\boldsymbol{e}_\beta} \boldsymbol{\sigma}\right) . \boldsymbol{e}^\beta + (\nabla_{\boldsymbol{e}_\gamma} \boldsymbol{\sigma}) . \boldsymbol{e}^\gamma \tag{7}$$

Denoting by $\operatorname{div}_2 \boldsymbol{\sigma}$ the sum of the first two terms, equation (7) may be rewritten as:

$$\operatorname{div} \sigma(\alpha,\beta,\gamma) - \operatorname{div}_2 \sigma + \frac{1}{h}\frac{\partial}{\partial\gamma}(\sigma.n) - \frac{1}{h}\sigma.\frac{\partial \boldsymbol{n}}{\partial\gamma} \tag{8}$$

On one hand, since $\boldsymbol{\sigma}$ and $\boldsymbol{\sigma}^t$ are in equilibrium with null body forces, one has:

$$0 = \operatorname{div}_2 \boldsymbol{\sigma} + \frac{1}{h}\frac{\partial}{\partial\gamma}(\boldsymbol{\sigma}.\boldsymbol{n}) - \frac{1}{h}\boldsymbol{\sigma}.\frac{\partial \boldsymbol{n}}{\partial\gamma} \tag{9}$$

$$\operatorname{div} \boldsymbol{\sigma}^m = \operatorname{div}_2 \boldsymbol{\sigma}^m - \frac{1}{h}\boldsymbol{\sigma}^m.\frac{\partial \boldsymbol{n}}{\partial\gamma} \tag{10}$$

$$0 = \operatorname{div}_2 \boldsymbol{\sigma}^t + \frac{1}{h}\frac{\partial}{\partial\gamma}\left(\boldsymbol{\sigma}^t.\boldsymbol{n}\right) - \frac{1}{h}\boldsymbol{\sigma}^t.\frac{\partial \boldsymbol{n}}{\partial\gamma} \tag{11}$$

where particular form of equation (10) is a direct consequence of equation (4). A linear combination of equations (9), (10) et (11) gives, thanks to equation (5) :

$$\frac{\partial}{\partial\gamma}(\boldsymbol{\sigma}.\boldsymbol{n})(\alpha,\beta,\gamma=t) = \frac{\partial}{\partial\gamma}\left(\boldsymbol{\sigma}^t.\boldsymbol{n}\right)(\alpha,\beta,\gamma=t) - h \operatorname{div} \boldsymbol{\sigma}^m(\alpha,\beta,\gamma=t) \tag{12}$$

One the other hand, simply differentiating equation (5) after having contracted at $\boldsymbol{n}$, one gets:

$$\frac{\partial}{\partial\gamma}(\boldsymbol{\sigma}.\boldsymbol{n})(\alpha,\beta,\gamma=t) = \left.\frac{\partial}{\partial t'}\right|_{t'=t}\left(\boldsymbol{\sigma}^{t'}.\boldsymbol{n}\right)(\alpha,\beta,t) + \frac{\partial}{\partial\gamma}\left(\boldsymbol{\sigma}^t.\boldsymbol{n}\right)(\alpha,\beta,\gamma=t) \tag{13}$$

Finally, a simple combination of equations (12) et (13) gives :

$$\forall t \in [0,T] \quad \left.\frac{\partial}{\partial t'}\right|_{t'=t}\left(\boldsymbol{\sigma}^{t'}.\boldsymbol{n}\right)(\alpha,\beta,t) = -h \operatorname{div} \boldsymbol{\sigma}^m(\alpha,\beta,\gamma=t) \tag{14}$$

Now, let $t$ be an arbitrary instant in $[0,T]$ and $t'$ in $[0,t]$. By definition, $\boldsymbol{\sigma}^{t'}(\alpha,\beta,\gamma)$ is the stress field induced in the elastic body $\Omega_{t'}$ by surface tractions $\boldsymbol{\sigma}(\alpha,\beta,t').\boldsymbol{n}(\alpha,\beta,t')$ on $S_{t'}$ and null tractions on $\partial\Omega_{t'} \setminus S_{t'}$. Therefore, $\boldsymbol{\sigma}^{t'}(\alpha,\beta,\gamma)$ is the stress field induced in $\Omega_t$ by surface tractions $\boldsymbol{\sigma}^{t'}(\alpha,\beta,t).\boldsymbol{n}(\alpha,\beta,t)$ on $S_t$ and null tractions on $\partial\Omega_t \setminus S_t$. Differentiating this field with respect to $t'$ (at constant domain $\Omega_t$), it is readily seen that $\left.\frac{\partial}{\partial t'}\right|_{t'=t}\boldsymbol{\sigma}^{t'}(\alpha,\beta,\gamma)$ is the stress field induced in $\Omega_t$ by surface tractions $\left.\frac{\partial}{\partial t'}\right|_{t'=t}\left(\boldsymbol{\sigma}^{t'}.\boldsymbol{n}\right)(\alpha,\beta,t)$ on $S_t$ and null tractions on $\partial\Omega_t \setminus S_t$. But, from equation (14), these surface tractions are seen to be nothing but $-h \operatorname{div} \boldsymbol{\sigma}^m(\alpha,\beta,t)$.

This is a Neumann elastic problem which admits a unique solution provided that the imposed surface tractions are self-equilibrated:

$$\forall t \in [0,T] \qquad \int_{S_t} h(\alpha,\beta,t)\,\mathrm{div}\,\underline{\underline{\sigma}}^m(\alpha,\beta,t)\,dS_t = 0 \tag{15}$$

$$\forall t \in [0,T] \qquad \int_{S_t} \underline{OM}(\alpha,\beta,t) \wedge \Big(h(\alpha,\beta,t)\,\mathrm{div}\,\underline{\underline{\sigma}}^m(\alpha,\beta,t)\Big)\,dS_t = 0 \tag{16}$$

where $O$ is an arbitrary origin and $M(\alpha,\beta,\gamma)$ the current integration point. In other terms, the field $\frac{\partial}{\partial t'}\Big|_{t'=t} \sigma^{t'}(\alpha,\beta,\gamma)$ is fully determined in $\Omega_0 \setminus \Omega_T$ by the data of the field $\boldsymbol{\sigma}^m$.

Then, existence and uniqueness of the field $\boldsymbol{\sigma}$ in $\Omega_0 \setminus \Omega_T$ under conditions (15) and (16) is a direct consequence of equation (5) which may be rewritten as :

$$\forall \gamma \in [0,T] \qquad \underline{\underline{\sigma}}(\alpha,\beta,\gamma) = \underline{\underline{\sigma}}^m(\alpha,\beta,\gamma) + \int_0^\gamma \frac{\partial}{\partial t'}\Big|_{t'=t} \underline{\underline{\sigma}}^{t'}(\alpha,\beta,\gamma)\,dt \tag{17}$$

where the integrant is fully determined by $\boldsymbol{\sigma}^m$.

This well-posedness proof is a general method to deal with the general problem. It leads to :

- Numerical methods. The matter removal is discretized in $N$ steps. For each step the field $\frac{\partial}{\partial t'}\Big|_{t'=t} \sigma^{t'}(\alpha,\beta,\gamma)$ is computed by a numerical solving of a Neumann elastic problem in $\Omega_t$ with surface traction $-h$ div $\boldsymbol{\sigma}^m$ on $S_t$. Then, a numerical integration in formula (17) gives a numerical approximation of the unknown field $\boldsymbol{\sigma}$ in $\Omega_0 \setminus \Omega_T$.
- Analytical methods. When the solving of the Neumann problem is explicit, because, for example, a closed-form Green function is known, one may derive explicit expression of the unknown field $\boldsymbol{\sigma}$ as a function of the field $\boldsymbol{\sigma}^m$. An example is displayed in next section.

## 4. The half-space problem

In this section, $\Omega_0$ is the half-space $z > 0$ in $\mathbb{R}^3$ with rectangular coordinate system $(O, \boldsymbol{e}_x, \boldsymbol{e}_y, \boldsymbol{e}_z)$. The matter removal process is the removal of infinite layers of uniform thickness. The thickness $h$ of the removed layer is taken as fictitious time parameter. It is readily seen that $\Omega_h$ is also a half-space and that $S_h$ is nothing but an infinite plane. Here, we add the supplementary hypothesis that the material is isotropic from the point of view of its elastic properties. $\nu$ denotes its Poisson coefficient.

First, a bidimensional problem is studied: it is assumed that $\boldsymbol{\sigma}$ and $\boldsymbol{\sigma}^m$ do not depend on the space variable $y$. $\boldsymbol{\sigma}^m$ is therefore a function of $x$ and

$h$ only. Equation (4) implies:

$$\forall x \ \forall h \quad \sigma^m_{xz}(x,h) = \sigma^m_{yz}(x,h) = \sigma^m_{zz}(x,h) = 0 \tag{18}$$

In this background, it may be shown (Ballard and Constantinescu, 1994c) that the following system holds:

$$\left\{ \begin{array}{lcl} \dfrac{\partial \sigma_{xx}}{\partial x} + \dfrac{\partial \sigma_{xz}}{\partial z} & = & 0 \\ \dfrac{\partial \sigma_{xy}}{\partial x} + \dfrac{\partial \sigma_{yz}}{\partial z} & = & 0 \\ \dfrac{\partial \sigma_{xz}}{\partial x} + \dfrac{\partial \sigma_{zz}}{\partial z} & = & 0 \\ \sigma_{xx}(x,z) & = & \sigma^m_{xx}(x,z) + \dfrac{2}{\pi}\ \text{P.v.} \displaystyle\int_{-\infty}^{+\infty} \dfrac{\sigma_{xz}(x',z)}{x-x'} dx' + \sigma_{zz}(x,z) \\ \sigma_{xy}(x,z) & = & \sigma^m_{xy}(x,z) + \dfrac{1}{\pi}\ \text{P.v.} \displaystyle\int_{-\infty}^{+\infty} \dfrac{\sigma_{yz}(x',z)}{x-x'} dx' \\ \sigma_{yy}(x,z) & = & \sigma^m_{yy}(x,z) + \dfrac{2\nu}{\pi}\ \text{P.v.} \displaystyle\int_{-\infty}^{+\infty} \dfrac{\sigma_{xz}(x',z)}{x-x'} dx' + 2\nu\sigma_{zz}(x,z) \end{array} \right. \tag{19}$$

with the following boundary conditions:

$$\sigma_{xz}(x,0) = 0, \qquad \sigma_{yz}(x,0) = 0, \qquad \sigma_{zz}(x,0) = 0, \tag{20}$$

which result from equation (2). The symbols P.v. in system (19) recall that the integrals are to be understood in the sense of Cauchy principal values. This system is therefore a singular integro-differential system with unknown $\boldsymbol{\sigma}$. Assuming, $\boldsymbol{\sigma}^m$ in $L^2$, it may be shown (Ballard and Constantinescu, 1994c) by use of Fourier transform in the sense of distributions that the unique distribution solution of system (19) is given by:

$$
\begin{aligned}
\sigma_{xx}(x,z) &= \sigma^m_{xx}(x,z) + \frac{2}{\pi}\int_0^z \int_{-\infty}^{+\infty} \frac{(x-x')^4 - 3(x-x')^2(z-z')^2}{[(x-x')^2+(z-z')^2]^3} \\
&\qquad \times \left[\sigma^m_{xx}(x',z') - \sigma^m_{xx}(x,z')\right] dx'dz' \\
\sigma_{xz}(x,z) &= \frac{4}{\pi}\int_0^z \int_{-\infty}^{+\infty} \frac{(x-x')^3(z-z') - (x-x')(z-z')^3}{[(x-x')^2+(z-z')^2]^3} \\
&\qquad \times \left[\sigma^m_{xx}(x',z') - \sigma^m_{xx}(x,z')\right] dx'dz' \\
\sigma_{zz}(x,z) &= \frac{2}{\pi}\int_0^z \int_{-\infty}^{+\infty} \frac{3(x-x')^2(z-z')^2 - (z-z')^4}{[(x-x')^2+(z-z')^2]^3} \\
&\qquad \times \left[\sigma^m_{xx}(x',z') - \sigma^m_{xx}(x,z')\right] dx'dz' \\
\sigma_{xy}(x,z) &= \sigma^m_{xy}(x,z) + \frac{1}{\pi}\int_0^z \int_{-\infty}^{+\infty} \frac{(x-x')^2 - (z-z')^2}{[(x-x')^2+(z-z')^2]^2} \\
&\qquad \times \left[\sigma^m_{xy}(x',z') - \sigma^m_{xy}(x,z')\right] dx'dz' \\
\sigma_{yy}(x,z) &= \sigma^m_{yy}(x,z) + \frac{2\nu}{\pi}\int_0^z \int_{-\infty}^{+\infty} \frac{(x-x')^2 - (z-z')^2}{[(x-x')^2+(z-z')^2]^2} \\
&\qquad \times \left[\sigma^m_{xx}(x',z') - \sigma^m_{xx}(x,z')\right] dx'dz' \\
\sigma_{yz}(x,z) &= \frac{2}{\pi}\int_0^z \int_{-\infty}^{+\infty} \frac{(x-x')(z-z')}{[(x-x')^2+(z-z')^2]^2} \\
&\qquad \times \left[\sigma^m_{xy}(x',z') - \sigma^m_{xy}(x,z')\right] dx'dz'
\end{aligned}
\tag{21}
$$

All the integrals in (21) are regular integrals. Such explicit inversion formulae may also be obtained in the three-dimensional case (i.e. with no *a priori* information on $\boldsymbol{\sigma}$. The corresponding explicit formulae may be found in (Ballard and Constantinescu, 1994c).

## References


Moore, M.G. and Evans, W.P. (1958) Mathematical correction for stress in removed layers in X-ray diffraction residual stress analysis, *SAE Trans.*, **Vol. no. 66**, pp 340–345.

Ballard, P. and Constantinescu, A. (1994a) Sur l'inversion d'un champ de contraintes résiduelles en fonction des contraintes mesurées en surface, *C. R. Acad. Sci. Paris, Série II*, **Vol. no. 318**, pp. 859–863

Ballard, P. and Constantinescu, A. (1994b) Reconstruction d'un champ de contraintes résiduelles à partir des contraintes mesurées sur des surfaces successives. Existence et unicité, *C. R. Acad. Sci. Paris, Série II*, **Vol. no. 319**, pp. 1117–1122

Ballard, P. and Constantinescu, A. (1994c) On the inversion of subsurface residual stresses from surface stress measurements, *J. Mech. Phys. Solids*, **Vol. no. 42**, pp. 1767–1787

# STRUCTURAL OPTIMIZATION BY THE HOMOGENIZATION METHOD

GRÉGOIRE ALLAIRE
*Commissariat à l'Energie Atomique*
*DRN/DMT/SERMA, CEA Saclay*
*91191 Gif sur Yvette, France*

AND

FRANÇOIS JOUVE
*Ecole Polytechnique*
*Centre de Mathématiques Appliquées*
*91128 Palaiseau, France*

**Abstract**

We discuss a method of structural optimization in the context of linear elasticity. We seek the optimal shape of an elastic body which is both of minimum weight and maximal stiffness under specified loadings. Mathematically, a weighted sum of the elastic compliance and of the weight is minimized among all possible shapes. This problem is known to be "ill-posed", namely there is generically no optimal shape and the solutions computed by classical numerical algorithms are highly sensitive to the initial guess and mesh-dependent. Our method is based on the homogenization theory which makes this problem well-posed by allowing microperforated composites as admissible designs. A new numerical algorithm is thus obtained which allows to capture an optimal shape on a fixed mesh. Such a procedure is called topology optimization since it places no explicit or implicit restriction on the topology of the optimal shape, i.e. on its number of holes or members.

## 1. Introduction.

We consider the following structural optimization problem : find the optimal shape that minimizes a weighted sum of its elastic compliance and weight. As usual the compliance (i.e. the work done by the load) is a global measure of the design's rigidity. No explicit or implicit restrictions are made on the



*P. Argoul et al. (eds.),*
*IUTAM Symposium on Variations of Domains and Free-Boundary Problems in Solid Mechanics*, 293–300.

shape's boundary. We work in a bounded reference domain $\Omega$, occupied by a linearly elastic material with isotropic Hooke's law $A_0$ (with bulk and shear moduli $\kappa$ and $\mu$), and loaded on its boundary by some given force $f$. An admissible structure $\omega$ is a subset of the reference domain $\Omega$ obtained by removing one or more holes (the new boundaries created this way are traction-free). The equations of elasticity for the resulting design are

$$\begin{cases} \sigma = A_0 e(u) & e(u) = \frac{1}{2}(\nabla u + \nabla^t u) \\ div\sigma = 0 & \text{in } \omega \\ \sigma \cdot n = f & \text{on } \partial\Omega \\ \sigma \cdot n = 0 & \text{on } \partial\omega \setminus \partial\Omega \end{cases} \tag{1}$$

where $u$ is the displacement vector and $\sigma$ the constraint matrix. The compliance of the structure $\omega$ is

$$c(\omega) = \int_{\partial\Omega} f \cdot u = \int_\omega A_0 e(u).e(u) = \int_\omega A_0^{-1}\sigma.\sigma. \tag{2}$$

Introducing a positive Lagrange multiplier $\lambda$, our structural optimization problem is to minimize, over all subsets $\omega \subset \Omega$, the objective function $E(\omega)$ equal to the weighted sum of the compliance and weight of $\omega$. In other words we want to compute minimizers of

$$\inf_{\omega\subset\Omega} \left(E(\omega) = c(\omega) + \lambda|\omega|\right). \tag{3}$$

The Lagrange multiplier $\lambda$ has the effect of balancing the two contradictory objectives of rigidity and lightness of the shape (increasing its value decreases the weight).

As is well known, in absence of any supplementary constraints on the admissible designs $\omega$ (namely, without smoothness of the boundary), the objective function $E(\omega)$ may have no minimizer, i.e. there is no optimal shape (see e.g. (Kohn *et al.*, 1986), (Lurie *et al.*, 1982), (Murat *et al.*, 1985)). The physical reason for this non-existence is that it is often advantageous to cut infinitely many small holes (rather than just a few big ones) in a given design in order to decrease the objective function. Thus, achieving the minimum may require a limiting procedure leading to a "generalized" design consisting of composite materials made by microperforation of the original material.

To cope with this physical behavior of nearly optimal shapes, we have to enlarge the space of admissible designs by permitting perforated composites from the start : this process is called *relaxation*. Such composite structures are determined by two functions $\theta(x)$ and $A(x)$ : $\theta$ is the local volume fraction of the original material, taking values between 0 and 1, and $A(x)$ is the effective Hooke's law determined by the microstructure of perforations.

Of course, we need to find an adequate definition of the relaxed objective function $\tilde{E}(\theta, A)$ which generalizes $E(\omega)$. This is done in the next section by using *homogenization theory* which gives optimal effective properties of composite materials. The ultimate goal is twofold : prove an existence theorem of relaxed or composite optimal design, and find a new numerical algorithm for computing optimal shapes.

The interested reader is referred to the pioneering works (Kohn *et al.*, 1986), (Lurie *et al.*, 1982), and (Murat *et al.*, 1985) for more details on the mathematical theory of relaxation by homogenization for structural design problems. The first numerical applications on meaningful problems have appeared more recently (see (Allaire *et al.*, 1997), (Allaire *et al.*, 1993), (Bendsoe *et al.*, 1988), (Jog *et al.*, 1994), (Suzuki *et al.*, 1991)).

## 2. Relaxed or homogenized formulation.

In this section we briefly recall the main results of (Allaire *et al.*, 1993), (Allaire *et al.*, 1997) concerning the relaxation procedure. We begin with a minimizing sequence for the objective function (3), i.e. a sequence of increasingly optimal shapes, $(\omega_\epsilon)_{\epsilon\to 0}$. This sequence can be regarded as finer and finer mixtures of the original material $A_0$ and void (holes) with lengthscale $\epsilon$ going to 0. Then, as a result of the homogenization theory, in the limit there exists an effective behavior of this fine mixture, i.e. a composite material of density $\theta(x) \in [0,1]$ and Hooke's law $A(x)$ such that the sequence of solutions of (1) converges to the solution of

$$\begin{cases} \sigma = A(x)e(u) & e(u) = \frac{1}{2}(\nabla u + \nabla^t u) \\ div\sigma = 0 & \text{in } \Omega \\ \sigma \cdot n = f & \text{on } \partial\Omega, \end{cases} \tag{4}$$

and the corresponding compliances also converge

$$\lim_{\epsilon\to 0} c(\omega_\epsilon) = \tilde{c}(\theta, A) = \int_{\partial\Omega} f \cdot u = \int_\Omega A(x)^{-1}\sigma.\sigma,$$

where the stress $\sigma$ is solution of the homogenized equation (4). Remark that, for a given value $\theta$ of the density, there are many different possible effective Hooke's law $A$ in a set $G_\theta$, the so-called $G$-closure set at volume fraction $\theta$, which is the set of all possible homogenized Hooke's law with density $\theta$. We can thus pass to the limit in the objective function and obtain the relaxed or homogenized functional

$$\lim_{\epsilon\to 0} E(\omega_\epsilon) = \min_{0\le\theta\le 1,\ A\in G_\theta} \tilde{E}(\theta, A),$$

where

$$\tilde{E}(\theta, A) = \tilde{c}(\theta, A) + \lambda \int_\Omega \theta(x).$$

The relaxed functional $\tilde{E}(\theta, A)$ has to be minimized over all admissible composite designes, i.e. over all density $\theta$ and effective Hooke's law $A \in G_\theta$. This problem is not entirely explicit since the precise characterization of the $G$-closure set $G_\theta$ is unknown ! However, by using the principle of complementary energy, we can restrict the set $G_\theta$ of admissible composites to a smaller set of optimal composites, namely the so-called sequential laminates which are explicitly known. Indeed, we rewrite the compliance as

$$\tilde{c}(\theta, A) = \min_{\substack{div\sigma=0 \text{ in } \Omega \\ \sigma\cdot n=f \text{ on } \partial\Omega}} \int_\Omega A(x)^{-1}\sigma.\sigma. \tag{5}$$

Then, the two minimizations, in $(\theta, A)$ and in $\sigma$, can be switched. Furthermore, a classical result of homogenization implies that the microstructure can be optimized pointwise in the domain. Therefore, the relaxed formulation becomes

$$\min_{\substack{div\sigma=0 \text{ in } \Omega \\ \sigma\cdot n=f \text{ on } \partial\Omega}} \int_\Omega \min_{0\leq\theta\leq 1,\ A\in G_\theta} \left(A^{-1}\sigma.\sigma + \lambda\theta\right). \tag{6}$$

For a fixed stress $\sigma$, the minimization of $A^{-1}\sigma.\sigma$ on $G_\theta$ is a classical problem in the theory of homogenization and composite materials. It amounts to find the most rigid composite of given density $\theta$ under the stress $\sigma$. In two dimensions, the result is

$$\min_{A\in G_\theta} A^{-1}\sigma.\sigma = A_0^{-1}\sigma.\sigma + \frac{(\kappa+\mu)(1-\theta)}{4\kappa\mu\theta}\left(|\sigma_1| + |\sigma_2|\right)^2 \tag{7}$$

where $\sigma_1$ and $\sigma_2$ are the eigenvalues of the 2 by 2 symmetric matrix $\sigma$. Furthermore, optimality in (7) is achieved for a so-called rank-2 sequential laminate aligned with the eigendirections of $\sigma$ (see (Allaire *et al.*, 1993), (Gibianski *et al.*, 1984) for details). In three dimensions, the result is more complicated, and we give it in the special case of Poisson's ratio equal to zero, i.e. $3\kappa = 2\mu$ (the general case is not much different in essence)

$$\begin{aligned} \min_{A\in G_\theta} A^{-1}\sigma.\sigma \quad &= \quad A_0^{-1}\sigma.\sigma + \\ + \begin{cases} \frac{(1-\theta)}{4\mu\theta}\left(|\sigma_1| + |\sigma_2| + |\sigma_3|\right)^2 & \text{if } |\sigma_3| \leq |\sigma_1| + |\sigma_2| \\ \frac{(1-\theta)}{2\mu\theta}\left((|\sigma_1| + |\sigma_2|)^2 + |\sigma_3|^2\right) & \text{if } |\sigma_3| \geq |\sigma_1| + |\sigma_2| \end{cases} \end{aligned} \tag{8}$$

where the eigenvalues of $\sigma$ are labeled in such a way that $|\sigma_1| \leq |\sigma_2| \leq |\sigma_3|$. Furthermore, optimality in the first regime of (8) is achieved by a rank-3 sequential laminate aligned with the eigendirections of $\sigma$, while in the second regime it is achieved by a rank-2 sequential laminate aligned with the two first eigendirections of $\sigma$.

After this crucial step, the minimization in $\theta$ can easily be done by hand, which completes the explicit calculation of the relaxed formulation. ¿From a mathematical point of view, one can prove that the relaxed formulation (6) admits a minimizer, that any minimizing sequence of the original problem (3) converges (in the sense of homogenization) to a minimizer of (6), and that the two infimum values of (6) and (3) are equal. However, in general there is no uniqueness of the minimizer.

## 3. Numerical algorithm.

At this point, using homogenization theory and introducing a relaxed formulation might appear to be just a trick for proving existence theorems. In truth its importance goes much further, and it is at the root of new numerical algorithms for computing optimal shapes. Indeed, it permits to separate the original minimization (3) in two different tasks : first, optimize locally the microstructure (the effective Hooke's law $A$) with explicit formula, second, minimize globally on the density $\theta(x)$. This has the effect of transforming a difficult "free-boundary" problem into a much easier "sizing" optimization problem in a fixed domain. It has many advantages : on the one hand, the computational cost is very low compared to traditional algorithms since the mesh is fixed (shapes are captured rather than tracked), on the other hand, it behaves as a topology optimizer (the final optimal shape may have a topology completely different from that of the initial guess). The key features of this type of algorithms have been first recognized by M. Bendsoe and N. Kikuchi (Bendsoe *et al.*, 1988).

Let us describe more precisely our algorithm. It is an alternate direction algorithm : we start with an initial design (usually full material everywhere), then, at each iteration, we compute the stress $\sigma$ solution of a linear elasticity problem with a Hooke's law corresponding to the previous design, and we update the design variables $\theta$ and $A$ in terms of $\sigma$ by using the explicit formula for the optimal laminated composite material in (7) or (8). We iterate this process until convergence which is detected when the density variation becomes smaller than some threshold. The elasticity problem is solved by finite elements (piecewise linear for the displacement and piecewise constant for the density and Hooke's law in (Allaire *et al.*, 1997)). There are some numerical difficulties associated to this choice of finite elements (checkerboard instabilities), but they can be easily cured (see (Allaire *et al.*, 1997), (Jog *et al.*, 1994), (Sigmund, 1994)).

This algorithm converges smoothly in a relatively small number of iterations (between 10 and 100, depending on the desired accuracy). Furthermore, in practice it is insensitive to the choice of initial guess and convergent under mesh refinement, suggesting that the numerical algorithm always

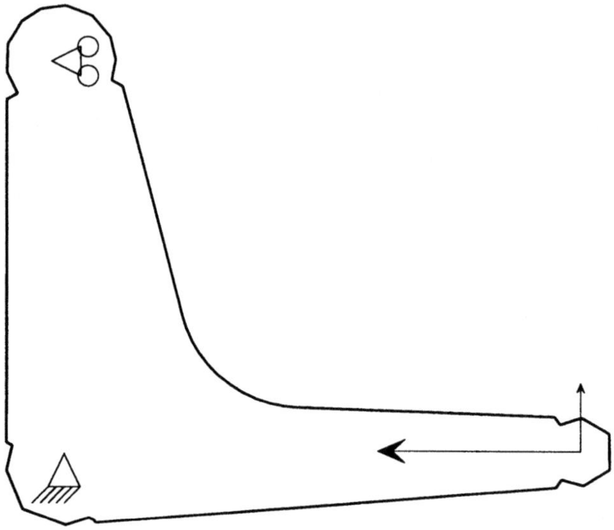

*Figure 1. Boundary conditions for a suspension triangle.*

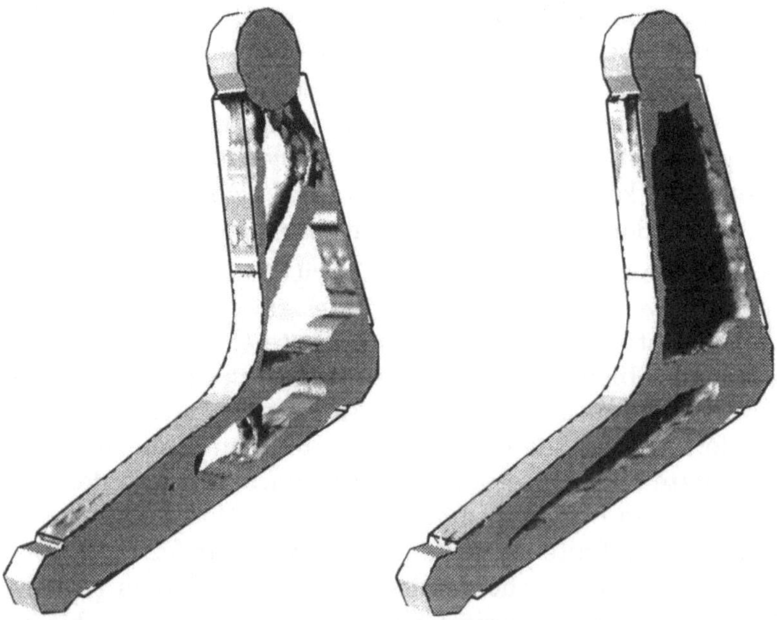

*Figure 2. Optimal suspension triangle; full 3-D solution (left) and plate reinforced solution (right).*

picks up the same global minimum. However, as expected, it usually produces homogenized optimal designs that include large region of composite materials with intermediate density. From a practical point of view, this is an undesirable feature since the primal goal is to find an optimal shape, i.e. a density taking only the values 0 or 1 ! The remedy is to introduce a *penalization* technique that will get rid of composite materials. The strategy is the following : after convergence has been reached on a homogenized optimal design, we run a few more iterations (around 10) of our algorithm during which we force the density to take values close to 0 or 1. More specifically, denoting by $\theta_{opt}$ the true optimal density, the penalization procedure amounts to update the density at the value $\theta_{pen} = \frac{1-\cos(\pi\theta_{opt})}{2}$. There is no specific reason to choose a cosinus-shape function for the penalized density, except that it works fine and yields surprisingly nice shapes featuring fine patterns instead of composite regions.

The success of this method is due to the fact that the relaxed design is characterized not only by a density $\theta$ but also by a microstructure $A$ which is hidden at the sub-mesh level. The penalization has the effect of reproducing this microstructure at the mesh level (see figure 1). Of course it is strongly mesh-dependent in the sense that the finer the mesh the more complicated the resulting "almost optimal" structure.

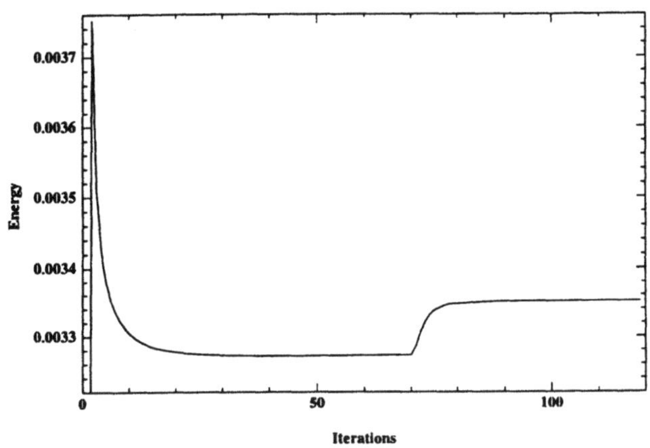


*Figure 3. Objective function versus iteration number.*

Figure 2 depicts a three-dimensional numerical example, namely a suspension triangle. The boundary conditions are indicated in a 2-D section of the 3-D mesh (see Figure 1). It is a multiple loads optimization, which means that the two arrows in Figure 1 correspond to two forces separately exerted. A full 3-D optimal structure has been computed and compared with a so-called "plate reinforced" optimal shape obtained by enforcing a middle section of the structure full of material. The full 3-D shape has 40% of material left from the original box (50% for the plate reinforced shape).

The process for computing such optimal shapes has been to compute first composite optimal shapes and then to apply the penalization procedure. The convergence is smooth and monotone, except when starting the penalization (see Figure 3).

## References

Allaire G., Belhachmi Z., Jouve F., (1996) The homogenization method for topology and shape optimization. Single and multiple loads case, *Revue Européenne des Eléments Finis* **5** pp. 649–672

Allaire G., Bonnetier E., Francfort G., Jouve F., (1997) Shape optimization by the homogenization method, *Numerische Mathematik* **76** pp. 27–68

Allaire G., Kohn R.V., (1993) Optimal design for minimum weight and compliance in plane stress using extremal microstructures, *Europ. J. Mech. A/Solids* **12** pp. 839–878

Bendsoe M., (1995) Methods for optimization of structural topology, shape and material, *Springer Verlag*

Bendsoe M., Kikuchi N., (1988) Generating Optimal Topologies in Structural Design Using a Homogenization Method, *Comp. Meth. Appl. Mech. Eng.* **71** pp. 197–224

Gibianski L., Cherkaev A., (1984) Design of composite plates of extremal rigidity, *Ioffe Physicotechnical Institute preprint* (in russian)

Jog C., Haber R., Bendsoe M., (1994) Topology design with optimized, self-adaptative materials, *Int. Journal for Numerical Methods in Engineering* **37** pp. 1323–1350

Kohn R., Strang G., (1986) Optimal Design and Relaxation of Variational Problems I-II-III, *Comm. Pure Appl. Math.* **39** pp. 113–182, 353–377

Lurie K., Cherkaev A., Fedorov A., (1982) Regularization of Optimal Design Problems for Bars and Plates I-II, *J. Optim. Th. Appl.* **37** pp. 499–521, 523–543

Murat F., Tartar L., (1985) Calcul des Variations et Homogénéisation, *Les Méthodes de l'Homogénéisation Théorie et Applications en Physique, Coll. Dir. Etudes et Recherches EDF, Eyrolles* pp. 319–369

Sigmund O., (1994) Design of material structures using topology optimization, *PhD thesis, report S 69, Dept. of Solid Mechanics, Technical University of Denmark*

Suzuki K., Kikuchi N., (1991) A homogenization method for shape and topology optimization, *Comp. Meth. Appl. Mech. Eng.* **93** pp. 291–318

# DEGENERATE SWEEPING PROCESSES

M. KUNZE
*Mathematische Institut der Universität Köln; Weyertal 86, D - 50931 Köln, Germany.*

AND

M.D.P. MONTEIRO MARQUES
*C.M.A.F. and Faculdade de Ciências da Universidade de Lisboa; Av. Prof. Gama Pinto 2, P-1699 Lisboa, Portugal.*

## 1. Introduction

The aim of this paper is to summarize some recent results concerning the evolutionary differential inclusions

$$-u'(t) \in \partial\delta_{C(t)}(Au(t)) \quad \text{a.e. in} \quad [0,T], \quad u(0) = u_0 \in D(A), \tag{1}$$

where $A$ is a maximal monotone and strongly monotone operator in a real Hilbert space $H$, and $t \mapsto C(t)$ is a set-valued mapping, cf. the precise assumptions below. Moreover, as usually, $\partial\delta_{C(t)}(v)$ denotes the cone of normals to the closed convex set $C(t)$ at the point $v \in C(t)$.

The investigation of inclusions of type (1) is motivated by problems in quasistatic elastoplasticity as follows: in some plastic flow problems one has to solve

$$-\frac{\partial q}{\partial t}(t,x) \in \partial\delta_{E(x)}\left(\frac{\partial W}{\partial q}(t,x,q(t,x))\right), \tag{2}$$

cf. e.g. [3]. In this formulation, every $E(x) \subset \mathbb{R}^n$ is a "rigidity set", $t \geq 0$ denotes process time, $x \in \Omega \subset \mathbb{R}^n$ is the material point, $q \in \mathbb{R}^n$ is an internal variable (like plastic strain or hardening), and $W$ denotes a (modified) stored energy function. For every fixed $x \in \Omega$, (2) has the general form

$$-q'(t) \in \partial\delta_\Gamma\Big(g(t,q(t))\Big) \tag{3}$$



*P. Argoul et al. (eds.),*
*IUTAM Symposium on Variations of Domains and Free-Boundary Problems in Solid Mechanics*, 301–307.

with $\Gamma = E(x)$ and $g(t,q) = \frac{\partial W}{\partial q}(t,x,q)$, and in fact (3) may easily be transformed into an inclusion of type (1), as explained in the following lemma.

**Lemma 1** *If $q : [0,T] \to H$ is a solution of (3) with initial value $q(0) = q_0$, then $u : [0,T] \to \mathbb{R}\times H =: \mathcal{H}$, defined by $u(t) = (t, q(t))$, is a solution of (1) with initial value $u_0 = (0, q_0)$ in the Hilbert space $\mathcal{H}$, where $C(t) := [t,\infty[\times\Gamma$ for $t \in [0,T]$ and $A(t,x) := (t, g(t,x))$ for $(t,x) \in \mathcal{H}$. Conversely, if $u = (\theta, q) : [0,T] \to \mathcal{H}$ solves (1), then $q : [0,T] \to H$ solves (3).*

Thus the study of inclusions such as (1) is of use in order to solve problems of type (3).

It is also worth mentioning that problems similar to (1) arise in the study of elliptic-parabolic variational (in-) equalities as were studied in [5] or [6], cf. also the references therein for additional information. To explain this, consider e.g. an elliptic-parabolic BVP arising in the modelling of flows in porous media, with obstacle at the boundary, which may depend on time, cf. [6]. For this, let $\Omega \subset \mathbb{R}^n$ be a bounded domain with smooth boundary $\partial\Omega$, and suppose that for each $t$ the boundary may be decomposed disjointly into $\partial\Omega = \overline{\Gamma_D(t)} \cup \overline{\Gamma_N(t)} \cup \overline{\Gamma_U(t)}$. Let $\rho : \mathbb{R} \to \mathbb{R}$ be Lipschitz continuous and nondecreasing, and let functions $r_D$, $q_N$, $r_U$, and $q_U$ on $\partial\Omega$ be given. Then the problem is to find a function $\theta = \theta(t,x)$ on $[0,T] \times \Omega$ such that

$$\begin{gathered}
\tfrac{\partial}{\partial t}[\rho(\theta)] - \Delta\theta = -f \quad \text{in} \quad ]0,T[\times\Omega\,, \quad \theta(0,\cdot) = \theta_0\,, \\
\theta = r_D \text{ on } \Sigma_D\,, \quad \tfrac{\partial}{\partial n}\theta = q_N \text{ on } \Sigma_N\,, \quad \theta \geq r_U\,,\ \tfrac{\partial}{\partial n}\theta \geq q_U\,, \\
(\theta - r_U)(\tfrac{\partial}{\partial n}\theta - q_U) = 0 \text{ on } \Sigma_U\,,
\end{gathered} \tag{4}$$

with $\Sigma_D = \bigcup_{0<t<T}(\{t\} \times \Gamma_D(t))$, and analogous for $\Sigma_N$ and $\Sigma_U$, cf. also { [1], [7], [8] }. Problem (4) may be reformulated in a suitable Hilbert space $H$ as

$$-\frac{d}{dt}[B\theta(t)] \in \partial\varphi^t(\theta(t)) + f(t)\,,$$

where every $\varphi^t$ is a convex and lower semicontinuous function, depending somehow regularly on $t \in [0,T]$. With the multivalued inverse $A = B^{-1}$, we arrive at

$$-u'(t) \in \partial\varphi^t(Au(t)) + f(t)\,. \tag{5}$$

Notice that we can obtain (1) from (5), by taking $f = 0$ (which is not important), and $\varphi^t = \delta_{C(t)}$, the indicator function of the set $C(t)$, i.e. $\delta_{C(t)}(x) = 0$ for $x \in C(t)$ and $\delta_{C(t)}(x) = \infty$ for $x \in H \setminus C(t)$.

Despite the similarity of (1) and (5), the problems are quite different, since in general with $\varphi^t = \delta_{C(t)}$ neither the sublevel sets $\{x \in H : \varphi^t(x) \leq$

$r\}$ are compact, nor $\varphi^t(x) \geq c\,|x|^2$ for some constant $c > 0$, and these were basic assumptions in [6], cf. (a1) and (a2) in that paper.

After this more or less heuristical introduction, we are going to explain the results we obtained for (1) resp. (3) in the following section. We are going to use only standard notations concerning lower semicontinuous function or maximal monotone operators, cf. e.g. { [2], [4]}.

## 2. Results

We start with the definition of solutions of (1) resp. (3).

**Definition 1** *(a) Let $u_0 \in D(A)$ be given such that $Au_0 \cap C(0) \neq \emptyset$. We call a pair $(u, v)$ a solution of (1), if $u \in W^{1,2}([0,T]; H)$, $v \in L^\infty([0,T]; H)$, $v(t) \in Au(t) \cap C(t)$ for a.e. $t \in [0,T]$, and if (1) is satisfied (with $v(t)$ inserted instead of $Au(t)$).*

*(b) Let $q_0 \in H$ be given such that $g(0, q_0) \in \Gamma$.*
*A function $q \in W^{1,2}([0,T]; H)$ is called a solution of (3), if $g(t, q(t)) \in \Gamma$ for a.e. $t \in [0,T]$, and if (3) is satisfied.*

Our general assumptions to prove the existence of solutions of (1) are:

(H1) $A : D(A) \to 2^H \setminus \{\emptyset\}$ is a maximal monotone operator (mmop, for short) such that $A = \partial\psi$ for some lsc, convex and proper $\psi : H \to \mathbb{R} \cup \{\infty\}$, and there exists $\beta > 0$ such that

$$\langle Ax - Ay,\, x - y\rangle \geq \beta\,|x - y|^2 \quad \text{for} \quad x, y \in D(A)\,. \tag{6}$$

(H2) For every $t \in [0,T]$, $\emptyset \neq C(t) \subset H$ is a closed convex set, and $t \mapsto C(t)$ is Lipschitz continuous, i.e. for some $L \geq 0$

$$\mathrm{d_H}(C(t), C(s)) \leq L|t - s| \quad \text{for} \quad t, s \in [0,T]\,, \tag{7}$$

with $\mathrm{d_H}$ being the Hausdorff-distance.

The following result was obtained in [10].

**Theorem 1** *Let $(H1)$ and $(H2)$ be satisfied. If in addition*

(H3a) *$C(0)$ is bounded, or*

(H3b) *there exists a function $M : [0,\infty[ \to [0,\infty[$ which maps bounded sets into bounded sets such that*

$$\|Ax\| = \sup\{|y| : y \in Ax\} \leq M(|x|) \quad \textit{for} \quad x \in D(A), \tag{8}$$

*and*

(H4a) *$D(A) \cap \overline{B}_R(0)$ is relatively compact for every $R > 0$, or*

(H4b) *$C(t) \cap \overline{B}_R(0)$ is compact for every $t \in [0,T]$ and $R > 0$,* *then (1) has a Lipschitz continuous solution, for every $u_0 \in D(A)$ with $Au_0 \cap C(0) \neq \emptyset$.*

In the bounded linear case we can do much better. Here we have

**Theorem 2** *Let $A : H \to H$ be linear, bounded and selfadjoint such that $\langle Ax, x\rangle \geq \beta |x|^2$ for $x \in H$. If (H2) holds for $C(\cdot)$ and if $Au_0 \in C(0)$, then (1) has a unique solution, and this solution is Lipschitz continuous.*

Simple examples show that this Theorem 2 (and hence also Theorem 1) might be wrong if $\beta = 0$ in (6).

In both theorems we had to assume that $A$ be a subdifferential, cf. $(H1)$. From [9] there is also available a version of Theorem 1 which exchanges this condition against a certain compactness assumption, formulated in terms of the Yosida-approximants $A_\mu$ of the mmop $A$.

**Theorem 3** *Let $(H1)$ be satisfied, without the assumption on $A$ to be a subdifferential, and let $(H2)$ hold. Assume in addition that $(H3b)$ and $(H4b)$ are fulfilled and also that*

*(H5) If $\mu_n \to 0^+$, $u_n \to u$ in $C([0,T]; H)$ and $v_n = A_{\mu_n} u_n \to v$ weakly in $L^2([0,T]; H)$, then even $v_n \to v$ in $L^2([0,T]; H)$.*

*Then (1) has a Lipschitz continuous solution, for every $u_0 \in D(A)$ such that $A^0 u_0 \in C(0)$, where $A^0$ is the minimal-norm selection of $A$.*

In particular this theorem is taylored to treat finite-dimensional problems, since both $(H4b)$ and $(H5)$ are compactness conditions. Via Lemma 1, Theorem 3 may be used to prove existence theorems for problems of the form (3), as shows the following result from [9].

**Theorem 4** *Let $g \in C(\mathbb{R} \times \mathbb{R}^n, \mathbb{R}^n)$ and $\Gamma \subset \mathbb{R}^n$ be closed and convex. Suppose that for some $\alpha > 0$ and $\beta \in ]-\infty, 1[$*

$$\langle g(t,x) - g(s,y),\, x - y\rangle \geq \alpha |x-y|^2 - \beta (t-s)^2 \quad \text{for} \quad t, s \in \mathbb{R},\ x, y \in \mathbb{R}^n. \tag{9}$$

*Then (3) has a solution in the sense of Definition 1 on every $[0,T]$ with $T > 0$.*

In the remaining part of this paper we will sketch the proof of Theorem 1 by means of a suitable discretization, following [10]. We remark that an alternative proof can be based on a double Yosida-Moreau regularization, by first replacing $A$ through $A_\mu$, and then also approximating the lower semicontinuous $\partial\varphi^t = \partial\delta_{C(t)}$ by $(\partial\varphi^t)_\lambda$. This very technical approach was followed in [6] for (5), and also in [9], where Theorem 1 (with assumptions $(H3b)$, $(H4)$ and $A^0 u_0 \in C(0)$) and Theorem 3 were proved by this method.

To describe the (implicit) discretization of (1), we note that for given $u_i$ this problem somehow consists in finding $u_{i+1} \in D(A)$ with $Au_{i+1} \in C(t_i)$ such that

$$-(u_{i+1} - u_i) = u_i - u_{i+1} \in \partial\delta_{C(t_{i+1})}(Au_{i+1}), \tag{10}$$

with a partition $0 = t_0 < t_1 < \ldots < t_k = T$ of $[0,T]$. This motivates the following definition.

**Definition 2** *Let $A$ be a mmop in $H$ such that* (6) *holds, and let $C \subset H$ be nonempty, closed, and convex. A map $D(A) \ni u \mapsto P_{A,C}(u) \in D(A)$ is called the approximation operator, if for every $u \in D(A)$ there exists a $P_{A,C}(u) = w \in D(A)$ such that $u - w \in N_C(Aw)$, i.e. $u - w \in N_C(v)$ for some $v \in Aw \cap C$.*

Because of (6), the element $P_{A,C}(u)$ is unique. It now turns out that the approximation operator $P_{A,C}$ exists, if $C$ is bounded. Moreover, there is also a good dependence of $P_{A,C}(\cdot)$ on $C$.

**Lemma 2** *Let $A$ be a mmop such that (6) holds. If $C \subset H$ is nonempty, closed, convex, and bounded, then $P_{A,C}(\cdot)$ exists. Moreover, if $C_1, C_2 \subset H$ are such that $P_{A,C_2}(\cdot)$ exists, and if $u_1 \in D(A)$ with $Au_1 \cap C_1 \neq \emptyset$, then*

$$|u_1 - P_{A,C_2}(u_1)| \leq \frac{1}{\beta} \mathrm{d_H}(C_1, C_2). \tag{11}$$

Although the sets $C(t)$ need not be bounded in case that $(H3b)$ holds in Theorem 1, we continue to argue with this additional assumption, since by intersection with increasing large balls this boundedness may be incorporated artificially.

If every $C(t)$ is bounded, in particular all the approximation operators $P_{A,C(t_{i+1})}(\cdot)$ exist, i.e. by definition of these operators we may solve (10) to obtain $u_{i+1} \in D(A)$ and some $v_{i+1} \in Au_{i+1} \cap C(t_{i+1})$. In addition, by (11) and (7) in $(H2)$ we find

$$|u_i - u_{i+1}| \leq \beta^{-1}\mathrm{d_H}(C(t_i), C(t_{i+1})) \leq \frac{L}{\beta} |t_i - t_{i+1}|. \tag{12}$$

Thus we can define a piecewise constant right-continuous approximation $u$ with values $u_1, \ldots, u_k$. Consequently, by (12) the variation of $u$ is bounded by $LT/\beta$. If we let shrink $\sup_i |t_i - t_{i+1}|$ to zero by inserting more and more partition points, we thus get a sequence of right-continuous approximants $u_n$ which are uniformly bounded in variation by $LT/\beta$, and they are also uniformly bounded in norm, since $u_n(0) = u_0$, the given initial value, for all $n \in \mathbb{N}$. This is enough to find a right-continuous $u : [0,T] \to H$

and a subsequence (again denoted by $u_n$) such that $u$ has variation $\leq LT/\beta$, and such that $u_n(t) \to u(t)$ weakly in $H$ for every $t \in [0, T]$, cf. [11, Theorem 0.2.1].

Thus it remains to show that $u$ is a solution of (1). For this the additional assumptions $(H3)$ and $(H4)$ will be used, whereas for the construction of the approximants only $(H1)$ and $(H2)$ were needed. Up to now we have obtained functions $u_n$ and $v_n$ such that $u_n(t) \in D(A)$, $v_n(t) \in Au_n(t)$ and $v_n(t)$ "almost in" $C(t)$ for $t \in [0, T]$. Then $(H3)$ is used to ensure that also $(v_n)_{n\in\mathbb{N}}$ is uniformly bounded, and thus w.l.o.g. $u_n \to u$ and $v_n \to v$ weakly in $L^2([0, T]; H)$ for some function $v$. But to conclude $v(t) \in Au(t)$ a.e., one of the weak convergences has to be improved to a strong convergence, and for this $(H4)$ is needed. Finally, $-u'(t) \in \partial\delta_{C(t)}(v(t))$ a.e. also follows from this improved convergence, and the Lipschitz continuity of $u$ is established by deriving suitable estimates of the type $|u_n(t) - u_n(s)| \leq c(|t - s| + 1/n)$ for the approximants, with $c > 0$ being independent of $n \in \mathbb{N}$. Hence the outline of the proof of Theorem 1 is complete, cf. [10] for the details. □

We also remark that our results remain true, if $t \mapsto C(t)$ is only assumed to be absolutely continuous, i.e. $d_H(C(t), C(s)) \leq |a(t) - a(s)|$ for some absolutely continuous (a.c.) real function $a$, the only difference being that the obtained solution is then a.c. instead of Lipschitz continuous.

## Acknowledgements

This research is supported by Deutscher Akademischer Austauschdienst (DAAD) and Conselho de Reitores das Universidades Portuguesas (CRUP), project A-4/96 "Acção Integrada Luso-Alemã, Inclusões Diferenciais em Sistemas Dinâmicos Não-Regulares", as well as by JNICT,FEDER and project PRAXIS/2/2.1/MAT/125/94.

## References

1. ALT H.W. & LUCKHAUS S.: Quasilinear elliptic-parabolic differential equations, *Math. Z.* **183**, 311-341 (1983)
2. BREZIS H.: *Opérateurs Maximaux Monotones*, North Holland Publ. Company, Amsterdam 1973
3. CARSTENSEN C. & MIELKE A.: A formulation of finite plasticity and an existence proof for one dimensional problems, preprint, IfAM Univ. Hannover
4. DEIMLING K.: *Nonlinear Functional Analysis*, Springer, Berlin 1985
5. DI BENEDETTO E. & SHOWALTER R.E.: Implicit degenerate evolution equations and applications, *SIAM J. Math. Anal.* **12**, 731-751 (1981)
6. KENMOCHI N. & PAWLOW I.: A class of nonlinear elliptic-parabolic equations with time-dependent constraints, *Nonlinear Anal.* **10**, 1181-1202 (1986)
7. KRÖNER D.: Parabolic regularization and behaviour of the free boundary for unsaturated flow in a porous medium, *J. Reine Angew. Math.* **348**, 180-196 (1984)
8. KRÖNER D. & RODRIGUES J.F.: Global behaviour for bounded solutions of a porous media equation of elliptic-parabolic type, *J. de Mathématique pures et ap-*

*pliquées* **64**, 105-120 (1985)
9. Kunze M. & Monteiro Marques M.D.P.: Existence of solutions for degenerate sweeping processes, to appear in *J. Convex Analysis*
10. Kunze M. & Monteiro Marques M.D.P.: On the discretization of degenerate sweeping processes, to appear in *Portugaliae Mathematica*
11. Monteiro Marques M.D.P.: *Differential inclusions in nonsmooth mechanical problems–shocks and dry friction*, Birkhäuser, Basel-Boston-Berlin, 1993

# ABOUT THE MATHEMATICAL MODELING OF IRREVERSIBILITY PROBLEMS

P. LABORDE
*Mathematiques pour l'Industrie et la Physique,*
*CNRS/UPS/INSA, UMR 5640*
*118, route de Narbonne*
*31062 TOULOUSE Cedex 04, FRANCE*

## 1. Introduction

We are concerned with the modeling of irreversible behaviors in Solid Mechanics and the numerical solution of the associated free boundary problem. A general enough framework is presented which allows us to represent typical stress/strain diagrams in the study of damage processes. We state some mathematical results in order to prove the coherence of the model with respect to the given diagrams. Theoretic tools are discussed for the solution of the corresponding discrete finite element problem.

## 2. A first approach to the problem

In order to introduce the method, we first begin by a one-dimensional analysis of the problem.

### 2.1. A ONE-DIMENSIONAL PROBLEM.

Let an orthotropic ceramic-ceramic composite material be given, which is subjected to a uniaxial tensile test into a direction of fibers. The main features of the corresponding *tensile force/longitudinal elongation* diagram $(D)$ are the following: $(a)$ the elastic stiffness is decreasing during the history of deformations; $(b)$ there are residual strains(or not); $(c)$ the elastic limit is increasing; $(d)$ a difference of behavior occurs according to the sign of the stress, precisely Young modulus is different in tension and in compression, moreover the behavior is purely elastic in compression.



*P. Argoul et al. (eds.),*
***IUTAM Symposium on Variations of Domains and Free-Boundary Problems in Solid Mechanics*, 309–316.**

In the present step, our aim is to find a relationship beetween the tensile force $\sigma \in \boldsymbol{R}$ and the longitudinal elongation $\varepsilon \in \boldsymbol{R}$ consistent with the diagram $(D)$. Let us study the following variational inequality of evolution with an internal parameter $a \in \boldsymbol{R}$ :

$$(1) \qquad \begin{cases} L(\sigma,a)\dot{\sigma}\,(\tau-\sigma)+M(a)\dot{a}\,(b-a)\geq\dot{\varepsilon}\,(\tau-\sigma) \\ \text{for all } (\tau,b) \text{ such that } \varphi(\tau,b)\leq 0; \quad \varphi(\sigma,a)\leq 0. \end{cases}$$

The yield function is written : $\varphi(\sigma,a)=\sigma-s_0-a$, where $s_0>0$. The meaning of the variable $a$ comes from the previous definition. For a given elongation/force pair $(\varepsilon,\sigma)$, let $\sigma_{max}$ be the elastic limit corresponding to the state $(\varepsilon,\sigma)$ on the diagram $(D)$. Then, the internal parameter is defined as $a=\sigma_{max}-s_0$, *see* section 2.2. In (1), we denote $L(\sigma,a)=(E(\sigma,a))^{-1}$ where

$$E(\sigma,a)=E(a) \quad \text{if } \sigma>0, \quad E_- \quad \text{otherwise.}$$

Young's modulus $E(a)$ in tension and $E_-$ in compression are obtained from the diagram $(D)$. Finally, the evolution of the internal parameter is governed in (1) using the quantity $M(a)=(E_T(a))^{-1}-(E(a))^{-1}$ computed from the tangent modulus $E_T(a)$.

## 2.2. STUDY OF THE MODEL.

Let $t$ be the evolution parameter. In order to prove the consistency of the model with respect to the diagram $(D)$, we are faced with the following question : "For a given strain path $\varepsilon(t)$, let $\sigma(t)$ be the stress response satisfying (1) and subjected to an admissible initial condition. Is $\sigma(t)$ unique?" The answer is well-known in the specific case $L(\sigma,a)=L_0>0$ (corresponding to the standard plasticity) where the problem is well-posed in the Sobolev space $H^1(0,T)$. In the general case, the same holds true at least after a smoothing of the elastic compliance $L(\sigma,a)$ in a neighbourhood of $\sigma=0$ (which does not significantly affect the representation of the given diagram). In the following proposition, we assume that $L(\sigma,a)$ is Lipschitz-continuous on the bounded subsets of $\boldsymbol{R}^2$. Moreover, $L(\sigma,a)$ is uniformly greater than a constant $\alpha>0$.

**Proposition 1.** *Under the above hypotheses, let $\varepsilon\in H^1(0,T)$ and $(\sigma_0,a_0)$ be such that $\varphi(\sigma_0,a_0)\leq 0$. Then, there exists a unique $(\sigma,a)\in(H^1(0,T))^2$ satisfying the variational inequality* (1) *for almost all $t\in(0,T)$, and such that $(\sigma(0),a(0))=(\sigma_0,a_0)$.*

The proposition, as also the following one, proceeds from more general results stated in section 3. A second question must be considered:"Let a state $(\sigma(t),a(t))$ and a strain rate $\dot{\varepsilon}(t)$ be given. Can we compute the stress rate $\dot{\sigma}(t)$? Moreover, is $\dot{\sigma}$ equal to the value derived from the given diagram

$(D)$?" In fact, the convenient notion of rate is defined by using the derivative to the right :

$$\frac{d^+\psi}{dt}(t) = \lim_{h\to 0, h>0} \frac{\psi(t+h) - \psi(t)}{h}.$$

**Proposition 2.** *At the time $t \in [0, T)$, let us assume that the given function $\varepsilon(.)$ is differentiable to the right. Then, the solution $\sigma(.)$ defined in Proposition 1 is differentiable to the right at the time $t$. Moreover, the generalized stress rate can be computed from the strain rate as follows.*
$(i)$ Either $\sigma - s_0 - a < 0$ (elastic range) or $(d^+\varepsilon/dt) \leq 0$ (elastic unloading):

$$(d^+\sigma/dt) = E(\sigma, a)\,(d^+\varepsilon/dt), \quad (d^+a/dt) = 0.$$

$(ii)$ Otherwise: $\sigma - s_0 - a = 0$ and $(d^+\varepsilon/dt) > 0$ (irreversible behavior), then

$$(d^+\sigma/dt) = E_T(a)\,(d^+\varepsilon/dt), \quad (d^+a/dt) = (d^+\sigma/dt).$$

The meaning of the variable $a$ can be deduced from the previous statement; it can be proved that

$$a(t) = sup\{(\sigma(t') - s_0)^+ : 0 \leq t' \leq t\}.$$

In conclusion, the variational inequality (1) is a convenient mathematical model to study the previous typical one-dimensional damage strain/stress diagram.

## 3. A framework of modeling.

The aim now is to define a general three-dimensional model taking into account some fondamental features of damage behaviors. Let $\varepsilon$ be the linearized strains tensor and $\sigma$ the stresses tensor in the space $S$ of the symmetrical 2nd-order tensors. The variational inequality (1) cannot be obtained from the standard models in irreversible processes Thermodynamics. So, let us look for a model adaptable to the present problem and which keeps the advantages of the Generalized Standard Materials (G.S.M.) (Halphen and Nguyen, 1975).

### 3.1. THE MODEL.

The following constitutive equations are a three-dimensional generalization of the variational inequality (1). In the spirit of the G.S.M., we denote by $s = (\sigma, a)$ the generalized stress, where $a = (a_k) \in \boldsymbol{R}^p$ are internal parameters, and $e = (\varepsilon, 0) \in S \times \boldsymbol{R}^p$ the generalized strain. We consider the following (strain $e$ / stress $s$) relationship:

(2) $\quad \varphi(s) \leq 0, \quad A(s)\dot{s}.(s^* - s) \geq \dot{e}.(s^* - s)$ for all $s^*$ s.t. $\varphi(s^*) \leq 0,$

where $\varphi(s)$ be a convex function and $s.s^*$ stands for the canonical inner product in $S \times \boldsymbol{R}^p$. The generalized elastic compliances operator is defined as

$$A(s)s^* = \begin{pmatrix} L(\sigma, a) & 0 \\ 0 & M(\sigma, a) \end{pmatrix} \begin{pmatrix} \sigma^* \\ a^* \end{pmatrix}.$$

In practice, the latter can be defined from experimental strain/stress diagrams characteristic of the behavior in the elastic range. In section 1, the diagram $(D)$ with damage shows a difference of behavior in tension and in compression. Such a diagram is nonlinear in the elastic range. So, we are led to consider the following preliminary problem of finding the constitutive equations of a given *elastic* material taking into account: $(a)$ the symmetry properties of the medium, $(b)$ a set of given (nonlinear) elastic diagrams corresponding to various loadings.

For instance, let us consider an isotropic material whose the behavior is known through the (three) diagrams obtained from a uniaxial tensile test and a simple shear test. A typical case to examine corresponds to the so-called bi-modulus materials where Young modulus depends on the sign of the tensile stress. So the elastic behavior cannot be linearized in infinitesimal strains. Such a constitutive nonlinearity, or more general ones, must be correctly taken into account using the canonical form of the elastic constitutive equations given by Rivlin-Ericksen theorem. A compatibility condition on the data is necessary to construct a continuous elastic response function. Moreover, the canonical form must be specified in a convenient way in order to determine the response function from the available informations, see (Laborde and Petitjean, 1996), (Laborde and Petitjean, 1997). In these references, the orthotropic case is also discussed.

**Remark.** If $A(s)$ does not depend on $s$, the model (2) corresponds to the G.S.M. with a quadratic free energy. But, in general, there does not exist a convex free energy associated to the model.

## 3.2. PROPERTIES OF THE MODEL.

Let $e^i$ be a orthonormed basis of $\boldsymbol{R}^3$. In the following, tensors $\tau = (\tau_{ij})$ will be represented in this basis. We assume that the elastic compliances $A(s)$ are uniformly coercive in $S \times \boldsymbol{R}^p$ and Lipschitz-continuous on the bounded sets. The following results extend the two previous propositions (Laborde and Michrafy, 1991).

**Proposition 3.** *Let a function $\varepsilon(t)$ be given such that $\varepsilon_{ij} \in H^1(0,T)$ and $(\sigma_0, a_0)$ an admissible initial data : $\varphi(\sigma_0, a_0) \leq 0$. Denoting $e = (\varepsilon, 0)$, there is a unique function $s(t) = (\sigma(t), a(t))$ such that $\sigma_{ij}, a_k \in H^1(0,T)$ and satisfying* (2) *for almost every $t \in (0,T)$, as also $(\sigma(0), a(0)) = (\sigma_0, a_0)$.*

The rate principle hereafter gives the (generalized) stress rate $(d^+s/dt)$, at the time $t$, as a function of the strain rate $(d^+e/dt)$ from the state $s$ and the purely elastic stress rate $(A(s))^{-1}$ $(d^+e/dt)$ (omitting the variable $t$). Let us denote by $K = \{\varphi \leq 0\}$ the convex set of admissible generalized stresses in $S \times \boldsymbol{R}^p$.

**Proposition 4.** *Under the hypotheses of Theorem 3, let a time $t \in [0,T)$ be given. If the the strain path $\varepsilon(.)$ is differentiable to the right at $t$, then the same holds true for the stress response $\sigma(.)$ and the internal parameter $a(.)$ defined in* Proposition 3. *Moreover $(d^+s/dt)$ is equal to the projection on the tangent cone to $K$ at the point $s$ of $(A(s))^{-1}$ $(d^+e/dt)$, in the sense of the elastic energy norm $s^* \to (A(s)s^*.s^*)^{1/2}$.*

**Remark.** In the specific case of the G.S.M., the coercivity assumption results from the strong convexity of the free energy. In the general framework (2), it is essential that the above hypotheses of convexity and coercivity are satisfied. This is a necessary condition to ensure that the specific models used in practice offer a convenient representation of given experimental diagrams.

3.3. APPLICATION : A SPECIFIC MODEL.

Let an orthotropic material be given, for which the available informations on the behavior are a strain/stress diagram of type $(D)$ into the orthotropy directions 1 and 2. Under the hypothesis of planes stresses into the direction 3, we can define a specific orthotropic model in the framework (2) which represents exactly the given behavior. In order to identify the model from the only previous informations, an assumption is necessary on the coupling of the damage between the directions 1 and 2 (in the definition of the directional Young moduli). Here, the basis $e^i$ is defined by the directions of orthotropy. In its simplest version, the specific model involves two internal parameters $(a_1, a_2) = a$ and the admissibility condition is written

$$\varphi_1 \leq 0 \quad \text{and} \quad \varphi_2 \leq 0 \quad \text{where} \quad \varphi_i = \sigma_{ii} - s_0 - a_i,$$

so that the yield function in (2) is defined as $\varphi = \max(\varphi_1, \varphi_2)$.

The unknown $a_i$ characterizes the state of irreversibility into the direction $i$. The corresponding generalized elastic compliances operator can be easily identified in a similar way to section 2. Let us note that $\varphi$ is a non-differentiable convex function. In consequence, the flow rule (2) defines the anelastic strain rate $\dot{\varepsilon}^{an} = \dot{\varepsilon} - L(\sigma, a)\,\dot{\sigma}$ and the internal parameters rate $\dot{a}$ by means of many (two) unknown multipliers:

$$\dot{\varepsilon}^{an} = \lambda_1 \frac{\partial \varphi_1}{\partial \sigma} + \lambda_2 \frac{\partial \varphi_2}{\partial \sigma} = \lambda_1\, e^1 \otimes e^1 + \lambda_2\, e^2 \otimes e^2,$$

$$-M(\sigma,a)\,\dot{a} = \lambda_1 \frac{\partial \varphi_1}{\partial a} + \lambda_2 \frac{\partial \varphi_2}{\partial a} = \lambda_1\, e^1 + \lambda_2\, e^2,$$

where $\otimes$ stands for the tensor product of two vectors. The rate principle stated in Proposition 4 enables us to compute these multipliers (Laborde and Michrafy, 1991).

Naturally, supplementary informations on the behavior under multiaxial loading would lead to more realistic models.

**Remark.** In the above example, the convexity of the admissibility function $\varphi$ proceeds from the definition of the internal variables. It is a feasible choice because the elastic limit is increasing in the course of the deformations history. This last property moreover leads to the coercivity of the generalized elastic compliances.

## 4. Numerical solution to the boundary value problem.

The previous general model allows us to consider efficient algorithms used in industrial finite element codes.

### 4.1. THE QUASI-STATIC PROBLEM.

At each time $t \in [0,T]$, the problem consists in finding the displacement field $u(t,x)$ and the stress field $\sigma(t,x)$ defined for all points $x$ of the deformable body $\Omega$ and satisfying the variational inequality (2) with $\dot{\varepsilon} = \varepsilon(\dot{u})$ in $\Omega$, as also the equilibrium equations and the usual boundary conditions, together with convenient initial conditions.

Let us consider a time discretization of the problem by using a (semi-) implicit scheme. The discretized constitutive equations at the step $n$ are written (compare to (2)):

$$(3)\qquad \begin{cases} s^n \in K, \qquad \text{and for all } s^* \in K: \\ A(s^{n-1})(s^n - s^{n-1}).(s^* - s^n) \geq (e^n - e^{n-1}).(s^* - s^n). \end{cases}$$

At the step $n$, the discrete generalized stress $s^n$ in (3) is the projection of the purely elastic discrete generalized stress $s^{n-1} + A(s^{n-1})^{-1}(e^n - e^{n-1})$ onto the convex set $K$, for the energy norm $s^* \to \left(A(s^{n-1})s^*.s^*\right)^{1/2}$ at the previous step.

This scheme is an extension of the radial return algorithm in Plasticity for a Von Mises criterium. In the present case, the expression of the projection is not explicit, and the computation of $s^n$ requires an iterative method. In some cases, we are able to express explicitly the projection on the admissibility set, *e.g.* for the yield function $\varphi = \max(\varphi_1, \varphi_2)$ introduced in section 2 (Petitjean, 1996).

Let $F(u) = 0$ be the equation obtained by applying a standard finite element method to the time-discretized problem. We denote by $u \in \boldsymbol{R}^N$ the vector of the degrees of freedom defining the finite element displacement increment, and $F(u) \in \boldsymbol{R}^N$ is the residual forces vector. In the expression of $F(.)$, the nonlinearity comes from the discrete generalized stress response at the step $n$, and therefore is reduced to the projection operator $P_K$ on the admissibility convex set, as above-mentioned.

4.2. COMPUTATIONAL ASPECTS.

The efficient algorithms in plasticity can be adapted in the present framework. Let us consider the class of Newton type methods for the solution of the finite element equation $F(u) = 0$. First of all, let us note that the mapping $F$ is not differentiable in general, because of the operator $P_K$. This fact is not proper to damage modeling; the same holds true in standard plasticity.

In order to define rigorously the wanted algorithms, we begin by recalling the differentiability properties of the projection operator $P_K$ on a convex set. It is known that such a mapping has directional derivatives at each point in the convex set. The derivative into a direction $d$ is equal to the projection of $d$ on the tangent cone to $K$ at this point. This general result allows us to compute the directional derivatives of $P_K$ from the expression of the yield function defining the set $K = \{\varphi \leq 0\}$. In (Laborde, 1992), we deal with the case where $\varphi$ is a differentiable function, such as a Von Mises type yield function. We also consider non-differentiable functions of the form $\varphi = \max(\varphi_1, \varphi_2)$ where $\varphi_k$ are differentiable convex functions.

In general, the projection operator is *not* differentiable, even for a smooth yield function. Let us now examine some usefull specific results. Suppose that $\varphi$ is a two times differentiable convex function such that $\varphi(0) < 0$. Then, $P_K$ is (Gateaux-)differentiable at each point which does not belong to the boundary $\{\varphi = 0\}$ of $K$. In the same way, if the functions $\varphi_k$ are two times differentiable convex functions, then a G-differentiability property can be stated in some points for the projection operator on $K = \{\varphi = \max(\varphi_1, \varphi_2) \leq 0\}$, and the derivative can be computed (Petitjean, 1996).

Now, we are in a position to give some indications on the use of generalized Newton algorithms for non-differentiable problems $F(u) = 0$ in Solid Mechanics. The generic iteration is written

$$R^{(i)}\,(u^{(i+1)} - u^{(i)}) = -F(u^{(i)}), \quad i = 1, 2, ...$$

where various choices are usually considered for the matrix $R^{(i)}$. The "tangent stiffness" method can be considered as a "linearization" of the continuous evolution problem. In fact, the strict definition of the algorithm requires the directional differentiability of $P_K$ at the value of the generalized stress

$s^{n-1} = (\sigma^{n-1}, a^{n-1})$ computed at the previous step. In particular, it may be noted that

$$R^{(i)} = R(u^{(i)}) \quad \text{such that} \quad R(u^{(i)})\, u^{(i)} = F'(0; u^{(i)})$$

the directional derivative of $F(u)$ at $u = 0$. In the general case, if $F$ is differentiable at 0, then the present procedure gives $R^{(i)} = F'(0)$.

A mathematical definition of the tangent stiffness algorithm for a differentiable yield function (Von Mises ...) will be found in (Laborde, 1992), and an application for a non-differentiable function $\varphi = \max(\varphi_1, \varphi_2)$ in (Michrafy, 1988). In the last reference, the "tangent operator" moreover is computed, and numerical simulations are given.

An alternative consists in the "consistent tangent stiffness" method which is close to the original Newton method (Simo *et al.*, 1985). Naturally, the practical difficulty lies in the computation of the tangent operator, and also in its strict definition since the nonlinearity is not differentiable. This efficient algorithm is studied from this point of view in (Petitjean, 1996) with an application to the solution of a damage problem in an industrial finite element code. Another application for a thermo-plasticity problem can be found in (Laborde *et al.*, 1997).

## References


Halphen, B. and Nguyen, Q.S. (1975) Sur les materiaux standard generalises, *J. Mecan.* **Vol. 14**, pp. 39–63

Laborde, P. (1992) Numerical solution of free boundary problems in Solids Mechanics, *Int. Ser. Numer. Analysis* **Vol. 107**, pp. 339–347

Laborde, P. and Michrafy, A. (1991) On general constitutive equations involving damage, *Eur. J. Mech., A/Solids* **Vol. 10 no. 2**, pp. 213–237

Laborde, P. and Petitjean, F. (1996) Symetries materielles et comportements non-lineaires en Elasticite, *C. R. Acad. Sci., Paris* **Vol. 323, Serie I**, pp. 1147–1152

Laborde, P. and Petitjean, F. (1997) Computation of form-invariant response functions in Elasticity, *Report MIP*

Laborde, P., Toson, B. and Pesque, J.J. (1997) On the consistent tangent operator algorithm for thermo-plastic problems, *Comp. Meth. Appl. Mech. Eng.*, to appear

Michrafy, A. (1988) Modelisation et calcul pour certains types de materiaux composites endommageables, *Thesis*, Bordeaux

Petitjean, F. (1996) Modelisation de comportements irreversibles en mecanique des structures, *Thesis*, Toulouse

Simo, J.C. and Taylor, R.L. (1985) Consistent tangent operators for rate independent elasto-plasticity, *Comp. Meth. Appl. Mech. Eng.* **Vol. 48**, pp. 101–118

# SHARP ENERGY ESTIMATES FOR FINITE ELEMENT APPROXIMATIONS OF NON-CONVEX PROBLEMS

M. CHIPOT
*Mathematische Institut, Universität Zürich*
*Winterthurstr. 190 - 8057 Zürich - Switzerland*

AND

S. MÜLLER
*MPI für Mathematik in den Naturwissenschaften*
*Inselstr. 22-26 - 04103 Leipzig - Germany*

## 1. Introduction

The goal of this note is to expose in a simple situation the key arguments which allow one to prove sharp energy estimates in the numerical analysis of problems with multiple-well energy in the calculus of variations. Let us recall that such problems arise naturally for instance in materials science. We refer the reader for this matter to [1], [5], [6], [7], [10]. Note also that our technique borrows widely from [8], [9] and can be extended to more general situations - see [3].

So, we will assume that we are in dimension 2 and $\Omega$ will denote the square $(0,1) \times (0,1)$ with boundary $\partial\Omega$. Setting

$$W_0^{1,\infty}(\Omega) = \{v : \Omega \to \mathbf{R} \mid v \text{ is Lipschitz continuous, } v = 0 \text{ on } \partial\Omega\}$$

we would like to consider the problem

$$\inf_{W_0^{1,\infty}(\Omega)} \int_\Omega W(\nabla u)\, dx\, dy = \inf_{W_0^{1,\infty}(\Omega)} \int_\Omega |u_x| + |\,|u_y| - 1|\, dx\, dy \tag{1.1}$$

where we have set

$$\nabla u = (u_x, u_y) \quad , \quad W(\xi_1, \xi_2) = |\xi_1| + |\,|\xi_2| - 1|. \tag{1.2}$$

More precisely we would like to address a discrete version of this problem. We denote by $\tau_h$ a regular (see [4]) family of triangulations of $\Omega$, with mesh



*P. Argoul et al. (eds.),*
*IUTAM Symposium on Variations of Domains and Free-Boundary Problems in Solid Mechanics*, 317–325.

size $h$, i.e.

$$h = \max\{\mathrm{diam(K)} | \mathrm{K} \in \tau_\mathrm{h}\}.$$

Then, if $P_1$ denotes the set of polynomials of degree one, we set

$$V_0^h = \{v : \Omega \to \mathbf{R} \mid v \text{ continuous}, v|_k \in P_1 \; \forall k \in \tau_h, v = 0 \text{ on } \partial\Omega\}$$

and we consider the discretized analogue of (1.1) :

$$\inf_{V_0^h} \int_\Omega W(\nabla u)\, dx\, dy. \tag{1.3}$$

The energy density $W$ is a nonnegative function that vanishes only on the two wells

$$\omega_+ = (0, 1) \quad , \quad \omega_- = (0, -1). \tag{1.4}$$

They play, of course, a crucial role in the minimization process. It is easy to establish that (1.1) does not admit a minimizer but that (1.3) does (see for instance [2]) thus the minimizers of (1.3) provide a minimizing sequence to (1.1). What we would like to establish here is:

**Theorem 1** *There exists a constant $C$ independent of $h \ll 1$ such that*

$$\inf_{V_0^h} \int_\Omega |u_x| + |\,|u_y| - 1|\, dx\, dy \leq C h^{\frac{1}{2}}. \tag{1.5}$$

*Moreover, this estimate is sharp in the sense that there exists a family of triangulation $\tau_h$, a constant $c$ independent of $h$ such that*

$$\inf_{V_0^h} \int_\Omega |u_x| + |\,|u_y| - 1|\, dx\, dy \geq c h^{\frac{1}{2}}. \tag{1.6}$$

The rest of the paper will be devoted to the proof of this theorem.

## 2. Proof of Theorem 1

### 2.1. THE PROOF OF (1.5)

Let us denote by $\alpha \in (0, 1)$ a real number that we will fix later on. Define

$$u(x, y) = \begin{cases} y & \text{if } \; y \in [0, h^\alpha], \\ 2h^\alpha - y & \text{if } \; y \in [h^\alpha, 2h^\alpha]. \end{cases} \tag{2.7}$$

Assume that $u$ is extended periodically - with period $2h^\alpha$ in the $y$ direction - on the whole $\mathbf{R}^2$. Clearly one has

$$\nabla u = \omega_+ \quad \text{or} \quad \omega_- \quad \text{a.e.} \quad \text{in } \Omega. \tag{2.8}$$

Since $u$ does not vanish on the boundary of $\Omega$ one sets

$$\hat{u} = u \wedge \text{dist}(x, \partial\Omega)$$

and then

$$u_h = \text{ the interpolate of } \hat{u} \text{ on } \tau_h, \tag{2.9}$$

i.e. $u_h$ denotes the function of $V_0^h$ that agrees with $\hat{u}$ at the nodes of the triangulation $\tau_h$. The two functions that are involved in the definition of $\hat{u}$ have a gradient that is uniformly bounded. Since the triangulation $\tau_h$ is regular, the gradient of $u_h$ is uniformly bounded independently of $h$ (see [4]). It then follows from (2.8) that one has

$$\int_\Omega W(\nabla u_h)\,dx\,dy = \int_{\{u_h \neq u\}} W(\nabla u_h)\,dx\,dy \leq C|\{u_h \neq u\}| \tag{2.10}$$

for some constant $C$, where $|\{u_h \neq u\}|$ denotes the measure of the set where $u_h$ is distinct from $u$. First one remarks that

$$0 \leq u \leq h^\alpha$$

(see (2.7)) and thus for $\text{dist}(x, \partial\Omega) > h^\alpha$ one has $\hat{u} = u$. So, for $\text{dist}(x, \partial\Omega) \geq h^\alpha + h$ the interpolate of $\hat{u}$ will be equal to the interpolate of $u$. Now, the interpolate of $u$ is $u$ itself except on a strip of size $2h$ around each of the lines $y = kh^\alpha, k \in \mathbf{N}$. Collecting this information one has clearly

$$|\{u_h \neq u\}| \leq 4(h^\alpha + h) + N\,2h \leq 8h^\alpha + (N+1)2h \tag{2.11}$$

where $N$ is the number of strips cutting $\Omega$. Note that for $h < 1$ and $\alpha \in (0,1)$ one has $h < h^\alpha$. Now, one has clearly

$$(N+1)h^\alpha \leq 1$$

so that by (2.10), (2.11) one gets

$$\int_\Omega W(\nabla u_h)\,dx\,dy \leq 8C(h^\alpha + h^{1-\alpha}).$$

Taking $\alpha = \frac{1}{2}$ leads to (1.5).

## 2.2. PROOF OF (1.6)

First remark that if we claim that (1.5) is sharp there are some particular triangulations - related to the wells - for which (1.5) could be improved. Indeed, choose for instance the triangulation of the (figure 1) and set

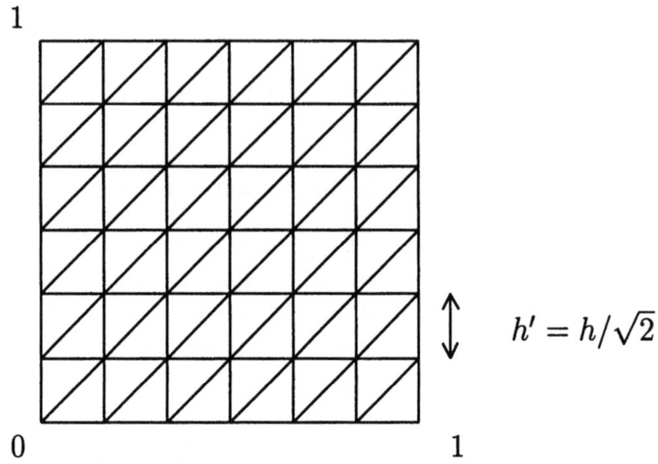


*Figure 1.*

$$u(x,y) = \begin{cases} y & \text{on } (0,h') \\ 2h'-y & \text{on } (h',2h'). \end{cases}$$

Extending $u$ periodically with $2h'$, and setting

$$u_h = \quad \text{the interpolate of } u \wedge \operatorname{dist}(x, \partial\Omega)$$

one clearly has

$$\inf_{V_0^h} \int_\Omega W(\nabla u)\, dx\, dy \le \int_\Omega W(\nabla u_h)\, dx\, dy \le ch \ll ch^{\frac{1}{2}}$$

for $h$ small. However, this estimate depends on the triangulation as we are about to see. Indeed consider now a family of triangulations as on (figure 2). For $x_0 = kh/2$, $k \in \mathbf{N}$ we are going to consider the strip $(x_0, x_0 + \frac{h}{2}) \times (0,1)$. Let $u \in V_0^h$ and $0 < \delta < \frac{1}{3}$. Recall that $\nabla u$ is constant on each of the triangles of $\tau_h$. Then for a given $u$ let us adopt the following definition:

**Definition 1** *Let $T \in \tau_h$ with one side at least of length $h$. We will say that $T$ is of*

$$\begin{array}{lll} \text{type } + & \text{if} & |u_x| \le \delta\,, \quad |u_y - 1| \le \delta\,, \\ \text{type } - & \text{if} & |u_x| \le \delta\,, \quad |u_y + 1| \le \delta\,, \\ \text{type } 0 & \text{else.} & \end{array}$$

In other words $T$ is of type $+$ or $-$ if $\nabla u$ is close to $\omega_+$ or $\omega_-$ respectively. Then on the strip $(x_0, x_0 + \frac{h}{2}) \times (0,1)$ we have triangles of various types. We denote by $N_0 = N_0(x_0)$ the number of triangles of type 0. We say that we have a change of phase along the line $x_0$ when two triangles change from type $+$ to type $-$ (see figure 3). Then we can prove,

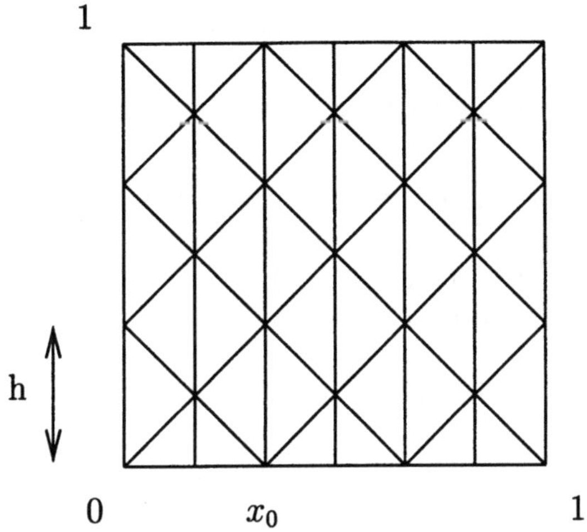


*Figure 2.*

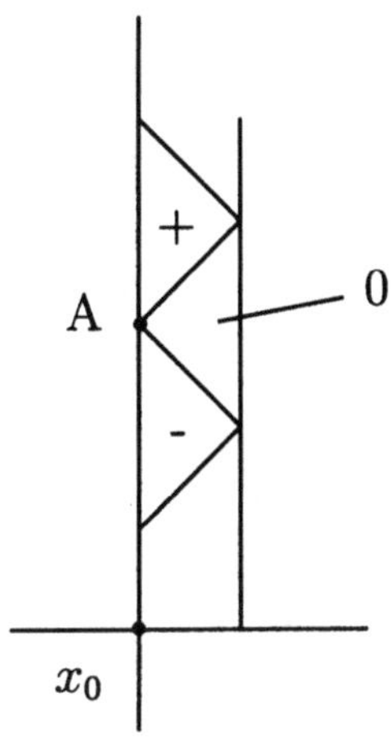


*Figure 3.*

**Lemma 1** *At each change of phase along the line $x = x_0$ the triangle having only one vertex on $x = x_0$ is of type* 0, *(see figure 3).*

**Proof.** Let us denote by $(u_x^+, u_y^+), (u_x^-, u_y^-)$ the gradients of $u$ in the triangle of type + and of type −, respectively. If $A = (x_0, y_0)$ one has

$$u(x_0 + \frac{h}{2}, y_0 + \frac{h}{2}) = u(x_0, y_0) + u_x^+ \frac{h}{2} + u_y^+ \frac{h}{2}$$

$$u(x_0 + \frac{h}{2}, y_0 - \frac{h}{2}) = u(x_0, y_0) + u_x^- \frac{h}{2} - u_y^- \frac{h}{2}.$$

So, in the triangle whose only vertex on $x = x_0$ is $A$ one has

$$u_y = \frac{1}{2}(u_x^+ - u_x^-) + \frac{1}{2}(u_y^+ + u_y^-).$$

Since $u_x^+, u_x^- \in [-\delta, \delta]$, $u_y^+ \in [1-\delta, 1+\delta]$, $u_y^- \in [-1-\delta, -1+\delta]$ one deduces that $|u_y| \leq 2\delta$ so that

$$|u_y \pm 1| \geq 1 - 2\delta > \delta.$$

Since $0 < \delta < \frac{1}{3}$. The considered triangle is thus of type 0.

**Lemma 2** *Let $u$ be a piecewise $\mathbf{C}^1$ function on $(a,b)$ such that $u'$ has a constant sign and $|u'| \geq C$. Then one has*

$$\int_a^b |u(z)|\,dz \geq C\left(\frac{(b-a)^2}{4}\right) \tag{2.12}$$

**Proof.** If $u$ does not vanish on $[a,b]$ then clearly

$$\int_a^b |u(z)|\,dz \geq \min\left(\int_a^b |u(z) - u(a)|\,dz, \int_a^b |u(z) - u(b)|\,dz\right)$$

So, considering possibly $u - u(a)$ or $u - u(b)$ instead of $u$ one can assume that $u$ vanishes at some point $\xi$. Since $u'$ has a constant sign one has

$$|u(z)| = \left|u(\xi) + \int_\xi^z u'(z')\,dz'\right| = \int_{[\xi,z]} |u'(z)|\,dz \geq C|\xi - z|.$$

Integration in $z$ on $(a,b)$ leads to (2.12).

**Lemma 3** *Assume that $(N_0 + 1)h \leq \frac{1}{2}$ then one has*

$$\int_0^1 |u(x_0, y)|\,dy \geq \frac{1-\delta}{16(N_0+1)}. \tag{2.13}$$

**Proof.** Consider a maximal chain of triangles of type + or −. In other words consider $(a_i, b_i)$ such that one has a change of phase or a "boundary point" at $(x_0, a_i), (x_0, b_i)$ and all the sides of triangles of the strip $(x_0, x_0 + \frac{h}{2}) \times (0,1)$ located on $x = x_0$ between $a_i$ and $b_i$ belong to triangles of the same type + or −. (A "boundary point" could be a vertex of one triangle of the boundary with all its sides strictly smaller than h .) Since $u_y$ has a constant sign and $|u_y| \geq 1-\delta$ one deduces from Lemma (2) that

$$\int_{a_i}^{b_i} |u(x_0,y)|\, dy \geq (1-\delta)\frac{(b_i \quad a_i)^2}{4}. \tag{2.14}$$

Denote by $N_1$ the number of maximal chains as above, i.e. $i = 1, ..., N_1$. Then from (2.14) one deduces

$$\sum_{i=1}^{N_1} \int_{a_i}^{b_i} |u(x_0,y)|\, dy \geq \frac{(1-\delta)N_1}{4} \sum_{i=1}^{N_1} \frac{(b_i-a_i)^2}{N_1}$$

and by a convexity argument

$$\int_0^1 |u(x_0,y)|\, dy \geq \sum_{i=1}^{N_1} \int_{a_i}^{b_i} |u(x_0,y)|\, dy \geq \frac{(1-\delta)N_1}{4} \left( \sum_{i=1}^{N_1} \frac{b_i-a_i}{N_1} \right)^2. \tag{2.15}$$

Due to Lemma 1 the sides of triangles on $x = x_0$ which do not belong to a chain belong to a triangle of type 0 or a small triangle of the boundary. So, one has

$$\sum_{i=1}^{N_1} b_i - a_i \geq 1-(N_0+1)h \geq \frac{1}{2}$$

and from (2.15) we deduce that

$$\int_0^1 |u(x_0,y)|\, dy \geq \frac{(1-\delta)}{16N_1}.$$

The result follows from the fact that by Lemma 1, $N_1 \leq N_0+1$.

**Lemma 4** *Let $u \in V_0^h$ then one has*

$$\int_0^1 |u(x_0,y)|\, dy \leq \int_\Omega W(\nabla u)\, dx\, dy. \tag{2.16}$$

**Proof.** One has

$$u(x_0,y) = u(0,y) + \int_0^{x_0} u_x(\xi,y)\, d\xi = \int_0^{x_0} u_x(\xi,y)\, dy.$$

Thus

$$|u(x_0,y)| \leq \int_0^1 |u_x(\xi,y)|\, d\xi \leq \int_0^1 W(\nabla u(\xi,y))\, d\xi$$

and the result follows by integration in $y$.

**Lemma 5** *(Estimate of the number of changes of phase). Let $u \in V_0^h$ such that*

$$\int_\Omega W(\nabla u)\,dx\,dy \le Ch^{\frac{1}{2}} \tag{2.17}$$

*where $C$ is a constant. If $h \le 64C^2$ then one has*

$$N_0 + 1 \ge \frac{1-\delta}{16C}h^{-\frac{1}{2}}. \tag{2.18}$$

*(Recall that $N_0$ is the number of triangles of type 0 in the strip $(x_0, x_0 + \frac{h}{2}) \times (0,1)$.)*

**Proof.** If (2.18) does not hold one has

$$(N_0+1)h < \frac{(1-\delta)h^{\frac{1}{2}}}{16C} < \frac{h^{\frac{1}{2}}}{16C} \le \frac{1}{2}.$$

Thus combining (2.13),(2.16) and (2.17) one obtains

$$\frac{1-\delta}{16(N_0+1)} \le \int_0^1 |u(x_0,y)|\,dy \le Ch^{\frac{1}{2}}$$

which leads to (2.18).

**End of the proof of Theorem 1**

Knowing that (1.5) holds consider a $u \in V_0^h$ such that

$$\int_\Omega W(\nabla u)\,dx\,dy \le Ch^{\frac{1}{2}} \tag{2.19}$$

On a triangle of type 0 for such an $u$ one has

$$|u_x| \ge \delta \text{ or } |\,|u_y - 1| \ge \delta,$$

i.e. $W(\nabla u) \ge \delta$. Thus if $S_{x_0} = (x_0, x_0 + \frac{h}{2}) \times (0,1)$

$$\int_{S_{x_0}} W(\nabla u)\,dx\,dy \ge N_0\delta\frac{h^2}{2}.$$

Assuming that $h$ is small - more precisely $\frac{(1-\delta)h^{-\frac{1}{2}}}{32C} > 1$ - one deduces from (2.18)

$$N_0 \ge \frac{1-\delta}{32C}h^{-\frac{1}{2}}$$

which leads to

$$\int_{S_{x_0}} W(\nabla u)\, dx\, dy \geq \frac{(1-\delta)\delta h^{\frac{3}{2}}}{64C}.$$

Since the number of stripes $S_{x_0}$ is $\frac{2}{h}$ one deduces that

$$\int_{\Omega} W(\nabla u)\, dx\, dy \geq \frac{(1-\delta)\delta h^{\frac{1}{2}}}{32C}$$

which completes the proof.

**Remark.** The exponent 1/2 appearing in (1.6), (1.5) is related to the linear growth of $W$. Other growth conditions would produce other exponents in (1.6), (1.5). Our argument extend also to general polygonal domains and more general functionals. For all this we refer to [3].

**Acknowledgements:** This paper was written when the first author was visiting the Universidade de Beira Interior in Covilhã. He would like to thank the mathematics department for its support. The second author would like also to thank the mathematics department in the University of Metz for its hospitality at the early stage of this research.

## References

1. J.M. Ball and R.D. James, Proposed experimental tests of a theory of fine microstructure and the two well problem, *Phil. Trans. R. Soc. London A*, **338**, (1992), 389–450.
2. M. Chipot, Numerical analysis of oscillations in nonconvex problems, *Numerische Mathematik*, **59**, (1991), 747–767.
3. M. Chipot and S. Müller, In preparation.
4. Ph. Ciarlet, *The finite element method for elliptic problems*. North-Holland, Amsterdam (1978).
5. J. Ericksen, Constitutive theory for some constrained elastic crystal, *Int. J. Solids and Structures*, **22** (1986), 951–964.
6. R. James and D. Kinderlehrer, Theory of diffusionless phase transitions, *Lecture Notes in Physics*, **344**, Springer Verlag (M. Rascle, D. Serre and M. Slemrod, Eds.), (1989), 51–84.
7. R. Kohn, Relaxation of a double-well energy, *Continuum Mech. Thermodyn.*, **3** (1991), 193–236.
8. R. Kohn and S. Müller, Branching of twins near an austenite-martensite interface, *Phil. Mag. A*, **66** (1992), 697–715.
9. R. Kohn and S. Müller, Surface energy and microstructure in coherent phase transitions, *CPAM*, **47** (1994), 405–435.
10. M. Luskin, On the computation of crystalline microstructure, *Acta Numerica*, (1996), 191–257.

# FINITE-ELEMENT ANALYSIS OF SOLID-FLUID, KINETICLY DRIVEN PHASE TRANSFORMATION

Y.M. LEROY AND J. GHOUSSOUB
*Laboratoire de Mécanique des Solides, Ecole Polytechnique, CNRS UMR 7649, Palaiseau 91128 Cedex, France*

**Abstract.** A finite-element scheme is proposed to analyze the solid-fluid phase transformation that occurs within grain boundaries and is characteristic of the deformation mechanism pressure solution.

## 1. Introduction

The deformation mechanism called pressure solution by Geologists is partly responsible for the destruction of primary intergranular porosity in the Earth's upper crust. That mechanism operates under temperature and stress conditions which are similar to those necessary for Coble creep with, however, the fundamental difference that a fluid phase is present at the interfaces between grains. A key issue to understand the actual deformation kinetics, which has not been resolved so far by experimentalists, is the structure of the grain contact which controls the transport of the desolved materials from the region of high stress concentration towards the pore free-fluid. That boundary has been thought to contain a thin film (Rutter, 1976) or to be made of a network of channels and islands (Lehner et Bataille, 1984/85). At the scale of the grain contact, that structure results from the evolution in time of a non-hydrostatically stress solid in thermodynamic disequilibrium with respect to the fluid in contact. This evolution is analyzed here as a phase transformation which is assumed to be controlled by a kinetic law. Diffusion is considered to be fast compared to the kinetic law characteristic time and the fluid to be in mechanical equilibrium at all times.

The objective of this paper is to present the numerical algorithm, based on the finite-element method, necessary to compute the change in mor-



*P. Argoul et al. (eds.),*
*IUTAM Symposium on Variations of Domains and Free-Boundary Problems in Solid Mechanics*, 327–335.

phology of a solid-fluid interface at the scale of the grain boundary. The essential part of the numerical scheme is a special finite-element which has $C_0$ continuity in the elastic solid and $C_1$ continuity along the fluid interface. The latter property is found necessary to capture the stabilizing influence of surface tension. Numerical results are presented for a rectangular grain having contact with a fluid on one of its sides and sustaining a plane strain compression. The initial evolution in time of the change in morphology of the interface due to the addition of a trigonometric perturbation is compared with the analytical predictions of a linear stability analysis to validate the numerical scheme. It is shown that the long time evolution of the trigonometric perturbation results in the formation of a cusp. That cusp corresponds to a dissolution pit which is present even if the fluid is supersaturated and results initially in a net deposition along the interface.

## 2. Governing equation and time discretization

Consider a solid occupying the volume $\Omega(t)$ and sustaining a loading on part of its boundary $\Gamma(t)$. Those volume and boundary are function of time and define the current configuration at time $t$. A section of $\Gamma(t)$, denoted by $\Gamma_I(t)$, is in contact with a fluid, Fig. 1. The migration of this boundary is controlled by a kinetic law and the new material is assumed to be added coherently to the solid and to have the same elastic properties as in the bulk solid.

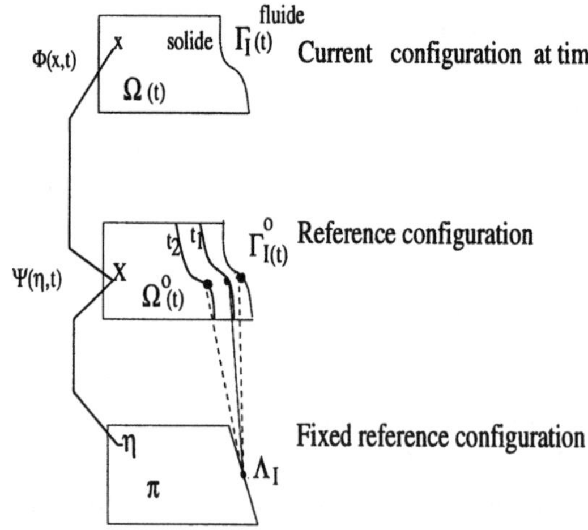


Fig.1 : The three configurations of the problem.

We aim to determine the position of the migrating interface $\Gamma_I(t)$ and for that purpose introduce two other configurations. The first, the reference configuration at time t, is occupied by the unstressed solid and denoted by $\Omega^o(t)$ with a fluid interface $\Gamma_I^o(t)$. The second, required to facilitate the linearization of the governing equations, is a fixed reference configuration $\Pi$ with boundary $\Lambda$ independent of time and such that a point on the boundary $\Gamma_I^o(t)$ is always the image of a point located on $\Lambda_I$, Fig. 1.

The problem is summarized by the search of two mappings, the first $\Phi(\mathbf{X}, t)$ from the reference configuration to the current configuration and the second $\Psi(\boldsymbol{\eta}, t)$ from the fixed reference configuration to the reference configuration. The first mapping $\Phi(\mathbf{X}, t)$ defines the material displacement due to loading of the elastic grain and the second $\Psi(\boldsymbol{\eta}, t)$ defines the evolution in geometry of the solid due to precipitation or dissolution. Note that a small strain approximation is made in the following and that no physical interpretation is proposed for the fixed reference configuration. It is often chosen as the current configuration to provide symmetric tangent operator upon linearization close to equilibrium (Pradeilles Duval, 1992). Here, it is found convenient to identify $\Pi$ as the computational domain used for the finite-element approximation.

## 2.1. GOVERNING EQUATIONS

The mechanical equilibrium of the stressed elastic solid is satisfied with the principle of virtual work written on the fixed reference configuration:

$$\int_{\Pi} \boldsymbol{\sigma} : \nabla^s_{\Omega}(\delta \mathbf{u}) J_{\Pi} d\Pi = \int_{\Lambda_I} (2\kappa\gamma - p)\mathbf{N} \cdot \delta\mathbf{u}\, J_{\Lambda} d\Lambda_I \,, \tag{1}$$

in which $\boldsymbol{\sigma}, \delta\mathbf{u}, 2\kappa, \gamma, p$ and $\mathbf{N}$ are the Cauchy stress, an admissible virtual displacement, the mean curvature, the capillarity of the solid-fluid contact, the fluid pressure and the normal vector on the solid fluid interface pointing towards the fluid, Fig. 1. In (1), the only external action presented is due to the fluid and is described by Laplace's condition. Note that $\nabla^s_{\Omega}$ denotes the symmetric part of the gradient with respect to the reference configuration of the quantity between parentheses. The scalars $J_{\Pi}$ and $J_{\Lambda}$ are the jacobian of the mapping between $\Pi$ and $\Omega^o$ and the interfaces $\Lambda_I$ and $\Gamma_I^o$, respectively.

The thermodynamic force $\chi$ that drives the migration of the solid-fluid interface is given by

$$\chi = h + \frac{p - 2\kappa\gamma}{\rho} - \mu \,, \tag{2}$$

in which $h, \rho$ and $\mu$ are the Helmholtz free energy of the solid, the current solid material density and the solute chemical potential in the fluid (Heidug and Leroy, 1994). The chemical potential, a function of the fluid pressure and the solute concentration, is assumed constant in what follows. A kinetic law is now proposed to relate the mass flux to this force $\chi$ so that the non-material velocity $U_N$ of the migrating interface (positive if directed towards the fluid) is given by

$$U_N = \frac{-\mathcal{L}(\chi)}{\rho_o}\,. \tag{3}$$

### 2.2. TIME DISCRETIZATION AND WEAK FORMULATION OF THE MIGRATION PROBLEM

Time is now discretized and we wish to determine the mapping $\mathbf{X}_{n+1}$ of a point $\boldsymbol{\eta}$ on $\Lambda_I$ at time $t_{n+1}$ from the knowledge of $\mathbf{X}_n$ at time $t_n$ and the non-material velocity vectors $\mathbf{U}_{n+1}$ and $\mathbf{U}_n$ oriented along the normal to the interface. For that purpose, a $\theta$-family of approximation is proposed as follows:

$$\mathbf{X}_{n+1} = \mathbf{X}_n + \Delta t[(1-\theta)\mathbf{U}_n - \theta\mathbf{U}_{n+1}] \equiv \mathbf{F}(\mathbf{U}_{n+1})\,. \tag{4}$$

A finite-element scheme will be introduced in the next section which approximate the value of $\mathbf{X}_{n+1}$ by $\mathbf{X}^h_{n+1}$. Denote the difference between the $\mathbf{F}(\mathbf{U}_{n+1})$ introduced above and $\mathbf{X}^h_{n+1}$ by $\mathbf{R}_{n+1}$ and let propose to weight this residual over the boundary to obtain a weak formulation of our evolution problem:

$$\int_{\Gamma^o_I(t)} \varphi \cdot \mathbf{R}_{n+1} d\Gamma^o_I(t) = 0\,, \tag{5}$$

in which $\varphi$ is a weighting function.

Solving equations (1) and (5) provides the solution to the elasticity problem as well as the new position of the migrating solid-fluid interface. The solution procedure requires a linearization if an implicit time integration is considered ($\theta \neq 0$). An explicit scheme results from the choice of ($\theta = 0$) which is made in the last section of this paper. Once the position of the interface is updated, it remains to adapt the mesh over the new domain following standard techniques discussed by Brackbill and Saltzman (1982).

## 3. Finite-element scheme

The choice of a spatial discretization is dictated by the role played by stresses and surface tension in the evolution of the solid-fluid interface which is best explained from the results of a stability analysis (Leroy and Heidug, 1994). A solid-fluid interface in thermodynamic equilibrium is destabilized if the stress component acting parallel to the interface differs from the fluid pressure. This action is due to the contribution of the Helmholtz free energy and the change in volume present in the driving force. The larger the difference between the stress component parallel to the interface and the fluid pressure the smallest is the wavelength of the dominant mode of instability.

There exists a dominant wavelength since surface tension penalizes the development of short wavelength modes leading to large changes in curvature. To capture those two competing roles, it is necessary to estimate the stress acting on the solid-fluid interface as well as the local curvature. A $C_0$ interpolation is sufficient for the first task but not for the second. Indeed, local discontinuities in the tangent at the element connections result in stress concentrations which have a destabilizing influence. However, the measure of curvature conducted within the element will not be affected by these discontinuities and thus will fail to activate the stabilizing role of surface tension. It is for that reason that a $C_1$ interface interpolation is necessary.

### 3.1. C1-INTERPOLATION ALONG THE INTERFACE

Consider the following interpolation along the solid-fluid interface

$$\mathbf{X} = \sum_{a=1}^{4} H_a \tilde{\mathbf{X}}_a \,, \tag{6}$$

in which $H_a$ are Hermite cubic polynomials of $\xi_1$ and $\tilde{\mathbf{X}}_a$ the following collection of nodal quantities: $(\mathbf{X}_3; \mathbf{X}_4; \lambda_1 \mathbf{T}_1; \lambda_2 \mathbf{T}_2)$, Fig. 2. The nodal positions and tangents are defined in Fig. 2 and it remains to explain the introduction of the two scalars $\lambda_1$ and $\lambda_2$ which weight the contribution of the tangent vectors in the interpolation (6). It is easily shown that those two scalars are the jacobian of the transformation between $\Lambda_I$ and $\Gamma_I^o(t)$ at node 3 and 4, respectively. Those two scalars need additional information to be computed locally. It is proposed to determine them from the knowledge of the position of an extra node, labeled node 5 in Fig. 2. For example, one could impose the curve passing trough $\mathbf{X}_3$ and $\mathbf{X}_4$ with the tangent $\mathbf{T}_1$ and $\mathbf{T}_2$ to intersect that extra node. In all cases, we consider that sufficient constraints are provided to compute these two scalars in terms of the collection of nodal quantities $\bar{\mathbf{X}}_b$ defined by $(\mathbf{X}_3; \mathbf{X}_4; \mathbf{X}_5; \mathbf{T}_1; \mathbf{T}_2)$.

### 3.2. A 2-D ISOPARAMETRIC ELEMENT WITH $C_1$ CONTINUITY ALONG ONE OF ITS SIDES

We need now to construct a 2-D finite-element which will have on one of its side the solid-fluid interface described above. The $C_1$ continuity condition is only required on that interface and can be relaxed in the bulk. In consequence, it is proposed to mix classical Lagrange polynomials and Hermite cubics to construct the 4-noded element shown in Fig. 2 in which the spatial interpolation is provided by:

$$\mathbf{X} = \sum_{a=1}^{4} N_a \mathbf{X}_a + \sum_{a=1}^{2} N_{a+4} \lambda_a \mathbf{T}_a \,, \tag{7}$$

in which the 6 shape functions $N_a$ are given by

$$N_1 = \tfrac{1}{4}(\xi_1 - 1)(\xi_2 - 1)\,, \quad N_4 = \tfrac{1}{2}H_2(\xi_1)(\xi_2 + 1)\,,$$
$$N_2 = \tfrac{-1}{4}(\xi_1 + 1)(\xi_2 - 1)\,, \quad N_5 = \tfrac{1}{2}H_3(\xi_1)(\xi_2 + 1)\,,$$
$$N_3 = \tfrac{1}{2}H_1(\xi_1)(\xi_2 + 1)\,, \quad N_6 = \tfrac{1}{2}H_4(\xi_1)(\xi_2 + 1)\,. \tag{8}$$

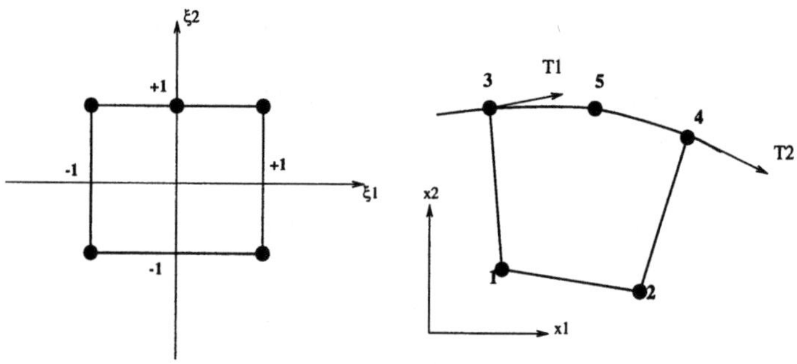


Fig.2 : The finite element with C1 continuity at the solid-fluid interface.

The next step in constructing this finite-element scheme is the material displacement interpolation. To obtain an isoparametric element, we propose to make use of the same interpolation for the displacement as in (7) in which the nodal unknowns are grouped in the collection $(\mathbf{u}_1; \mathbf{u}_2; \mathbf{u}_3; \mathbf{u}_4; (\nabla_\Gamma \mathbf{u})_1; (\nabla_\Gamma \mathbf{u})_2)$. The first 4 unknowns are the material displacements of nodes 1 to 4 and the last two correspond to the surface gradient of the material displacement at nodes 3 and 4. $\nabla_\Gamma$ denotes the surface gradient with respect to the fixed reference configuration $\Lambda_I$

### 3.3. CONVERGENCE OF THE FINITE-ELEMENT SCHEME

The convergence of the finite-element scheme proposed above relies on the proof of completeness and conformity. The second property is clearly satisfied since no displacement discontinuities can be generated at the interface between two elements. To assess the element completeness, we recourse to the classical argument that a rigid body motion or a constant strain field imposed by adequate nodal displacements to a finite element will result in a uniform strain state within that element. The proof is easily constructed and is omitted here for sake of brevity.

## 4. Results

The example chosen to illustrate the potential of the finite-element scheme proposed is a rectangular grain in contact with a fluid on one of its sides, Fig. 3. The solid sustains a lateral compression $\sigma_o$ and the pressure $p$ of the fluid is set to 30 $MPa$. The elastic solid is characterized by an elastic modulus of 10 $GPa$ and a Poisson's ratio of 0.2. The solid reference density is 2000 $kg/m^3$. A linear kinetic law is chosen such that the expression $\mathcal{L}(\chi)$ in (3) is replaced by $L\chi$ with a value for the phenomenological parameter $L$ set to 1.

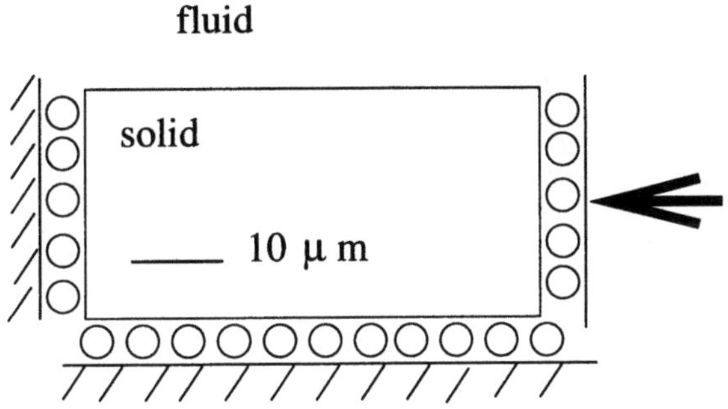


Fig.3 : The rectangular grain under a plane strain compressive loading.

The kinetics law provides the only time scale of this problem and time is normalized in the following by a characteristic time $t_c$ set to $10^{+5}s$. From the definition of the driving force in (2) it is clear that a thermodynamic equilibrium exists for the whole flat interface if the appropriate chemical potential is selected.

That equilibrium could be unstable depending on the difference between the lateral compressive stress and the fluid pressure. This conclusion is inferred from the findings of a linear stability analysis which stipulate that the rate of growth or decay of a trigonometric perturbation with wavenumber $\omega$ added to the interface position is

$$\lambda_\omega \frac{\rho_o^2 H}{LG} = -\frac{\gamma H}{G}(\omega H)^2[1-(1-2\nu)\frac{(\sigma_o+p)}{2G}\frac{\cosh(\omega H)\sinh(\omega H)}{\cosh(\omega H)\sinh(\omega H)+\omega H}] \qquad (9)$$
$$+(\frac{\sigma_o+p}{G})^2\omega H(1-\nu)\frac{\cosh^2(\omega H)}{\cosh(\omega H)\sinh(\omega H)+\omega H},$$

in which $H$ and $G$ are the grain thickness and the modulus of elasticity in shear. Note from (9) that the first term of the right hand side is always negative and results from the effects of surface tension $\gamma$. The second term is proportional to the square of the sum $\sigma_o + p$. If the lateral stress $\sigma_o$ is equal to $-p$ then the interface is stable to any perturbation since $\lambda_\omega$ is then always negative. The sum $\sigma_o + p$ needs to be non-zero to have a positive $\lambda_\omega$ and thus an unstable interface. for large enough value of $\omega H$.

These analytical results are now exploited to validate the numerical scheme proposed. The thickness $H$ of the grain is chosen to be 50 $\mu m$ and the dimensionless wavenumber $\omega H$ is 1. The amplitude of the perturbation chosen is 2% of the thickness. For a compressive stress $\sigma_o$ equal to $-p$ the rate of growth is negative and compares favorably with the results of the finite-element method presented in Fig. 4. In the same figure, results are also presented for a $\sigma_o$ of $-5/3p$ for which the initial rate of growth of the instability is identical to the analytical prediction.

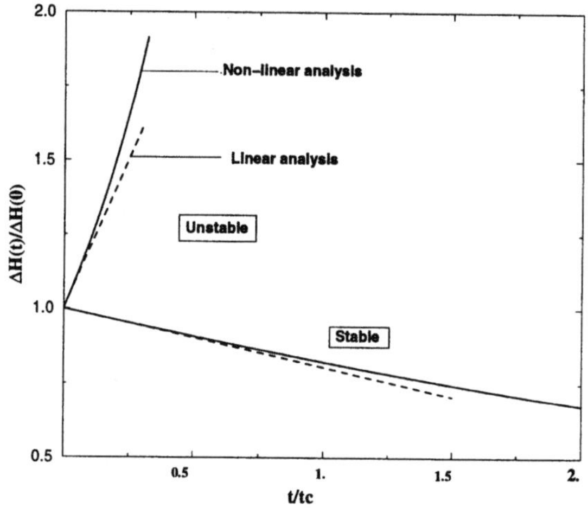


Fig.4 : Comparison between the finite element solutions and the predictions of the linear stability analysis.

The long term evolution of the interface morphology is now discussed for a chemical potential larger than the one necessary for equilibrium. The development of the instability is similar to the one which occurs from an equilibrium with the difference that there is in this case a general thickening of the grain due to deposition of the solute, Fig. 5.

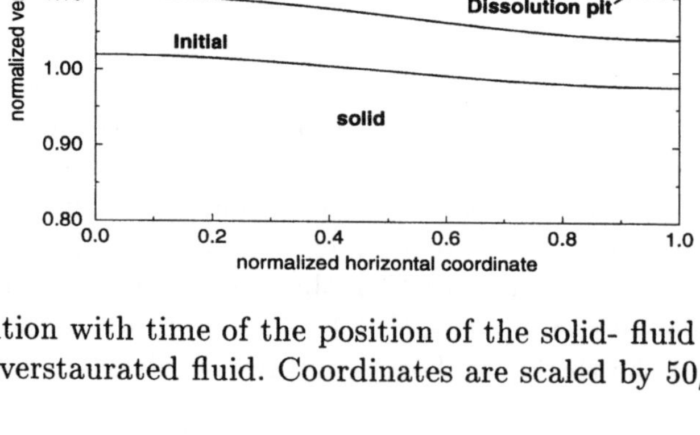


Fig.5 : Evolution with time of the position of the solid- fluid interface for an overstaurated fluid. Coordinates are scaled by $50\mu m$.

In that figure, we observe that the interface migrates due to the deposition process but that the thinner section of the grain is not as fast to migrate as the thicker part. This difference results in an increase in the amplitude of the perturbation added to the interface. However, the shape of the perturbation does not remain close to a trigonometric function and it is observed, already at time $t/t_c$ of 0.32, that the convex part of the interface is longer than the concave section due to a shift to the right of the inflexion point. This shift accelerates with time due to the built up of stress concentration along the right hand boundary of the grain which leads ultimately to a change in sign of the driving force in that region. At that time $t/t_c$ of 0.62 , there is now a dissolution pit, despite the initial supersaturated conditions, and a cusp is formed. This dissolution resembles the under-cutting mechanism often discussed in the literature and proposed to explain the migration of fluid from the pore to the grain boundary.

**Acknowledgments:** The ideas presented here were discussed repeatedly with W.K. Heidug (Shell Research, Holland) and F.K. Lehner (University of Bonn, Germany). This research was partly supported by the "Groupement de Recherche, Géomécanique des Roches Profondes".

## References

Brackbill, J.U and Saltzman, J.S. (1982) Adaptive zoning for singular problems in two dimensions, *J. Comput. Phys.*, **46**, pp. 342-368.

Lehner, F.K and Bataille, J.(1984-85) Non-equilibrium thermodynamics of pressure solution, *PAGEOPH*, **122**, pp. 53-85.

Leroy, Y.M and Heidug, W.K. (1994) Geometrical evolution of stressed and curved solid-fluid phase boundaries. Part 2- Stability of cylindrical pores, *J. Geophys. Res.*, **99**, B1, pp. 517-530.

Heidug, W.K and Leroy, Y.M. (1994) Geometrical evolution of stressed and curved solid-fluid phase boundaries. Part 1- Transformation kinetics, *J. Geophys. Res.*, **99**, B1, pp. 505-515.

Pradeilles Duval, R.M. (1992) Évolution de systèmes avec des surfaces de discontinuités mobiles: application au délaminage, doctoral thesis, École Polytechnique, France.

Rutter, E.H. (1976) The kinetics of rock deformation by pressure solution *Phil. Trans. R. Soc. Lond.* A. **283**, pp. 203-219.

Spiers, C.J, Schutjens, P.M.T.M, Brzesowsky, R.H, Peach, C.J, Liezenberg, J.L and Zwart, H.J.(1990) Experimental determination of constitutive parameters governing creep of rocksalt by pressure solution, *in :* Knipe, R and Rutter, E.H.(eds) *Deformation mechanisms, Rheology and Tectonics. Jl. geol. Soc. Spec.Publ.*, **54**, pp. 215-227.

# AUTHOR INDEX

# Mechanics

## *SOLID* MECHANICS AND ITS APPLICATIONS

*Series Editor:* G.M.L. Gladwell

20. W. Schiehlen (ed.): *Advanced Multibody System Dynamics.* Simulation and Software Tools. 1993 ISBN 0-7923-2192-8
21. C.-W. Lee: *Vibration Analysis of Rotors.* 1993 ISBN 0-7923-2300-9
22. D.R. Smith: *An Introduction to Continuum Mechanics.* 1993 ISBN 0-7923-2454-4
23. G.M.L. Gladwell: *Inverse Problems in Scattering.* An Introduction. 1993 ISBN 0-7923-2478-1
24. G. Prathap: *The Finite Element Method in Structural Mechanics.* 1993 ISBN 0-7923-2492-7
25. J. Herskovits (ed.): *Advances in Structural Optimization.* 1995 ISBN 0-7923-2510-9
26. M.A. González-Palacios and J. Angeles: *Cam Synthesis.* 1993 ISBN 0-7923-2536-2
27. W.S. Hall: *The Boundary Element Method.* 1993 ISBN 0-7923-2580-X
28. J. Angeles, G. Hommel and P. Kovács (eds.): *Computational Kinematics.* 1993 ISBN 0-7923-2585-0
29. A. Curnier: *Computational Methods in Solid Mechanics.* 1994 ISBN 0-7923-2761-6
30. D.A. Hills and D. Nowell: *Mechanics of Fretting Fatigue.* 1994 ISBN 0-7923-2866-3
31. B. Tabarrok and F.P.J. Rimrott: *Variational Methods and Complementary Formulations in Dynamics.* 1994 ISBN 0-7923-2923-6
32. E.H. Dowell (ed.), E.F. Crawley, H.C. Curtiss Jr., D.A. Peters, R. H. Scanlan and F. Sisto: *A Modern Course in Aeroelasticity.* Third Revised and Enlarged Edition. 1995 ISBN 0-7923-2788-8; Pb: 0-7923-2789-6
33. A. Preumont: *Random Vibration and Spectral Analysis.* 1994 ISBN 0-7923-3036-6
34. J.N. Reddy (ed.): *Mechanics of Composite Materials.* Selected works of Nicholas J. Pagano. 1994 ISBN 0-7923-3041-2
35. A.P.S. Selvadurai (ed.): *Mechanics of Poroelastic Media.* 1996 ISBN 0-7923-3329-2
36. Z. Mróz, D. Weichert, S. Dorosz (eds.): *Inelastic Behaviour of Structures under Variable Loads.* 1995 ISBN 0-7923-3397-7
37. R. Pyrz (ed.): *IUTAM Symposium on Microstructure-Property Interactions in Composite Materials.* Proceedings of the IUTAM Symposium held in Aalborg, Denmark. 1995 ISBN 0-7923-3427-2
38. M.I. Friswell and J.E. Mottershead: *Finite Element Model Updating in Structural Dynamics.* 1995 ISBN 0-7923-3431-0
39. D.F. Parker and A.H. England (eds.): *IUTAM Symposium on Anisotropy, Inhomogeneity and Nonlinearity in Solid Mechanics.* Proceedings of the IUTAM Symposium held in Nottingham, U.K. 1995 ISBN 0-7923-3594-5
40. J.-P. Merlet and B. Ravani (eds.): *Computational Kinematics '95.* 1995 ISBN 0-7923-3673-9
41. L.P. Lebedev, I.I. Vorovich and G.M.L. Gladwell: *Functional Analysis.* Applications in Mechanics and Inverse Problems. 1996 ISBN 0-7923-3849-9
42. J. Menčik: *Mechanics of Components with Treated or Coated Surfaces.* 1996 ISBN 0-7923-3700-X
43. D. Bestle and W. Schiehlen (eds.): *IUTAM Symposium on Optimization of Mechanical Systems.* Proceedings of the IUTAM Symposium held in Stuttgart, Germany. 1996 ISBN 0-7923-3830-8

Kluwer Academic Publishers – Dordrecht / Boston / London

# Mechanics

## *SOLID* MECHANICS AND ITS APPLICATIONS

*Series Editor:* G.M.L. Gladwell

44. D.A. Hills, P.A. Kelly, D.N. Dai and A.M. Korsunsky: *Solution of Crack Problems*. The Distributed Dislocation Technique. 1996 ISBN 0-7923-3848-0
45. V.A. Squire, R.J. Hosking, A.D. Kerr and P.J. Langhorne: *Moving Loads on Ice Plates*. 1996 ISBN 0-7923-3953-3
46. A. Pineau and A. Zaoui (eds.): *IUTAM Symposium on Micromechanics of Plasticity and Damage of Multiphase Materials*. Proceedings of the IUTAM Symposium held in Sèvres, Paris, France. 1996 ISBN 0-7923-4188-0
47. A. Naess and S. Krenk (eds.): *IUTAM Symposium on Advances in Nonlinear Stochastic Mechanics*. Proceedings of the IUTAM Symposium held in Trondheim, Norway. 1996 ISBN 0-7923-4193-7
48. D. Ieşan and A. Scalia: *Thermoelastic Deformations*. 1996 ISBN 0-7923-4230-5
49. J. R. Willis (ed.): *IUTAM Symposium on Nonlinear Analysis of Fracture*. Proceedings of the IUTAM Symposium held in Cambridge, U.K. 1997 ISBN 0-7923-4378-6
50. A. Preumont: *Vibration Control of Active Structures*. An Introduction. 1997 ISBN 0-7923-4392-1
51. G.P. Cherepanov: *Methods of Fracture Mechanics: Solid Matter Physics*. 1997 ISBN 0-7923-4408-1
52. D.H. van Campen (ed.): *IUTAM Symposium on Interaction between Dynamics and Control in Advanced Mechanical Systems*. Proceedings of the IUTAM Symposium held in Eindhoven, The Netherlands. 1997 ISBN 0-7923-4429-4
53. N.A. Fleck and A.C.F. Cocks (eds.): *IUTAM Symposium on Mechanics of Granular and Porous Materials*. Proceedings of the IUTAM Symposium held in Cambridge, U.K. 1997 ISBN 0-7923-4553-3
54. J. Roorda and N.K. Srivastava (eds.): *Trends in Structural Mechanics*. Theory, Practice, Education. 1997 ISBN 0-7923-4603-3
55. Yu. A. Mitropolskii and N. Van Dao: *Applied Asymptotic Methods in Nonlinear Oscillations*. 1997 ISBN 0-7923-4605-X
56. C. Guedes Soares (ed.): *Probabilistic Methods for Structural Design*. 1997 ISBN 0-7923-4670-X
57. D. François, A. Pineau and A. Zaoui: *Mechanical Behaviour of Materials*. Volume I: Elasticity and Plasticity. 1998 ISBN 0-7923-4894-X
58. D. François, A. Pineau and A. Zaoui: *Mechanical Behaviour of Materials*. Volume II: Viscoplasticity, Damage, Fracture and Contact Mechanics. 1998 ISBN 0-7923-4895-8
59. L. T. Tenek and J. Argyris: *Finite Element Analysis for Composite Structures*. 1998 ISBN 0-7923-4899-0
60. Y.A. Bahei-El-Din and G.J. Dvorak (eds.): *IUTAM Symposium on Transformation Problems in Composite and Active Materials*. Proceedings of the IUTAM Symposium held in Cairo, Egypt. 1998 ISBN 0-7923-5122-3
61. I.G. Goryacheva: *Contact Mechanics in Tribology*. 1998 ISBN 0-7923-5257-2
62. O.T. Bruhns and E. Stein (eds.): *IUTAM Symposium on Micro- and Macrostructural Aspects of Thermoplasticity*. Proceedings of the IUTAM Symposium held in Bochum, Germany. 1999 ISBN 0-7923-5265-3

Kluwer Academic Publishers – Dordrecht / Boston / London

MIX
Papier aus verantwortungsvollen Quellen
Paper from responsible sources
FSC® C105338

If you have any concerns about our products,
you can contact us on
**ProductSafety@springernature.com**

In case Publisher is established outside the EU,
the EU authorized representative is:
**Springer Nature Customer Service Center GmbH**
**Europaplatz 3, 69115 Heidelberg, Germany**

Printed by Libri Plureos GmbH
in Hamburg, Germany